Mathematical Models
in
Water Pollution Control

Mathematical Models in Water Pollution Control

Edited by

A. James
Head of Division of Public Health Engineering
University of Newcastle upon Tyne

A Wiley–Interscience Publication

JOHN WILEY & SONS

Chichester · New York · Brisbane · Toronto

Library of Congress Cataloging in Publication Data:
Main entry under title:

Mathematical models in water pollution control.

'A Wiley-interscience publication.'
'Based upon papers presented at a conference,
"The use of mathematical modelling in water
pollution control," held at the University of
Newcastle upon Tyne in September, 1973.'
Bibliography: p.
1. Water-Pollution—Mathematical models—
Congress. 2. Water quality management—
Mathematical models—Congresses. 3. Sewage—
Purification—Mathematical models—Congress.
I. James, A.
TD423.M35   363.6'1   77-7214

ISBN 0 471 99471 5

Photosetting by Thomson Press (India) Limited, New Delhi and
printed in Great Britain by the Pitman Press, Bath, Avon.

List of Contributors

J. F. ANDREWS	*Civil and Environmental Engineering Department, University of Houston, Texas.*
M. B. BECK	*University Engineering Department, Cambridge. Ernest Cook Trust Research Fellow in Environmental Sciences.*
K. BOWDEN	*Directorate of Resource Planning, Thames Water Authority.*
L. GALLAGHER	*ICI Petrochemicals Division, Billingham, Teesside.*
J. H. N. GARLAND	*Rivers Division, Water Research Centre, Stevenage.*
N. M. D. GREEN	*Department of Civil Engineering, University of Newcastle upon Tyne.*
M. J. HAMLIN	*Department of Civil Engineering, University of Birmingham.*
G. D. HOBBS	*ICI Central Management Services, Wilmslow, Cheshire.*
K. J. IVES	*Department of Civil and Municipal Engineering, University College, London.*
A. JAMES	*Department of Civil Engineering, University of Newcastle upon Tyne.*
D. G. JAMIESON	*Directorate of Resource Planning, Thames Water Authority.*
J. N. R. JEFFERS	*Institute of Terrestrial Ecology, Grange-over-Sands.*
G. L. JONES	*Microbiology Division, Water Research Centre, Stevenage.*
J. KANPTON	*Department of Civil Engineering, University of Newcastle upon Tyne.*
O. LINDHOLM	*Norwegian Institute of Water Research, Oslo.*
C. J. S. PETRIE	*Department of Engineering Mathematics, University of Newcastle upon Tyne.*
J. A. STEEL	*Metropolitan Water Division, Thames Water Authority.*
T. H. Y. TEBBUTT	*Department of Civil Engineering, University of Birmingham.*
A. E. WARN	*Sub-Directorate of Water Resource Planning, Anglian Water Authority.*
D. E. WRIGHT	*Sir William Halcrow & Partners, London.*

Contents

I TECHNIQUES

1 Application of some Systems Engineering Concepts and Tools to Water Pollution Control Systems   **3**
J. F. Andrews

1.1   Introduction   3
1.2   System definition and structure   4
1.3   Materials, energy, and information   9
1.4   Dynamic behaviour   12
1.5   Alternate means of accomplishing objectives   12
1.6   Factors for judging the value of a system   13
1.7   System optimization   15
1.8   Some systems engineering tools   16
1.9   Summary   35
1.10   References   36

2 The Modelling of Engineering Systems—Mathematical and Computational Techniques **39**
C. J. S. Petrie

2.1   Introduction   39
2.2   Case study 1—A biological reactor   40
2.3   Case study 2—Effluent dispersion in a river   44
2.4   Case study 3—The trickling filter   50
2.5   Discussion   53
2.6   References   54
2.7   Appendix: Some mathematical detail   54
2.8   Bibliography   64

3 Statistical Techniques in the Field of Water Pollution Control **67**
N. M. D. Green

3.1   Introduction   67
3.2   Design of experiments for point estimation of parameters   ..   68
3.3   Descriptive statistics   71
3.4   Statistical synthesis and simulation for the purposes of prediction   77
3.5   Conclusion   79
3.6   References   80

4 Optimization and its Application to a Unit Process Design Problem   **81**

J. Knapton

4.1   Introduction   81

4.2   Classification   81

4.3   Linear mathematical programming   82

4.4   Nonlinear iterative methods   84

4.5   Nonlinear programming   92

4.6   Application of optimization methods   96

4.7   References   101

4.8   Appendix: Simplex method fortran program   103

II APPLICATION TO POLLUTED ENVIRONMENTS

5 Reservoir Algal Productivity   **107**

J. A. Steel

5.1   Introduction   107

5.2   Model structure   108

5.3   Gross photosynthesis and maximum biomass   115

5.4   Net photosynthesis   123

5.5   Assimilation 'efficiency'   125

5.6   Biomass charges and nutrients   126

5.7   Algal concentrations   130

5.8   Conclusion   134

5.9   References   135

6 Modelling of Dissolved Oxygen in a Non-Tidal Stream **137**

M. B. Beck

6.1   Introduction   137

6.2   Mathematical models for DO−BOD interaction   138

6.3   The development of internally descriptive models—a review   144

6.4   Model objectives, identification, parameter estimation, and validation   150

6.5   A case study: the River Cam   153

6.6   Conclusions   163

6.7   References   164

7 Nitrification in the River Trent **167**

J. H. N. Garland

7.1   Introduction   167

7.2   Features of the Trent basin   167

7.3   The oxidation of ammoniacal nitrogen and its effect on the mass flow of oxygen in the Trent   168

7.4   Application of Michaelis−Menten and Monod growth relationships to results of BOD measurements   175

7.5   An intuitive model of the nitrification process in a river   ..   180

7.6 Results of intensive nitrification surveys conducted on a 23 km long reach in the mid-Trent in South Derbyshire between Willington and Shardlow   184
7.7 Summary and conclusions..   189
7.8 References   191

8 Estuarine Dispersion   193
L. Gallagher and G. D. Hobbs
8.1 Introduction   193
8.2 The dispersion process   193
8.3 Turbulent diffusion   195
8.4 Longitudinal dispersion   198
8.5 Dispersion in practice   200
8.6 Conclusion   203
8.7 References   204

9 The Modelling of Marine Pollution   207
A. James
9.1 Introduction   207
9.2 Dispersion models   207
9.3 Biological models in marine pollution   215
9.4 References   222

III APPLICATION TO WASTE TREATMENT

10 Modelling of Sewerage Systems   227
O. Lindholm
10.1 Introduction   227
10.2 Program configuration   227
10.3 Program descriptions   228
10.4 Optimization technique   233
10.5 Assumptions   234
10.6 Examples and demonstrations   235
10.7 Summary   245
10.8 References   246

11 Sedimentation   247
M. J. Hamlin and T. H. Y. Tebbutt
11.1 Introduction   247
11.2 Sedimentation theory..   247
11.3 Nature of suspensions   248
11.4 Sedimentation in practice   249
11.5 The Birmingham pilot plant   250
11.6 Mathematical modelling of the sedimentation process   251
11.7 Deterministic prediction models   251

11.8 Hydraulic modelling 254
11.9 Performance curves 255
11.10 Stochastic models 257
11.11 Conclusion 261
11.12 References 262

12 A Mathematical Model for Bacterial Growth and Substrate
  Utilization in the Activated-Sludge Process.. 265
    G. L. Jones
    12.1 Introduction 265
    12.2 Limitations of the equations 266
    12.3 Bacterial growth 269
    12.4 Model for bacterial growth and substrate utilization in waste
      treatment 270
    12.5 Application of the model 271
    12.6 Predictions of the model 272
    12.7 Nitrification.. 276
    12.8 Discussion 276
    12.9 References 277

13 The Development of a Dynamic Model and Control Strategies for the
  Anaerobic Digestion Process.. 281
    J. F. Andrews
    13.1 Introduction 281
    13.2 Model development 282
    13.3 Model verification 290
    13.4 Process stability 293
    13.5 Control strategies 296
    13.6 Summary 301
    13.7 References 302

14 An Ecological Model of Percolating Filters.. 303
    A. James
    14.1 Introduction 303
    14.2 The environment 303
    14.3 The community 309
    14.4 Filter model.. 314
    14.5 References 318

15 CIRIA Model for Cost-effective Wastewater Treatment 319
    K. Bowden and D. E. Wright
    15.1 Introduction 319
    15.2 Study objectives 320
    15.3 Limits and assumption of the prototype model 320
    15.4 General concepts 322
    15.5 Performance relationships 328

15.6   Costs   329
15.7   Conclusion   334
15.8   References   337

16 Optimization Model for Tertiary Treatment Rapid Filtration **339**
    K. J. Ives
16.1   Introduction   339
16.2   Mathematical models..   339
16.3   Operational optimum for uniform filters   342
16.4   Operational optimum for graded-media filters   345
16.5   Economic optimum design   348
16.6   References   351

IV APPLICATION TO WATER RESOURCES

17 The Trent Mathematical Model **355**
    A. E. Warn
17.1   Introduction   355
17.2   The structure of the model   355
17.3   The river model   356
17.4   Optimization in the river model   360
17.5   The allocation model   365
17.6   The investment model   368
17.7   Results   368
17.8   Future work   372
17.9   References   373

18 A Hierarchical Approach to Water-Quality Modelling **377**
    D. G. Jamieson
18.1   Introduction   377
18.2   Planning models   379
18.3   Design models   380
18.4   Operation models   386
18.5   Data requirements   387
18.6   References   388

19 Systems Analysis and Modelling Strategies in Ecology **389**
    J. N. R. Jeffers
19.1   Introduction   389
19.2   The role of models in ecology   391
19.3   The necessary stages in a modelling strategy   395
19.4   The inter-relationship between scientists and decision-makers   398
19.5   Some examples   400
19.6   References   409

Subject Index   **411**

Preface

The technology of water pollution control has developed over the last hundred years in an evolutionary manner. Empirical ideas have been tested and the successful ones incorporated into treatment practice. An early example of such developments was the evolution of percolating filters from land treatment. The main difficulties encountered with land treatment were clogging of the pores, so the initial experiments examined the drainage characteristics of other material such as crushed rock. Beds of stone known as contact beds were successfully used on a fill-and-draw basis and then, with the introduction of continuous inflow and removal, developed into percolating filters. Similarly, with the development of the activated sludge process, theory often lagged behind practice.

Theoretical developments have been slow because of the great degree of variation in natural waters and treatment plants. The action of random biological populations on wastes of variable composition produces data from which it is difficult to discern any general principles. Even in processes like sedimentation which have a physico-chemical rather than a biological basis the complicating effects of flocculation have obstructed the development of a coherent hydraulic description. It is in such a situation that the techniques of mathematical modelling can offer considerable assistance. For example simulation can be used to mathematically reconstruct a chain of causal connections where the individual relationships contain nonlinear elements, or are interdependent, or contain some other type of complexity which prevents a classical solution. Numerical methods and computing may then be used to find answers to the sets of equations which would be insoluble by analytical methods.

In this way the behaviour of complex natural systems like estuaries or a treatment process like percolating filters can be studied theoretically and alternative management policies can be explored. Other techniques like optimization may then be used to find the best combination of management strategies for combinations of treatment units or for water resource networks. Mathematical modelling is, therefore, an extremely valuable aid in speeding up the theoretical understanding of water pollution and its control.

This book is based upon papers presented at a conference, 'The Use of Mathematical Modelling in Water Pollution Control', held at the University of Newcastle upon Tyne in September, 1973. Since then interest in modelling techniques for pollution control has increased and it was felt that the publication of a revised version of the proceedings would usefully summarize the present state of knowledge.

It is noteworthy that authors of the chapters have a wide spectrum of backgrounds—engineers, mathematicians, chemists, and biologists. The rapid growth of modelling has resulted from an interdisciplinary approach. It is the aim of this book to encourage such cooperation because only in this way can the possibilities and limitations of modelling be properly appreciated.

A. James

'Though reared upon the base of outward things These are chiefly such structures as the mind builds for itself'.

W. Wordsworth

I

Techniques

1

Application of some Systems Engineering Concepts and Tools to Water Pollution Control Systems

JOHN F. ANDREWS

1.1 INTRODUCTION

The word 'systems' is one of the most popular words of our time and has pervaded all fields of science and engineering as well as popular thinking and the mass media. Professions and job titles have appeared in recent years under names such as systems science, systems theory, systems analysis, systems engineering, and others. However, very few people, even among scientists and engineers, can give a concise, accurate definition of these terms. This confusion, which is both natural and not very serious, has been one of many factors contributing to a reluctance on the part of some engineers to adopt and use some of the basic concepts and tools of systems engineering.

The word 'systems' would lead one to think that there should be as many kinds of systems engineers as there are systems and this is indeed the case. Since water pollution control systems have been around for many years, the question might fairly be asked 'Why include the word systems?' The author can answer this question only by stating that he has attempted to adopt some of the key concepts and analytical techniques found in the systems engineering literature and apply these to the analysis, design, and operation of wastewater treatment plants. Included among these concepts and analytical techniques are:

(1) Looking at a system as an integrated whole yet with recognition of the interactions between the elements in a system and between the system and its environment. A good example is the activated sludge process in which the interactions between the aeration basin and the secondary settler (elements) must be clearly defined in order for the process (system) to function adequately. The activated sludge process is also influenced by the temperature of the environment in which it is operated.
(2) Recognition of the universality of characteristics among systems. Even though two systems may appear vastly different, there are some basic common characteristics which can be used to obtain a better understanding of the systems. For example, biological processes used in wastewater

treatment have many common characteristics since the basic principles of microbiology apply to all of these processes.

(3) An increased awareness of the importance of dynamic behaviour, information handling needs, and an orderly examination of alternate ways of accomplishing objectives including establishment of those factors of major importance in comparing the value of different alternatives. An example is the use of control systems, which involve information handling, to improve the dynamic behaviour of wastewater treatment plants. This would be an alternate to increasing the size of the plant as is usually the case in conventional plant design.

(4) A team or interdisciplinary approach to the analysis, design, or operation of a system. From the viewpoint of the author, the most important member of this team is that person who is intimately familiar and experienced with the system to be analysed, designed, or operated. However, the talents of a wide variety of disciplines are needed in the analysis, design, or operation of water pollution control systems and included among these are most branches of engineering and the sciences.

(5) The engineer involved in a systems study has at his disposal a wide variety of relatively new analytical tools. Included among these are:
  (a) Mathematical modelling
  (b) Computer simulation
  (c) Transient response analysis
  (d) Control theory
  (e) Optimization techniques

The author is a relatively new user of systems engineering concepts and techniques and makes no claim to be an expert in any of the techniques illustrated in this chapter. Moreover, his experience with these concepts and techniques has been primarily restricted to the analysis, design, and operation of wastewater treatment plants and most of the examples presented herein are therefore drawn primarily from this field. For a more detailed discussion of systems theory and systems engineering as applied to a variety of fields, the reader is referred to the books of von Bertalanffy (1968) and Chestnut (1965). Motard (1966) has presented an excellent discussion of how systems engineering concepts and techniques are being incorporated into engineering curricula in U.S. Universities.

1.2 SYSTEM DEFINITION AND STRUCTURE

There are many different types of water pollution control systems ranging from large complex regional systems for river basins down to septic tanks for individual dwellings. For analysis, design, or operation of these systems it is first necessary to delineate their boundaries. The more complex systems are then structured by identification of individual elements or subsystems. These subdivisions are made in a fashion such that the important inputs and outputs

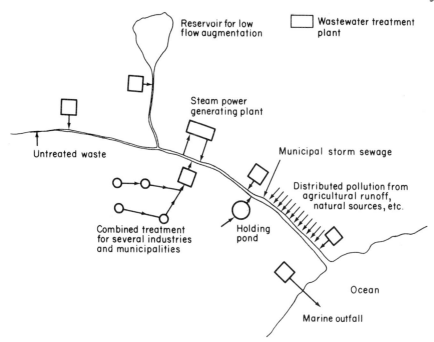

Figure 1.1 River basin pollution control system

as well as the environment in which the subsystem must operate can be reasonably well defined. Each subsystem should also have well-defined objectives which contribute to the overall objective of the system.

The breakdown of a system into its subsystems is illustrated in Figures 1.1, 1.2, and 1.3. One of the subsystems for the river basin shown in Figure 1.1 is the water and wastewater system for a municipality as presented in Figure 1.2. The wastewater treatment plant shown in Figure 1.3, the system with which the author is most familiar, is a subsystem for the municipality or could be a subsystem for the water and wastewater system of an industry. The treatment plant, in its turn, can be subdivided into individual processes such as primary sedimentation, anaerobic digestion, and so on. However, in discussing these individual processes, one must always be aware that they are part of a larger system, the treatment plant, and be conscious of the interactions between the individual processes as well as the contribution of the individual processes to the overall objectives of the plant.

The process of subdividing a system into its components and reconstituting the system from these components may be compared with viewing an American football game through the zoom lens of a camera. This is a sport in which both team effort and a high level of individual performance are essential for success. Focusing on an individual player (component) enables one to throughly analyse the performance of that player; however, the objective of the football team (system) is to score touchdowns or prevent the other team from scoring

6

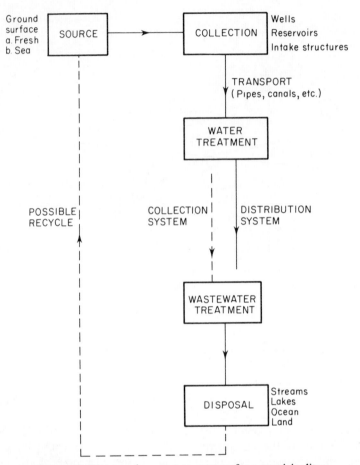

Figure 1.2 Water and wastewater system for a municipality

touchdowns and one occasionally suffers the frustration of the camera being focused on one player while another scores the touchdown. Stated in another and more old fashioned way, 'you can't see the forest for the trees.'

When the camera is focused on the entire playing field, the team objective and some of the interactions between the players become much clearer; however, one loses much of the detail of the game, such as the individual performance of some of the players and close observation of the interactions between individual players. A play may fail or succeed due to the performance of a single player and in order to improve the team (system) performance, a coach needs to know the strengths and weaknesses of each player (component). Another old saying is that 'a chain is no stronger than its weakest link'.

The systems engineering approach has sometimes failed to give reasonable predictions concerning the behaviour of a system because of inadequate attention to description of the individual components. This type of superficial treatment of a system has caused some people to oppose the systems engineering

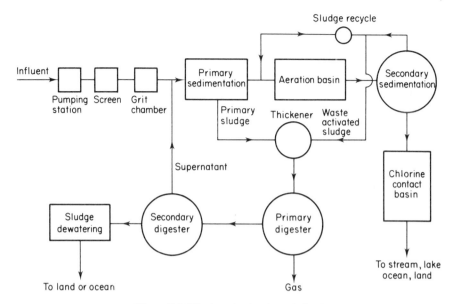

Figure 1.3 Wastewater treatment plant

approach. On the other hand, faulty predictions have also resulted from the work of those who have become so engrossed in detailed analysis of the components that they have not given adequate attention to interactions between components and the function of the system as an integrated whole.

Once the system boundaries have been defined and the system has been structured into its components, the inputs and outputs for the system must be determined as illustrated in Figure 1.4. The inputs and outputs may consist of materials, energy, or information and may be further classified as controllable or uncontrollable. In establishing these inputs, it must be remembered that they may not come from other well-defined systems such as the sewer coming into a plant or the flow from a reservoir but may be such inputs as temperature, wind, political decisions, and so on, from the environment in which the system operates. Some typical inputs which must be considered for the three systems shown in Figures 1.1, 1.2, and 1.3 are:

(1) Transfer of water or wastewater to or from other river basins.
(2) Flow of seawater into or out of an estuary by tidal action.
(3) Oxygen input to a stream by algal photosynthesis.
(4) Thermal energy input to a stream from a power plant.
(5) A political decision to extend a water distribution system to areas outside the municipal limits.
(6) Infiltration of groundwater into a sewerage system.
(7) Seasonal changes in temperature for biological processes.
(8) A signal from a flowmeter in a sewer for actuation of a bypass valve in a wastewater treatment plant.

8

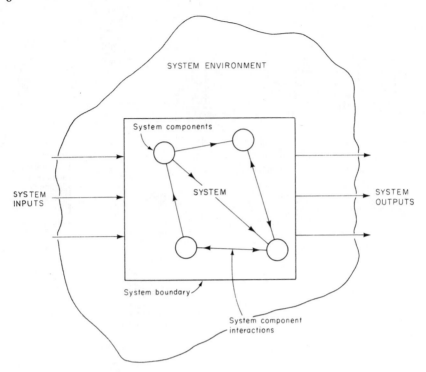

SYSTEM ENVIRONMENT

System components

SYSTEM INPUTS

SYSTEM

SYSTEM OUTPUTS

System boundary

System component interactions

Figure 1.4 General concept of a system

The outputs of the system must also be defined in order to determine if the objectives of the system have been satisfied. They also serve as inputs to other systems. In the case of water pollution control systems, numerical values for these outputs (suspended solids, biochemical oxygen demand, phosphate concentration, and so forth) are frequently specified by governmental agencies and are known as standards. Important outputs which can affect other systems, such as the generation of solid wastes and air pollutants in wastewater treatment plants, must also be considered. The creation of an unsightly condition in a stream (output for the stream) can serve as a strong input to the people using that stream. This input to the people may feedback to the plant via the political route thus resulting in another input to the plant.

Judgement must be used in establishing the inputs and outputs for a system. Time and money in either analysis, design, or operation can be saved by eliminating from consideration those inputs which have little or no effect on the performance of the system. However, lack of consideration of an important input can have serious consequences. Examples might be lack of consideration of the effect of a large flow of an industrial waste on the performance of a wastewater treatment plant or the effects of nitrogenous oxygen demand on the oxygen resources of the receiving body of water.

1.3 MATERIALS, ENERGY, AND INFORMATION

Environmental engineers are familiar with the theory and technology involved in the collection, transportation, processing, and distribution of materials and energy. However, they are not as accustomed to consciously thinking of information in the same terms although this is of equal or greater importance. One reason for this may be that only in recent years has an adequate aid to information processing, the digital computer, become available. Greater familiarity with the quantitative handling of information should result in improved water pollution control systems. Information is needed by management for decision making, by the consulting engineer for plant design, by the operator for evaluation of plant performance, and by automatic control systems for appropriate control action. Examples of important materials, energy, and information for water pollution control systems are:

Materials

(1) Dissolved oxygen in a stream.
(2) Materials for a pipeline.
(3) Specific pollutants discharging to a wastewater treatment plant.
(4) Chemicals used in water or wastewater treatment.

Energy

(1) Energy for transporting water to a treatment plant.
(2) Thermal pollution from power plants.
(3) Utilization of energy for wastewater treatment including energy generation in the plant.
(4) Comparison of alternate processes with respect to energy requirements.

Information

(1) Stream flow records for determining treatment plant location or required efficiency of treatment plants.
(2) Wastewater characteristics for treatment plant design.
(3) Water level in a distribution system storage reservoir for automatic pump activation.
(4) Efficiency of wastewater treatment plants for use in plant operations or by regulatory agencies.
(5) Industrial wastewater quantities and characteristics for development of a regional wastewater management system.

The technology involved in information handling is of more recent vintage than that for materials and energy; however, many of the same concepts are applicable. The handling of materials, energy, and information all involve collection, transportation, processing, storing, and distribution. Flow diagrams are used in the handling of materials, energy, and information, and examples of information flow diagrams are given in Figure 1.5 where the temperature of a process is to be controlled either manually or automatically. The temperature of the process is changed from its desired or reference value by some input disturbance such as a change in environmental temperature or heat input to the process. This change is measured by a sensor such as a thermometer. In a manual control system (Figure 1.5a), the measured temperature is transmitted to the man in the control loop by visual observation. The man processes this information by mentally comparing the observed temperature with the desired temperature, and adjusts the heat input to the process in an attempt to bring the temperature back to its desired value. Several iterations of this procedure

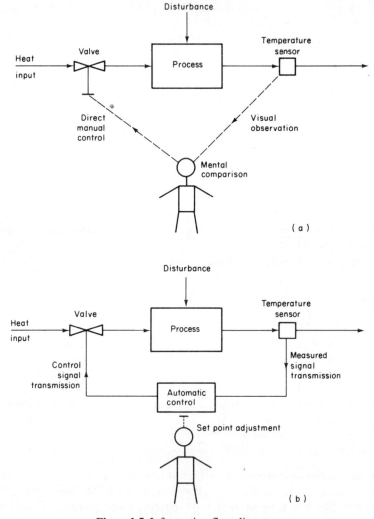

Figure 1.5 Information flow diagrams

may be needed before the desired temperature is attained. The man has 'closed the loop' by 'feedback' of information from the process output to the process input.

In the automatic system, the man is replaced in the feedback loop by a controller. The temperature sensor transmits a signal to the controller. An additional device, a transducer, may be needed between the sensor and controller for amplification or changing the form of the signal so that it can be transmitted to or understood by the controller. The controller compares the signal with a stored reference signal or set point to determine if an error exists. If an error does exist, the controller computes, by means of a control algorithm, the amount of control action needed. It then transmits a signal to a final control element, the valve in this instance, to adjust the heat input to the process. A transducer may also be needed between the controller and the final control element. The automatic system is also iterative since the computed adjustment of the control valve may not give the desired temperature. It should be noted that the man continues to participate in the feedback loop on an intermittent basis since he must select the set point value on the basis of this judgement and experience.

The accuracy and timeliness of information are of key importance and must be selected to accomplish the objectives of the information handling system. Fine control of the process temperature will require frequent and accurate measurement of temperature, rapid transmission of signals, and sophisticated automatic controllers. Manual control using a simple thermometer may be quite adequate for coarse temperature control. Another example of the need to consider the timeliness of information is provided by the biochemical oxygen demand (BOD) test which is the most widely used means of determining treatment plant efficiency. This test requires five days to complete and therefore represents a significant information time delay. From the viewpoint of the regulatory agency, this time delay is not normally serious since the information is transmitted to the agency at monthly intervals and an even longer period of time may be required to process the data and take legal or other action. However, from the viewpoint of the treatment plant operator, this is a serious time delay since corrective action for low process efficiency must be taken well before five days have elapsed. Information obtained from the BOD test may therefore be of considerable value to the regulatory agency but of much less value to the plant operator.

The handling of information costs time and money just as for the handling of materials and energy and it is usually necessary to accept some uncertainty in the information. Materials, energy, and information are interdependent and trade-offs are possible in both design and operation. For example, the collection of a large amount of information concerning wastewater flow rates and characteristics for the design of a small wastewater treatment plant is usually not justified since it would be more economical to oversize the plant thus increasing the expenditures of materials and energy but decreasing the cost of information. However, this would not be the case for large treatment plants and it can therefore be seen that the time and money spent in collecting information for plant design should be a function of size or capital cost, of the plant. The same is true for plant operation and it is frequently possible to decrease both treatment plant size and operational costs by increasing and making better use of information obtained by instrumentation and more frequent laboratory analyses. However, the value of this information should always be balanced against its cost.

1.4 DYNAMIC BEHAVIOUR

Changes are always taking place in the inputs, outputs, or environment of a system as well as in the characteristics of the system itself. It is important to identify the nature of these changes and the rates at which they occur. Some may be so slow that they need not be considered for a particular system while others may be so rapid and of such short duration that they also have no appreciable effect on the system. However, there are many transients which do affect the behaviour of a system and these should be considered in analysis, design, and operation. Transients are of special importance in the development of management and control systems since the very need for these systems is caused by dynamic or transient behaviour. Some examples of important transients in water pollution control systems are:

(1) Fluctuations in stream flow rate.
(2) Changes in land use patterns in a river basin or municipality.
(3) Water demand changes in a water distribution system.
(4) Start-up of an anaerobic digester.
(5) Changes in regulatory agency standards.
(6) Fluctuations in influent wastewater flow rate and composition for a wastewater treatment plant.
(7) Temperature variations in water and wastewater.

Little attention has been paid to the dynamic behaviour of wastewater treatment plants even though there are wide fluctuations in the plant inputs and environment. It is not surprising therefore, that there are also wide fluctuations in plant efficiency as well as occasional gross failures. There are many reasons for this but among the major reasons have been inadequate consideration of dynamic behaviour during both plant design and operation as well as a lack of attention to the interactions between design and operation. Almost all design formulae are based on the assumption of steady state and use average, or at best maxima and minima, values for the inputs. Under such conditions, it is relatively easy for a design engineer to slip into the habit of 'steady state thinking'.

1.5 ALTERNATE MEANS OF ACCOMPLISHING OBJECTIVES

Even under the best of conditions, there are still many unknowns which cannot be accurately evaluated for water pollution control systems. Different designers or operators will therefore arrive at different solutions for the accomplishment of objectives because of the uncertainty in information available to them and their differences in background, experience, or philosophy. Some of the alternate solutions may be readily shown to not be feasible; however, the other alternates should be examined in as quantitative a fashion as possible and that solution accepted which most adequately accomplishes the objectives. Some of the tools which may be used for this examination are opinion polls, economic analysis, modelling and simulation, observation of existing similar systems, and so forth. Some possible alternatives which have been considered for water pollution control systems are:

(1) Collection and transport of wastewaters generated in a region to a central treatment plant with disposal at one point in the stream or the use of a number of smaller plants with disposal at several points in the stream.
(2) Use of low flow augmentation, detention of wastewaters in holding ponds, or increase of treatment efficiency during periods of low stream flow.
(3) Types and arrangements of processes used in wastewater treatment plants.
(4) Treatment of industrial wastes separately or mixed with municipal wastes.

1.6 FACTORS FOR JUDGING THE VALUE OF A SYSTEM

In comparing alternate systems for the accomplishment of objectives, it is necessary to establish those factors which will be of major importance in judging the value of a system. These factors should be expressed in as quantitative a fashion as possible and must reflect the desires of the users of the system. What is the value of clean water to the sportsman? Industry? The public as a whole? A list of factors commonly used in judging the value of a system is given below. This list is merely representative and is not meant to be all inclusive. Furthermore, these factors and their relative importance will probably change with time, perhaps during the life of the system.

(1) *Performance.* This is one of the most common bases for system evaluation. Prediction of the performance of the system may be made from physical or mathematical models, experience, or observation of similar systems. Examples of performance requirements are those of regulatory agencies which may, for example, specify that the performance of a river basin pollution control system must be adequate to maintain the dissolved oxygen concentration in a stream above 4·0mg/l or that the BOD removal in a wastewater treatment plant be equal to or greater than 85%. One of the disadvantages of fixed standards such as these is that they do not consider the inherent variability in the system or its environment. A standard which does consider this variability is that for coliform organisms in water supplies which permits a variation in the observed results in accordance with a statistical model of the test procedure.
(2) *Cost.* This is always an important consideration and unfortunately in the

past has been weighed more heavily than performance in evaluating water pollution control systems. Accurate evaluation of costs requires the development of an economic model which may include such factors as the design life of the system, capital and operating costs, and taxation policies. Since there is usually no profit motive involved in water pollution control systems, profitability studies are usually not required. However, an attempt to quantitatively evaluate the benefits of having clean water must frequently be made and this is a difficult task.

(3) *Reliability.* Reliability is the probability that the system or its subsystems will perform in the manner intended. There are strong interactions between reliability, performance, and cost. The desired performance will not be obtained if certain parts of the system fail. Redundant units can be incorporated in the system to avoid failure; however, this increases the cost. Inadequate attention has been paid to reliability in the design of wastewater treatment plants as evidenced by the large number of occurrences of plant bypassing due to process or equipment failure. In river basin systems, the effects on the stream of failure to maintain the desired characteristics either continuously or intermittently must be considered. A higher degree of reliability should be required for a treatment plant discharging into a stream with no excess assimilative capacity than for one discharging into a stream with considerable excess capacity. This additional reliability can be obtained through either design or improved operation. One system in which great attention is paid to reliability is the chlorination system for public water supplies where failure can have disastrous consequences.

(4) *Time.* The word 'time' can have different meanings and one of these, consideration of the dynamic or time dependent behaviour of the system, has already been discussed. Another important but ill-defined aspect of time, is establishment of the design period or expected useful life of the system. Considerable uncertainty is involved in establishing design periods since this may involve such items as future projections of river basin development or estimates of future municipal populations. A distinction must also be made between the time required to make the system and the time required for the system itself to accomplish its objectives. One aspect of system operation time is the time required to process wastewater per unit of plant capacity. Decreased processing times can result in lower costs as well as lower space requirements which can be of considerable importance in larger cities where space is at a premium.

(5) *Maintainability.* Quick and simple maintenance is usually of considerable value to the user of a system. As would be expected there is a strong interaction between maintainability and reliability since more reliable systems may require less maintenance and the reliability of systems can be improved by better maintenance. Maintainability also illustrates the importance of devoting adequate attention in the design phase to the

manner in which the system will be operated. Instrument packages located in a remote area for monitoring the condition of a stream must be highly reliable and easily maintained. Unskilled personnel are used for the operation of many water and wastewater treatment plants and simple maintenance is therefore needed. Installation of complex control systems in situations where adequate maintenance is not available can be a waste of money and lead to poor performance.

(6) *Flexibility.* Flexibility is needed so that the system can be adjusted to accommodate uncertainties or changes with respect to time. Water and wastewater treatment plants must be designed with sufficient flexibility to accommodate future expansions. Management systems must have enough flexibility to accommodate to policy changes such as stricter enforcement of stream standards. Water and wastewater treatment plants should be flexible enough to accommodate plant modifications as technological advances occur.

1.7 SYSTEM OPTIMIZATION

Once the factors for judging the value of a system have been selected, it is necessary to combine them in a weighted fashion to optimize the system. This is illustrated in Figure 1.6 where the outputs of the system are the judgment factors. Each factor is assigned a relative weight and they are then combined to give a single function known as the objective function which is to be optimized. Inherent in the optimization process is the concept of constraints or limits for the judgment factors. For example, it may be desired to operate a wastewater treatment plant at minimum cost subject to the constraint that the BOD removal efficiency be equal to or greater than 85%. There are many types of

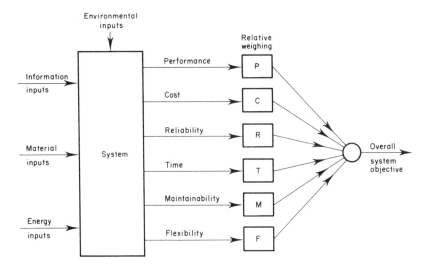

Figure 1.6 Combination of judgment factors for system optimization

constraints and some, such as sociological and political constraints, may be difficult to express in a quantitative fashion. Since there may be disagreements as to what factors and constraints should be used as well as uncertainty in the numerical values of the constraints and relative weights for each factor, it is not surprising that there will also be disagreement as to what comprises an optimum system. Some optimization problems of importance for water pollution control systems are:

(1) Allocation of costs to users of water resources in a river basin.
(2) Operational strategy for a water distribution system.
(3) Efficiency of treatment to be provided at different points along the length of a river.
(4) Operational strategy for a wastewater treatment plant.
(5) Types, sizes, and arrangement of processes in a wastewater treatment plant.
(6) Organizational plan for a river basin authority.

It will be noted from the types of problems mentioned, that the results of system optimization studies can be used for the operation of existing systems or in comparing alternate designs for new systems. In either case, considerable human judgment is required for systems as complex as most water pollution control systems.

Various analytical techniques such as trade-offs, the differential calculus, statistical methods, gradient techniques, and linear and dynamic programming are available to assist in solving optimization problems and will not be further discussed herein. The details of these techniques may be found in books by Zahradnik (1971), Smith, and others (1970), and Hillier and Lieberman (1967), among others.

1.8 SOME SYSTEMS ENGINEERING TOOLS

A wide variety of relatively new analytical tools are available to the engineer concerned with the analysis, design, or operation of water pollution control systems. Included among these are network analysis, transient response analysis, stochastic processes, stability analysis, graph theory, game theory, mathematical modelling, computer simulation, linear programming, control theory, and so forth. The author makes no claim to be an expert in the use of any of these techniques and further discussion will be restricted to those tools with which the author is most familiar, these being: (1) mathematical modelling, (2) computer simulation, (3) transient response analysis, and (4) control theory. The author is planning to use some of the optimization techniques in studies on optimal control strategies for wastewater treatment plants. However, this work will be preceded by the field verification of the dynamic models and control actions to be used in the optimization studies.

Modelling and simulation

Modelling and simulation are techniques frequently used (and sometimes abused!) in today's scientific and engineering investigations. Neither modelling nor simulation is new since scaled-down physical models have long been used for simulation in such diverse areas as astronomy (planetariums), hydraulic engineering (river models), architecture (building models), and chemical engineering (pilot plants). Biologists, chemists, and sociologists have studied model organisms, model compounds, and model communities, respectively. Even the hypothesis which is formulated in applying the scientific method can be considered as a verbal model of a system, and simulation is frequently used to test the validity of the hypothesis when it is not feasible to test the real system.

A model may be thought of as being a representation of a system in a form suitable for demonstrating the way the system behaves, while simulation involves subjecting models to various changes in such a way as to explore the possible effects of these changes on the real system. When properly applied, modelling and simulation can result in considerable savings in both time and money. For example, it is usually much less expensive to construct a pilot plant (physical model) and conduct experiments (physical simulation) on the treatment of a specific industrial wastewater than it is to build and experiment with a full scale plant. Modelling and simulation may also be used to explore the effects of changes on a system when it is impractical to experiment with the real system. The use of a scaled-down physical model of a river for studying the effects of flooding or mathematical modelling and computer simulation to determine the appropriate location of a large wastewater treatment plant along the length of a stream are examples of this. Another advantage of modelling and simulation is that they usually involve considerable analysis of the system and can therefore result in an increase in fundamental knowledge about the system.

There are many types of models and the type selected depends primarily upon the purpose for which it is to be used. For example, the schematic diagrams given in Figure 1.1, 1.2, and 1.3 for a river basin, water and wastewater system, and wastewater treatment plant, respectively, are adequate as pictorial models to illustrate the components and describe some of the interactions for their respective systems. However, they would not be adequate for a quantitative description of system behaviour. Another common type of model is a procedural model which describes in a sequential fashion the steps involved in a procedure such as the planning of a river basin project, construction of a wastewater treatment plant, or the start-up of a particular process such as the anaerobic digester. A procedural model may be nothing more than a written and ordered list of tasks to be accomplished for a project or may be in the form of a flow chart involving logical operations as commonly prepared in the development of computer programs. The addition of time to a procedural model can result in a schedule for completing a project or the more sophisticated flow charting method known as PERT (Program Evaluation Review Technique) which can be used to follow the progress of the project as well as to determine the critical path or longest time required to complete the project. Still other models of interest are reliability models which use probability theory to evaluate the reliability of a system, and economic models which focus on the cost aspects of a system.

Mathematical models are commonly used for a more quantitative description of system performance and consist of one or more equations relating the important inputs, outputs, and characteristics of the system. To those not familiar with systems engineering terminology, the term 'mathematical model' is sometimes frightening since it may bring to mind large sets of complex equations. However, this need not be the case since most of the common engineering design formulae may also be called mathematical models. In fact, as simple an expression as $y = mx + b$ can be considered as a mathematical model where the system output (y) is related to the system input (x) by the system parameters (m, b). However, for more complex systems or more adequate description of system performance it may be necessary to use larger numbers of equations and more powerful mathematical tools such as differential and partial differential equations, difference equations, probability theory, and others.

Mathematical models may be classified in many different ways and one of the most important for wastewater treatment processes is the distinction between dynamic and steady state models. The basic principles involved can be illustrated by the development of a model for the continuous addition of an inert substance or tracer to a continuous-flow stirred tank reactor (CFSTR) as illustrated in Figure 1.7. The model is developed by making a material balance on the inert tracer. The general form for a material balance is presented in equation 1.1 and the material balance on the tracer is given in equation 1.2. Since the reactor is stirred and the contents are homogeneous, the concentration of tracer in the reactor and in the reactor effluent are identical. The reaction term is zero since the tracer is inert and does not participate in any reactions. If the inert tracer had been added as a pulse, instead of continuously, term one would also become zero immediately after addition of the tracer.

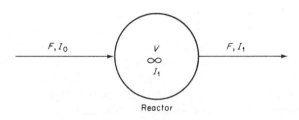

$$V = \text{Reactor volume}$$
$$F = \text{Flow rate}$$
$$I_0 = \text{Influent tracer concentration}$$
$$I_1 = \text{Effluent tracer concentration}$$

Figure 1.7 CFSTR with inert tracer input

$$\begin{array}{ccccccc}
\text{Rate of} & & \text{Rate of} & & \text{Rate of} & & \text{Rate of} \\
\text{Material} & + & \text{Appearance or} & = & \text{Material} & + & \text{Accumulation} \\
\text{Flow into} & & \text{Disappearance} & & \text{Flow Out} & & \text{of Material} \\
\text{Reactor} & & \text{of Material due} & & \text{of Reactor} & & \text{in Reactor} \\
& & \text{to Reaction} & & & &
\end{array} \quad (1.1)$$

$$\begin{array}{ccccccc}
(1) & & (2) & & (3) & & (4)
\end{array}$$

$$F I_0 \quad + \quad 0 \quad = \quad F I_1 \quad + \quad V \frac{\mathrm{d} I_1}{\mathrm{d} t} \qquad (1 \cdot 2)$$

Where:

F = flow rate
V = reactor volume
I = tracer concentration
0 = subscript denoting reactor influent
1 = subscript denoting reactor effluent

Most models currently in use are based on the assumption of steady state which is attained by setting term four, the derivative term in the material balance, equal to zero. This has the effect of reducing the differential equation to an algebraic equation thus simplifying the model. For the example presented, such a simplification is unnecessary since an analytical solution is easily obtained. However, prior to the advent of computer simulation, such simplifications were frequently the only means of obtaining solutions for even slightly more complex models. Consequently, most models currently used for describing the performance of water pollution control systems are steady state models. Steady state models have proven their value on a qualitative basis by indicating needed changes in process design. However, in most instances they are not adequate to describe process operation since the inputs are far from constant and there is considerable variation in effluent quality with respect to time. Dynamic models are needed for the description of process operation and would also be of value in design by permitting a comparison of different processes with respect to stability.

Another important classification of mathematical models is as deterministic or stochastic. Deterministic models are those in which the inputs, outputs, and system parameters can be assigned a definite fixed number, or series of fixed numbers, for any given set of conditions. In contrast, the principle of uncertainty is introduced in stochastic or probabilistic models and statistical techniques must be used to express the model in a mathematical form. An example would be a model to express the probability of low flows occurring in a stream as a function of time between low flow periods. This type of model is frequently used to determine the stream flow when evaluating the effect of a wastewater treatment plant discharge on a stream. It should also be noted that models may have both deterministic and stochastic features. For example, the wastewater flow into a treatment plant may have a deterministic component expressed as a daily cycle but superimposed on this may be random or stochastic changes in the flow rate.

Mathematical models may be quite complex and consist of large numbers of equations using many mathematical tools. However, Occam's razor applies here as well as to verbal hypotheses in that the simplest possible expressions should be used. The purpose for which the model is to be used must be defined so that one can be developed which is adequate for the intended use. Mathematical elegance should be secondary and it should be remembered that a model which is too complex may be subject to either misuse or disuse. In some instances a coarse or approximate model, using rough estimates for numerical values and relating only the most important inputs, outputs, and system parameters, may be adequate. In other instances a fine or highly quantitative model with highly accurate numerical values and including many more factors may be necessary. An analogue may be drawn here between qualitative and quantitative analytical chemistry.

In developing models, it must be realized that they are evolutionary in nature and will change as more knowledge is gained about the system. The degree of accuracy required for representation of the system by the model may also change with time. A model which is quite adequate as a first approximation may be replaced at a later date by a more exact model with better estimates of the coefficients, fewer empirical relationships, and inclusion of more variables. This evolutionary nature of models is not always recognized and can lead to reluctance on the part of an investigator to either modify or discard a model in the same fashion that investigators in past years have sometimes been reluctant to modify or discard verbal hypotheses.

Whenever possible, model development should be based on scientific knowledge about the fundamental biological, chemical, and physical phenomena which govern the system. Some of the more important phenomena for wastewater treatment systems are stoichiometry, equilibrium relationships, reaction kinetics, gas laws, diffusion theory, and conservation equations (mass, momentum, and energy balances). A knowledge of information theory is useful in the development of models for control systems. Models based on fundamental principles give more insight into system behaviour and may be more reliably extrapolated to different designs or operational strategies. In addition, the use of fundamental principles enables one to search for similar characteristics of other systems and thus draw on existing knowledge in other branches of engineering or science.

When systems are complex and poorly understood, as are most wastewater treatment systems, empirical relationships may be necessary for all or parts of the model. Models based on such relationships have been used throughout the history of engineering and may be perfectly satisfactory for a given purpose as long as their limitations are recognized. One of the principal limitations of empirical models is that they cannot be extended with any degree of confidence to similar systems or beyond the range over which data has been collected. The general tendency is for models to become more fundamental and less empirical as more knowledge is gained about the system. However, the obtaining of fundamental knowledge can be expensive as well as time consuming and from a practical viewpoint, it is expected that a significant portion of the relationships in water pollution control systems will be empirical for some time.

After a mathematical model has been developed for a system, the equations which comprise the model must be solved in order to predict the behaviour of the process. This procedure is known as simulation and can be defined as the use of a model to explore the effects of changing conditions on the real system. Obviously, the model must be a reasonable representation of the real system in order for the results to be meaningful since the simulation results can be no better than the mathematical model and data on which they are based.

Although simulation is not new since physical models have long been used for the simulation of systems, computer simulation is relatively new and a brief discussion of this is in order. Prior to the advent of computers, a computational or simulation bottleneck existed and efforts at mathematical modelling were frequently of no practical value since the equations could not be solved. However, computers have largely eliminated this bottleneck with analog, digital, or hybrid computers being used for the simulations. In some instances, the advantage of the computer is in high speed numerical analysis of equations for which analytical solutions are not available. In other instances, the advantage of the computer is in its ability to handle large numbers of equations thus permitting the simulation of larger and more complex systems.

The early use of computers was largely restricted to specialists since a considerable amount of time was required to learn to program the computer. However, the advent of general purpose computer languages such as FORTRAN and ALGOL with the availability of special purpose subroutines which can be easily called in, has greatly reduced both the time needed to learn to use the computer and to write programs to solve specific problems. However, an even more exciting development has been the recent availability of continuous system modelling languages (Franks, 1967; IBM, 1968) for use on the digital computer. These languages are heavily user oriented, thus permitting the engineer or scientist to concentrate on model development and simulation results rather than on the details of the computations. They are especially delightful for those not experienced in using the computer since they can be learned in a relatively short period of time. For example, students with no prior computer experience are usually able to write their own programs for significant simulations after less than four hours of instruction. The practising engineer can obtain access to these simulation languages through computer time-sharing services using a remote terminal in his own office.

One of the most commonly used simulation languages is CSMP/360 (IBM, 1968). This program or language may be thought of as being one level above such languages as FORTRAN since CSMP statements resemble ordinary mathematics much more closely and are automatically translated into FORTRAN by the computer. Unlike most other computer languages, CSMP statements, with few exceptions, can be written in any order and are automatically sorted by the computer to establish the correct order of information flow. CSMP provides a number of standard functions such as integrators, comparators, limiters, etc., which are used to represent the mathematical model. These standard functions are augmented by the usual FORTRAN functions such as square root, sine, and logarithms. The basic arithmetic operator symbols such as (*) for multiplication and (/) for division are the same as those used in FORTRAN. An easily used and fixed format is provided for tabular and graphic output of selected variables at selected increments of time.

The writing of a CSMP program is best illustrated by example and a program for the dynamic model of a CFSTR with continuous addition of an inert tracer (equation 1.2) is given below. The asterisk in column one of the punched card denotes a comments card which can be used to insert

```
*           CFSTR WITH INTER TRACER INPUT

*           V = REACTOR VOLUME (L), F = FLOW RATE (L/HR)
*           I0,I1 = INFLUENT,EFFLUENT TRACER CONCENTRA-
            TION(MG/L)

PARAM     V = 10.0,  F = 2.0,  I0 = 100.0
INCON     I1IT = 0.0

DYNAM
          I1DOT = F*I0/V-F*I1/V
          I1 = INTGRL(I1IT,I1 DOT)

METHOD    RKSFX
TIMER     DELT = 0.01,  FINTIM = 20.0,  PRDEL = 0.5,  OUTDEL = 0.5
PRINT     I1
PRITPLT   I1
TITLE     CFSTR WITH INERT TRACER INPUT
LABEL     CFSTR WITH INERT TRACER INPUT
END
STOP
ENDJOB
```

appropriate comments or explanations in the program. The PARAM and INCON cards are used to insert numerical values, or data, for parameters, constants, and initial conditions. The DYNAM section contains the mathematical model and the CSMP functions needed to solve the model. In this instance, INTGRL is the only CSMP function used with I1DOT representing the differential equation to be integrated and I1IT representing the initial condition for this equation. The other cards are control statements and include METHOD which specifies the numerical integration technique to be used (six are available), the integration interval or step size (DELT), the length of time over which the variables are to be examined (FINTIM), the time intervals at which tabular or graphical outputs are to be given (PRDEL, OUTDEL), and the variables which are to be printed or plotted (PRINT, PRPLT). An example of the tabular output is presented below. Standard E notation is used to indicate the power of 10 to which the numbers in the columns must be raised.

CFSTR WITH INERT TRACER INPUT

TIME	I1
0.0	0.0
5.0000E−01	9.5120E 00
1.0000E 00	1.8211E 01
1.5000E 00	2.5923E 01

The example presented is a very simple example and makes use of only a small fraction of the power of CSMP. A total of 34 functions are available and only one of these (INTGRL) has been used. Regular FORTRAN statements can be used in the program and a TERMIN section can be added after the DYNAM section to control subsequent runs as might be needed in optimization problems. The language can handle sizeable numbers of equations since 300 INTGRL blocks are provided.

Computer simulation has many of the same advantages—and disadvantages—as physical simulation. A great deal of knowledge can be gained about a system through the development of a mathematical model and the subsequent computer simulations using this model. Sensitivity analysis, or the response of the model to changes in specific inputs or system parameters, can be used for model improvement by indicating those variables which are most significant. Sensitivity analysis can also be used for model simplification by indicating those variables which have little effect on the outputs. Considerable monetary savings are frequently realized by using simulation since experimentation on the computer is usually less expensive than construction of a full scale plant or physical model with subsequent experimentation. Simulation permits examination of very large systems such as river basins where physical experimentation may not be possible. Time can be compressed on the computer with simulations being conducted in minutes. This is especially important for systems such as wastewater treatment plants where rates are relatively slow and physical experimentation may require days, weeks, or even months.

There are, of course, disadvantages to computer simulation. The results of the simulations are no better than the mathematical model and data on which they are based, and many first simulations may give inaccurate or even completely misleading results. Simulation using physical models does not have this disadvantage; however, it does present the problem of scaling up from the model to the real system. This disadvantage of computer simulation can be overcome by iterating between model development, computer simulation, experimentation with physical models, and field observations, since these complement one another. Knowledge gained in simulation is useful for modifying the mathematical model, guiding physical experimentation, and establishing the type and frequency of field observations needed. This iterative technique also points out another important aspect of modelling and simulation, this being the need for model verification. The case and speed with which computer simulations can frequently be made may lead to a neglect of this very important portion of model development and in the extreme, can result in one becoming so enamoured with the techniques that the purpose for using them is almost forgotten. This can lead to the generation of large quantities of worthless results if the model is not a reasonable representation of the real system. In addition to physical experimentation and field observations, other techniques which may be used for validation of models are common sense, information from the literature, and comparison of simulation results using other available mathematical models.

For a more detailed discussion of modelling and simulation the reader is referred to the books of Chestnut (1965), Denbigh and Turner (1971), Himmelblau and Bischoff (1968), Himmelblau (1970), Franks (1967) and McLeod (1968). Examples of some articles with specific reference to the modelling and simulation of wastewater treatment systems are those by Andrews (1969, 1971a), Andrews and Graef (1971), Knowles (1970), Smith (1969), Lawrence and McCarty (1970), Bryant and colleagues (1971) and Andrews and Lee (1973).

Transient response analysis

After a dynamic mathematical model has been developed for a system and a simulation technique selected, the time varying behaviour of the outputs can be predicted from the inputs. This procedure is called transient response analysis. The techniques used in transient response analysis are also of value in developing models. This is accomplished by subjecting the system to standardized inputs or signals, known as forcing functions, and comparing the output response with those for standardized models. Examples commonly encountered in water pollution control practice are the injection of dyes into a sedimentation basin or stream to evaluate the hydraulic regime of the basin or stream. The forcing functions used may be periodic or non-periodic and deterministic or stochastic.

Transient response analysis can be illustrated using the dynamic model for a CFSTR with inert tracer input and the CSMP program for this model. The model (equation 1.2) is usually recast into the form shown in equation 1.3 where V/F is called the time constant (τ) of the system

$$\tau\frac{dI_1}{dt} + I_1 = I_0 \tag{1.3}$$

and is a measure of the time required for the process output to respond to a disturbance in the input. I_0 is the forcing function and I_1 is the response. Four non-periodic forcing functions commonly used are the step, impulse, pulse, and ramp functions. These functions, and the response of the model to these functions, are presented in Figure 1.8. The CSMP program gives the response of the reactor at time ≥ 0, to a step increase of I_0 from 0·0 to 100·0 mg/litre.

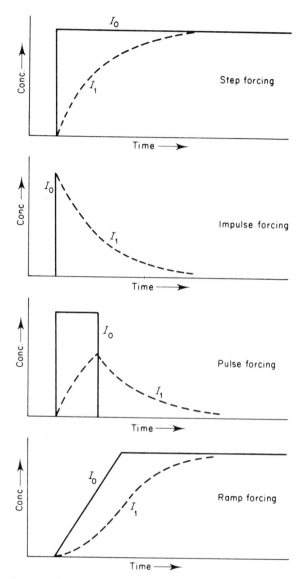

Figure 1.8 Response of CFSTR to non-periodic inputs of
an inert tracer

This was accomplished by setting I1IT on the INCON card equal to 0·0 and
I0 on the PARAM card equal to 100·0. The response to an impulse is obtained
by the reverse, i.e. setting I1IT equal to 100·0 and I0 equal to 0·0. Responses to
pulses and ramps can be easily obtained by using other CSMP functions.

The time constant (τ) indicates how rapidly a system will respond to a change
in the input and is therefore an important measure of the dynamic characteristic
of a system. For the example presented, it is numerically equal to the time

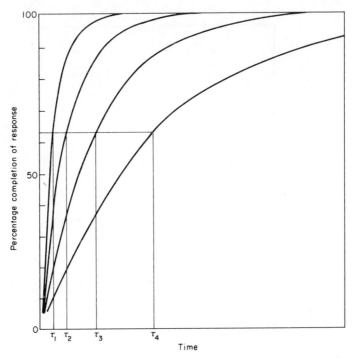

Figure 1.9 Effect of time constant on the response of a CFSTR to the step input of an inert tracer

required for the output concentration (I_1) to equal 63.2% of the input or steady state concentration (I_0) for a step forcing of the reactor. The effect of different time constants on the rate of response is illustrated in Figure 1.9. The time constant can be varied by changing either F or V. In interpreting the effect of the time constant, it should be remembered that this illustration is for the addition of an inert tracer to a CFSTR. The addition of a reacting substance to the reactor would necessitate the inclusion of a reaction term in the model and could substantially change the value of the time constant as demonstrated by Perlmutter (1965).

Two common periodic inputs are the sinusoid and the pulse train. The response to a pulse train would be the same as illustrated in Figure 1.8, except repeated and perhaps without time for the process to come to steady state between pulses. The periodic feeding of an anaerobic digester or the dosing of a trickling filter can be represented by pulse trains. Analysis of the response of a system to sinusoidal forcing is known as frequency response analysis and sinusoids are frequently used as first approximations to represent the daily cycle of influent flow rate variations for a wastewater treatment plant. For more quantitative evaluations, inputs corresponding to those actually observed in the field can be obtained by using the function generator elements in CSMP. The effect of random variations can be superimposed on these deterministic inputs by using the noise or random number generators in CSMP.

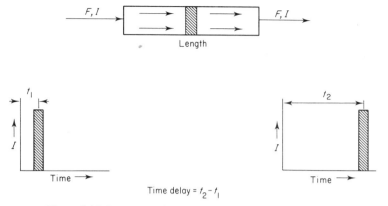

Figure 1.10 Response of a plug flow reactor to a pulse input

The transient responses which have been presented for the standard forcing functions are for a linear, first order system (CFSTR with inert tracer input) and it would be expected that the responses would be different for other systems. A plug flow reactor is an example of a zero order system and the response of such a reactor to a pulse input of an inert tracer is shown in Figure 1.10. No mixing is assumed to occur in a plug flow reactor and the pulse input appears unchanged in shape in the reactor effluent but delayed by a time equal to the time constant or residence time (V/F) of the reactor. This time delay is called dead time or pure time delay and is frequently encountered in wastewater treatment plants and their associated control systems. It would be of considerable importance when attempts are made to calculate process efficiency based on grab samples of the influent and effluent taken at the same time. Second order systems (systems represented by second order differential equations) are also encountered. Two coefficients, a time constant (τ) and a damping coefficient (ψ) are required to represent a second order system. Such systems can exhibit oscillatory behaviour which is greatly influenced by the numerical value of the damping coefficient.

For a more detailed discussion of the theory of dynamic behaviour, the reader is referred to the books of Cooper and McGillem (1967), Shilling (1963), and Perlmutter (1965). Thoman's (1972) book illustrates the application of dynamic analysis to the behaviour of natural systems such as rivers and estuaries. Bryant (1969) has applied frequency response analysis to the dynamic hydraulic behaviour of wastewater treatment plants. Andrews and Graef (1971) have used CSMP to study the dynamic behaviour of the anaerobic digester. Grieves (1972) has simulated the dynamic behaviour of the rotating disc biological filter on the analog computer. Busby (1973) has used the hybrid computer for dynamic analysis of the several versions of the activated sludge process.

Control strategies

The development of a control strategy is intimately related to the dynamic behaviour of a system since the need for control, whether it be manual or

28

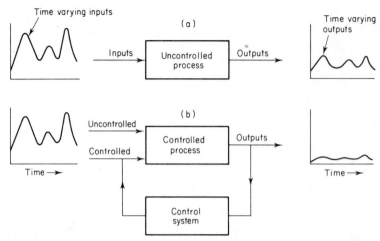

Figure 1.11 Time varying inputs and outputs for controlled and uncontrolled processes

automatic, is brought about by dynamic or transient behaviour. As shown in Figure 1.11, control systems frequently represent an economical means for improving the performance of a treatment plant and should receive more consideration for both existing plants and in the design of new plants. When the characteristics of the plant to be controlled are fixed, as for existing plants, the design of the control system must be established in much the same iterative fashion as for plant design. However, in the design of a new plant, it is possible to strike a better balance between the effort and expense devoted to the plant and that devoted to the control system. The major item of significance is that both the control system and the controlled plant affect the plant outputs and the two should therefore be designed as an integrated system. This additional freedom to modify both the plant design and the control system design to enhance overall performance is one of the key differences between the systems engineering approach and the traditional control engineering approach.

Figure 1.11 is somewhat misleading in that it illustrates only one of the many potential benefits from the incorporation of modern control systems into treatment plant design. Other benefits are possible and included among these are the possibility of decreasing plant size by better control, improved reliability and process stability, minimization of the need for operating personnel, decreased operational costs, and faster plant start-up. For a more detailed discussion of the potential behefits to be obtained from modern control systems the articles by Andrews (1972, 1973) should be consulted. The economic feasibility of installing such systems in wastewater treatment plants has been addressed by Andrews (1971*b*) and Smith (1971).

Regardless of whether control is to be manual, automatic, or a mixture of the two, some of the same basic questions must be answered in development of a control strategy. Included among these are:

(1) What measurements should be made for initiation of the control strategy?

Measurements might be made on the system influent or effluent, the system environment, or could be internal to the system. Associated items of importance are the required accuracy and frequency, time required for making the measurements, and the availability of instruments. The dynamic behaviour of the process and its associated control system must be considered in establishing information needs.

(2) What control actions should be taken? This is of particular importance in wastewater treatment plants since many plants have been designed without adequate consideration of dynamic behaviour or operational characteristics. In most plants, possible control actions are therefore very limited.

(3) How should the information be transmitted from the sensor to the controller and from the controller to the final control element? This is an elementary question; however, in many wastewater treatment plants, information which has been collected is not used for initiation of control but is simply filed for the record. The control loop is therefore broken.

(4) How should the information be processed in order to determine the type and amount of control action needed? A variety of control algorithms involving such calculations as conversion of units, comparison with desired values, averaging, multiplication, integration, statistical analysis, and so on, may be required.

Dynamic modelling, computer simulation, and transient response analysis can be of considerable value in answering questions one and two. However, of equal or greater importance is an intimate knowledge of the system to be controlled. In wastewater treatment the answers to these questions are highly dependent upon the wastewater to be processed, process characteristics, and environment in which the plant will be operated. Much additional research will be needed to quantify the answers to questions one and two.

A sizeable body of literature relative to questions three and four is available in the field of control engineering. Some of the more pertinent material will be briefly discussed herein; for more details, the reader should consult the books of Perlmutter (1965), Shilling (1963), Hougen (1972), Tucker and Wills (1962), and Lloyd and Anderson (1971).

Once the measurement and control action have been established for a control loop, it is customary to prepare a block diagram showing the information flow and a diagram for a simplified feedback control loop is given in Figure 1.12. The output signal from the block representing the system goes to a comparator where it is subtracted from the set point or desired value of the signal. The resultant signal from the comparator is called the error signal and goes to the controller where various mathematical operations called control algorithms or control modes are performed on the error signal to calculate the required value of the signal to be transmitted to the final control element. The final control element then modifies one of the input variables to the process so as to bring the process output signal closer to the desired value.

In simulating the process and its associated control system, it is necessary to

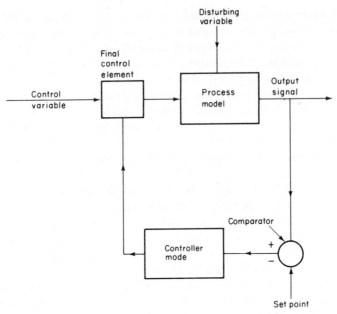

Figure 1.12 Block diagram of a simplified feedback control loop

consider the dynamic characteristics of all of the components of the total system. The blocks given in Figure 1.12 can be considered to represent dynamic mathematical models or transfer functions of the process and controller. It may also be necessary to include blocks representing the dynamic characteristics of such elements as the sensor, transducer, transmission lines, and final control element. Examples are the time delay involved in pumping a sample for analysis to a residual chlorine analyser and the relationship between valve stem position and flow rate through a control valve.

The simplest form of a control algorithm is two-position control which means that the final control element is either in the completely open or completely closed position. There are several variations of this with on–off control being the most common. In on–off control, as soon as the measured value differs from the desired value, the final control element is either completely closed or completely opened depending upon whether the error signal is positive or negative. This type of control is frequently used in air conditioning and refrigeration systems. On–off control may also be coupled with a time cycle controller and this is frequently used to control the density of sludge pumped from a primary sedimentation basin. For example, the timer may turn the sludge pump on once each hour and the pump will run until it receives a signal from a sludge density meter indicating that the sludge density has fallen below some desired or set point level.

A very commonly used control algorithm is that provided by the three-mode controller equation which is also known as PID control. This equation, given below, allows calculation of the amount of control action as a function of four terms. The first term is called the bias

$$C_A = K_B + K_p e + K_I \int_0^t e \, dt + K_D \frac{de}{dt} \qquad (1.4)$$

$$(1) \quad (2) \qquad (3) \qquad (4)$$

where:

C_A = amount of control action
K_B = bias control coefficient
K_p = proportional control coefficient
K_I = integral control coefficient
K_D = derivative control coefficient

and is provided to allow the amount of control action to be set at approximately 50% of the full value whenever the measured value exceeds the set point. The second term provides proportional control in which the amount of control action is proportional to the error signal. Proportional control is sometimes called 'throttling control' and indicates any type of control system in which the final control element, i.e. a valve, is purposefully maintained in some intermediate position between fully open and fully closed. The third term gives control action proportional to the integral of the error signal and therefore reflects the length of time that the error has persisted. Integral control is also called reset control since proportional control alone can result in offset, i.e. a difference at steady state between the values of the measured value and the set point. The addition of integral control to proportional control eliminates offset or resets the control system. The fourth term in the three-mode controller equation is proportional to the derivative of the error signal and thus reflects the rate of change of the error. To a certain extent, derivative control anticipates the amount of needed control action and can bring the measured variable back to its set point more quickly. However, it can also result in poor control if the process contains a significant amount of dead time or if the measured signal is noisy.

The number of terms to be used in controlling the system and the values selected for their coefficients or 'gains' (K_p, K_I, K_D) is dependent upon the dynamic behaviour of the process. Determination of the values for the coefficients is known as 'controller tuning' and is primarily a trial and error process although there are optimization programs available. Modelling and simulation can be of considerable value in this respect.

On–off and PID controllers are widely used and have proven their value in a variety of control applications. However, there are many situations in which process performance could be further improved by the use of more advanced control modes. The simpler of these could be the addition of circuits to take into account such items as dead time in sensors and sampling apparatus, and filters for removing high frequency components of the signal which one might not want to consider in initiation of a control action. More sophisticated control algorithms are being increasingly used and included among these are multiple variable control, feedforward control, various optimal control algorithms, and computer control.

Two common types of multiple variable control are ratio and cascade control. Ratio control simply means maintaining one variable, the controlled variable, in a preset ratio to a second variable, the disturbing or uncontrolled variable. Cascade control involves two controllers where a primary or master controller is used to adjust the set point of a secondary or slave controller. A good example of ratio control is found in the activated sludge process where the recycled sludge flow rate is maintained as a set fraction of the wastewater flow rate to the aeration basin (Figure 1.13). This type of control is usually initiated in an attempt to maintain a more constant concentration of mixed liquor suspended solids (MLSS) in the aeration basin. It would be classified as an open loop control system since there is no feedback from the key variable of interest, the MLSS. Closed loop control, using MLSS as the measured variable and recycle flow rate as the controlled variable, is also illustrated in Figure 1.13.

Feedforward control is similar to ratio control in that it is open loop control. Information for feedforward control is obtained by measuring the inputs to the process instead of the outputs as in feedback control. The amount and type of control action needed is then predicted using a dynamic mathematical model. Feedforward control is theoretically capable of perfect control since no error need exist, as for feedback control, before the control action is initiated. This can be of special importance for processes with large time constants as commonly encountered in wastewater treatment. However, considerable knowledge of the system is required to develop the necessary mathematical models and it is expected that feedforward control strategies for wastewater treatment systems will have to incorporate a considerable amount of feedback control until their dynamic behaviour is more clearly defined.

Among books and articles which may be consulted for information on conventional control systems for water and wastewater treatment plants are those by Ryder (1969), Austin (1967), and Babcock (1968). Brouzes (1969)

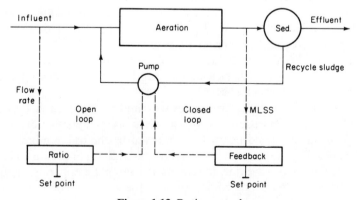

Figure 1.13 Ratio control

has employed a multiple variable control system using a dynamic mathematical model for the activated sludge process. Andrews and Lee (1973) and Busby (1973) have presented possible dynamic models and control strategies for several versions of the activated sludge process and Graef and Andrews (1972) have done the same for the anaerobic digestion process.

Control systems involving many control loops, or using advanced control algorithms, require considerable computing power and it is only logical that digital computers are being increasingly used for process control. By mid-1968, 1700 process control computers were either installed or placed on order for a variety of applications in industries throughout the United States and 5900 installations are forecast by 1975 (U.S. Department of Labor, 1970). A report on the feasibility of computer control for wastewater treatment plants has been prepared by the American Public Works Association (1970) and computers are now being installed in several plants throughout the world. The author is aware of at least 30 plants throughout the world which have either installed or placed process control computers on order. Included among these are plants located in Norwich, Stockholm, Paris, Tokyo, Melbourne, Philadelphia, Los Angeles, New York, Chicago, Detroit, and Atlanta. Only a brief discussion of computer control can be presented herein. The books of Lee, and others (1968), Savas (1965), Lowe and Hidden (1971), and Smith (1972) should be consulted for more information.

A block diagram of a fully developed computer controlled plant is presented in Figure 1.14. However, computer control can be installed in stages and it is important to realize that a digital computer can be of substantial value for information handling even when a human operator must be used to completely or partially replace any of the information links shown in Figure 1.14. The first step taken toward computer control is usually data processing and moni-

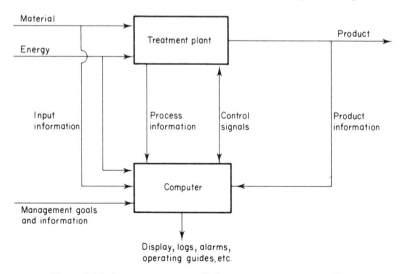

Figure 1.14 A computer controlled wastewater treatment plant

toring. In this mode, the computer collects data, processes it into a more meaningful form, and displays it to the operator. There may also be substantial manual entry of data. The operator then adjusts the set points of his automatic controllers. In this mode of operation, the operator is still in the feedback loop and the computer only assists him in running the plant. Typical functions which may be performed by the computer are:

(1) Scan process sensing instruments, check for instrument malfunctions, and convert raw data into engineering units.

(2) Process data into a more meaningful form for the operator by such operations as smoothing, curve fitting, integration, differentiation, and statistical analysis.

(3) Monitor and report on the status of process equipment. For example, the computer can monitor the on−off status of pumps, values, motors, and compressors. It can check for overheating of bearings and excessive vibration of motors and compressors.

(4) Compare process variables against high−low limits and sound alarms.

(5) Determine normally unmeasurable variables by computation from measured variables. An example would be calculation of oxygen utilization rate in the activated sludge process from a mass balance on oxygen.

(6) Prepare operating logs and display information to the operator. This can be in tabular form, plotted trend charts of particular variables, or graphical displays on a cathode ray tube.

(7) Store maintenance schedules and notify operator of needed maintenance.

(8) Furnish the operator with an operating guide upon request. For example, procedures to be followed in case of a process upset.

(9) Furnish data for other computers or provide reduced operating data to higher management. Examples might be the furnishing of data on costs to the business computer in an accounting section or monthly reports on operation to a State pollution control agency.

The next stage in computer control is to replace the operator in at least some of the control loops. In this mode, the computer calculates the amount of control action needed and directly initiates the control action. The control algorithms used by the computer may range from simple on−off control up to some form of optimal control. However, only the simpler algorithms are being used in wastewater treatment since these systems are poorly defined from a dynamic point of view and control strategies are still very empirical. This points out one of the major advantages of using a digital computer for control; the control strategy can be easily changed by reprogramming (software changes) the computer whereas conventional controllers normally require replacement or rewiring (hardware changes). Still another advantage is that the computing power and storage capabilities of the digital computer enable it to handle complex interactions between variables and implement the more advanced control strategies such as feedforward and optimal control.

An objection frequently raised to the incorporation of computer control in wastewater treatment systems is that reliable on-line sensors are not available for many important measurements. However, this problem is not as serious as might be expected since many changes in treatment processes occur relatively slowly and there is frequently adequate time for a man to participate in the data collection pathway by performing analyses on-site or in the plant laboratory with data input to the computer via remote terminals such as simple card readers or teletypes. An example might be the performance of wet chemical analyses using an automated wet chemical analyser such as the Technicon Autonanlyzer. The important thing is that the information be entered into the control strategy instead of being simply filed for the record.

Prospective users should recognize that although computer control offers many potential benefits, the adaptation of a process to computer control can be a difficult and time consuming task. A substantial amount of technical manpower is required to implement computer process control and it has been reported (U.S. Department of Labor, 1970) that from 2 to 21 man-years have been required to complete computer control projects in industrial plants. Bailey (1969) has discussed some of the problems which have been encountered in adapting industrial processes to computer control. It is expected that considerable development work will be required to take full advantage of computer control for wastewater treatment systems since these systems are complex with poorly understood behaviour.

1.9 SUMMARY

Water pollution control is accomplished by many different systems varying in both size and complexity. Each system requires individual consideration; however, there are some basic concepts of systems engineering which are applicable to all. Among these concepts are:

(1) Looking at a system as an integrated whole yet with recognition of the interactions between the elements in a system and between the system and its environment.
(2) Recognition of the universality of characteristics among systems.
(3) A team or interdisciplinary approach to the analysis, design, or operation of a system.

In the analysis, design, or operation of a system it is first necessary to determine the objectives of the system and delineate the system boundaries and internal structure. The inputs, outputs, environment, and internal characteristics should be expressed in terms of materials, energy, and information handling and it should be recognized that these are dynamic and will therefore change with respect to time. A search should be made for alternate means of accomplishing the system objectives and those which appear most feasible should be compared on the basis of those factors which are of major importance in judging the value of the system. Some of the most commonly used factors are; (1) performance, (2) cost, (3) reliability, (4) time, (5) maintainability, and (6) flexibility.

36

These factors are then combined in a weighted fashion to yield a single function for optimization of the system.

Several powerful analytical tools are available for use in the analysis, design, or operation of water pollution control systems. A discussion of four of these, mathematical modelling, computer simulation, transient response analysis, and control theory, is presented. Stress is placed on the development of dynamic mathematical models from fundamental principles and solution of the equations which comprise the model by a continuous system simulation language. Solution of the model, or prediction of the time varying behaviour of the outputs of the model, is known as transient response analysis. A simple reactor, a CFSTR with an inert tracer input, is used to illustrate the basic principles of dynamic modelling, computer simulation, and transient response analysis.

Control systems are primarily involved with the handling of information. This may be done manually or by automatic control systems. However, in either case, some of the same basic questions must be answered and included among these are: (1) what information should be collected? (2) how should the information be transmitted? (3) how should the information be processed? (4) what control actions should be taken? Once these questions are answered, it is usually found that control systems frequently represent an economical means for improving plant performance.

A wide variety of control algorithms are available and included among these are two-position control, PID, ratio, cascade, and feedforward control. Many of these algorithms can be implemented using conventional control hardware; however, the more sophisticated controller models are best implemented by a process control computer. Computer control systems also offer many other advantages and are being increasingly used in wastewater treatment plant operations.

1.10 REFERENCES

Andrews, J. F. (1973). 'Dynamic Models and Control Strategies for Wastewater Treatment Systems', *Proceedings 8th Annual Workshop on Mathematical Modelling in Environmental Engineering* (Keinath, T. M., and Wanielista, M. P., Eds), Association of Environmental Engineering Professors, Clemson University, Clemson, S. C., 303.

Andrews, J. F. (1972). 'Control Systems for Wastewater Treatment Plants', *Water Research*, **6**, 575.

Andrews, J. F. (1971a). 'Kinetic Models of Biological Waste Treatment Processes', *Biological Waste Treatment* (Canale, R. P., Ed.), John Wiley & Sons, New York, 5.

Andrews, J. F. (1971b). 'Control of Wastewater Treatment Plants—The Engineer as an Operator', *Water and Sewage Works*, **118**, (1), 26.

Andrews, J. F. (1969). 'Dynamic Model of the Anaerobic Digestion Process', *Journal Sanitary Engineering Division, American Society of Civil Engineers*, **95**(SA1), 95.

Andrews, J. F., and Graef, S. P. (1971). 'Dynamic Modeling and Simulation of the Anaerobic Digestion Process', *Anaerobic Biological Treatment Processes, Advances in Chemistry Series No. 105*, American Chemical Society, Washington, D.C., 126.

Andrews, J. F., and Lee, C. R. (1973). 'Dynamics and Control of a Multi-Stage Biological Process', *Proceedings IVth International Fermentation Symposium* (Terui, G., Ed.), Society of Fermentation Technology, Japan, Yamada-Kami, Suita-shi, Osaka, Japan, 55.

American Public Works Association (1970). *Feasibility of Computer Control of Wastewater Treatment*, American Public Works Association, Chicago.

Austin, J. H., (Ed.) (1967). *Proceedings 9th Sanitary Engineering Conference on Instrumentation, Control, and Automation for Water Supply and Wastewater Treatment Systems*, University of Illinois, Urbana.

Babcock, R. H. (1968). *Instrumentation and Control in Water Supply and Waste Disposal*, Reuben H. Donnelley Corp., New York.

Bailey, S. J. (1969). 'On-Line Computer Users Polled', *Control Engineering*, **16**, (1), 86.

Bertalanffy, L. Von (1968). *General System Theory: Foundations, Development, Applications*, George Braziller, New York.

Brouzes, P. (1969). 'Automated Activated Sludge Plants with Respiratory Metabolism Control', *Advances in Water Pollution Research* (Jenkins, S. H., Ed.), Pergamon Press, London.

Bryant, J. O. (1969). 'Transient and Frequency Response Analysis of Water and Wastewater Treatment Systems', *Proceedings 4th Annual Workshop on Applications of Systems Analysis in Sanitary Engineering* (Andrews, J. F., Ed.), American Association of Professors in Sanitary Engineering, Clemson University, Clemson, S. C.

Bryant, J. O., Wilcox, L. C., and Andrews, J. F. (1971). '*Continuous Time Simulation of Wastewater Treatment Plants*', Presented at the 69th National American Institute of Chemical Engineers Meeting, Cincinnati, Ohio.

Busby, J. B. (1973). 'Dynamic Models and Control Strategies for the Step Feed Activated Sludge Process', *Ph.D. Dissertation*, Clemson University, Clemson, S. C.

Chestnut, H. (1965). *Systems Engineering Tools*, John Wiley & Sons, New York.

Cooper, G. R., and McGillem, C. D. (1967). *Methods of Signal and System Analysis*, Holt, Rinehart, and Winston, New York.

Denbigh, K. G., and Turner, J. C. R. (1971). *Chemical Reactor Theory*, 2nd ed., Cambridge University Press, London.

Franks, R. G. E. (1967). *Mathematical Modeling in Chemical Engineering*, John Wiley & Sons, New York.

Graef, S. P., and Andrews, J. F. (1972). *Process Stability and Control Strategies for the Anaerobic Digester*. Presented at the 45th Annual Conference, Water Pollution Control Federation, Atlanta, Georgia (In Press).

Grieves, C. G. (1972). 'Dynamic and Steady State Models for the Rotating Disc Biological Reactor', *Ph.D. Dissertation*, Clemson University, Clemson, S. C.

Hiller, F. S., and Lieberman, G. J. (1967). *Introduction to Operations Research*, Holden-Day, San Francisco.

Himmelblau, D. M. (1970). *Process Analysis by Statistical Methods*, John Wiley & Sons, New York.

Himmelblau, D. M., and Bischoff, K. B. (1968). *Process Analysis and Simulation: Deterministic Systems*, John Wiley & Sons, New York.

Hougen, J. O. (1972). *Measurements and Control Applications for Practising Engineers*, Cahners Books, Boston.

International Business Machine Corporation (1968). *System 360 Continuous System Modeling Program (360A-CX-16X) User's Manual*, International Business Machine Corp., New York.

Knowles, G. (1970). 'Mathematical Models as an Aid in the Solution of Water Pollution Control Problems', *Chemistry and Industry*, 697.

Lawrence, A. W., and McCarty, P. L. (1970). 'Unified Basis for Biological Treatment Design and Operation', *Journal Sanitary Engineering Division, American Society of Civil Engineers*, **96**, (SA3), 757.

Lee, T. H., Adams, G. E., and Gaines, W. M. (1968). *Computer Process Control: Modeling and Optimization*, John Wiley & Sons, New York.

Lloyd, S. G., and Anderson, G. D. (1971). *Industrial Process Control*, Fisher Controls Co., Marshalltown, Iowa.

Lowe, E. I., and Hidden, A. E. (1971). *Computer Control in Process Industries*, Peter Peregrinus Ltd., London (1971).

McLeod, J. (1968). *Simulation*, McGraw-Hill, New York.

Motard, R. L. (1966). 'Systems Engineering: Engineering Come of Age and Its Academic Image', *Journal Engineering Education*, **56**(6), 198.

Perlmutter, D. D. (1965). *Introduction to Chemical Process Control*, John Wiley & Sons, New York.

Ryder, R. A. (1969). 'Dissolved Oxygen Control in Activated Sludge', *Proceedings 24th Industrial Waste Conference*, Purdue University, Lafayette, Indiana, 238.

Savas, E. S. (1965). *Computer Control of Industrial Processes*, McGraw-Hill, New York.

Shilling, G. D. (1963). *Process Dynamics and Control*, Holt, Rinehart, and Winston, New York.

Smith, C. L. (1972). *Digital Computer Process Control*, Intext Educational Publishers, San Francisco.

Smith, C. L., Pike, R. W., and Murrill, P. W. (1970). *Formulation and Optimization of Mathematical Models*, International Textbook, Scranton, Pa.

Smith, R. (1969). 'Preliminary Design of Wastewater Treatment Systems', *Journal Sanitary Engineering Division, American Society of Civil Engineers*, **95** (SA1), 117.

Smith, R. (1971). *Wastewater Treatment Plant Control*, Presented at the Joint Automatic Control Conference, Washington University, St. Louis, Mo.

Thoman, R. V. (1972). *Systems Analysis and Water Quality Management*, Environmental Research & Applications, New York.

Tucker, G. K., and Wills, D. M. (1962). *A Simplified Technique of Control System Engineering*, Honeywell, Inc., Fort Washington, Pa.

U. S. Department of Labor (1970). *Outlook for Computer Process Control*, U. S. Government Printing Office, Bulletin No. 1658, Washington, D. C.

Zahradnik, R. L. (1971). *Theory and Techniques of Optimization for Practising Engineers*, Barnes & Noble, New York.

2

The Modelling of Engineering Systems— Mathematical and Computational Techniques

C. J. S. PETRIE

2.1 INTRODUCTION

I do not propose to try and teach mathematical techniques, or even give a catalogue of useful methods and tricks of the trade. Since there are many excellent textbooks at all levels, it seems to me more important to express some personal opinions on mathematics in engineering, starting from the view that mathematics is essentially redundant and in that sense trivial. What I mean by this is that once a complete and properly formulated mathematical model of a process has been obtained, the deductions from this model, in terms of predictions for plants not yet built or for operating conditions not yet experienced, follow necessarily by a process of logical inference. In the engineering context mathematical techniques are concerned with making these logical deductions in a limited time and at a limited cost, since the problems are usually posed by engineers of finite lifespan and strictly limited means. This leaves plenty of scope for the ingenuity and low cunning which Bellman (1961) identifies as the typical attributes of the mathematician, so I have not yet talked myself out of a job.

It is because of these cost and time constraints, as well as because of our incomplete understanding of the physical world and our incomplete data, that there is no such thing as a complete and properly formulated mathematical model of any real process. Therefore, the mathematical tools available play a large part in determining the size and structure of the mathematical models which are constructed, and mathematical techniques can aid physical intuition in making suitable valid approximations to a model. A further, and perhaps more important aspect than the incompleteness of actual mathematical models is the question of proper formulation of the model, and here mathematical principles can be invaluable in ensuring that a model is logically complete and that its equations do have solutions (and preferably unique ones).

As a further limitation of a very large topic, I small not get involved in the latest (and most exciting—according to their evangelists) branches of mathematics since I am not competent to present these well, nor should I do so fairly, since I hold the opinion that the unfamiliar language of, say, functional analysis widens even further the gap between the mathematician and the practising engineer. No doubt in time, and once the language barrier is overcome, the clarity with which concepts and mathematical principles can be presented and the generality of the 'modern' approach will give us the possibility of significant advances in engineering science.

What I have chosen to do is to illustrate the interaction between the engineer and the relatively old-fashioned mathematician by means of three case studies. These are presented to show how engineering problems look to a mathematician, what mathematicians can do and, a little of how he does it. Details of some of the techniques are given in the Appendix. The case studies have all grown out of problems brought to me by members of staff of engineering departments in the University of Newcastle upon Tyne and, while they may have grown out of all recognition in the ivory tower of Engineering Mathematics, I must record my thanks to the originators of the problems—J. L. Woods of the Department of Agricultural Economics for case study 1, and Dr. A. James of the Department of Civil Engineering for case studies 2 and 3. Let the extent to which the problems are unrecognizable be a warning against the enthusiasm of the mathematician.

2.2 CASE STUDY 1—A BIOLOGICAL REACTOR

We consider a constant volume stirred tank reactor, with volume V, whose state is completely described by the concentrations of substrate (S) and of microorganisms (X) in the reactor. There is a supply of substrate in the feed stream which has concentration S_f and volumetric flow rate v. The stream leaving the reactor must also have volumetric flow rate v and concentrations S and X of substrate and microorganisms. Mass balances for the two components give differential equations for X and S as functions of time:

$$V dX/dt = V\mu X - vX \tag{2.1}$$

$$V dS/dt = V\mu X/Y + vS_f - vS \tag{2.2}$$

Here μ is the specific growth rate of the microorganisms, which describes the reaction kinetics, and Y is the yield coefficient which depends on the stoichiometry. If we write θ for the mean residence time (or retention time) in the reactor $(\theta = V/v)$ and assume a familiar form of dependence of growth rate on substrate concentration,

$$\mu = \hat{\mu} S/(K_s + S) \tag{2.3}$$

(Monod, 1949) we obtain the coupled pair of nonlinear ordinary differential equations

$$dX/dt = \hat{\mu} S/(K_s + S) - 1/\theta X \tag{2.4}$$

$$dS/dt = \hat{\mu} S X/[Y(K_s + S)] + (1/\theta)(S_f - S) \tag{2.5}$$

as the mathematical model of our idealized reactor.

This is typical of what we may call the 'black box' system model, with inputs (concentrations S_f and zero) and outputs (S and X) related by some sort of mathematical operation which models the system response as seen from outside without going into details of what happens inside the system. In this case the nonlinear terms arising from the kinetics lead to two disadvantages. We cannot use the approach of Danckwarts (1953) of using quantities like residence time distribution (which we might measure by tracer experiments) to improve the model and take account of imperfect mixing. The reason is essentially that where the kinetics are not first order (i.e. they are nonlinear) the extent of reaction in any small element of fluid depends on the detailed history of its environment during its passage through the reactor, and not just on an averaged effect of this detail. Also, fairly obviously, the standard techniques of solution in terms of exponentials or by means of Laplace transforms (so popular in control engineering—the home of the black box approach) are not available. This is less serious, since there is no difficulty in principle (and probably none in practice) in solving the system differential equations numerically for any given value of the parameters $\hat{\mu}$, K_s, Y, θ, and S_f if we are given starting values for X and S.

The basic choice for the engineer here seems to be between a model which describes the kinetics accurately but idealizes the reactor (and whose equations must be solved numerically) and a model which gives a very rough description of the kinetics but a much better picture of the effect of the reactor (and also requires the experimental determination of residence time distributions). This is something of an oversimplification, and there is an obvious third choice (if the necessary time and money are available) of trying to model in detail what goes on inside the reactor. This approach is exemplified (partially, and in another context) by case study 3.

Here I want to look at the first choice and qualify the statement that the equations must be solved numerically by showing how we can find out a good deal about the solutions of a pair of equations like (2.4) and (2.5) without actually solving them. The contrast is between finding out everything about a few particular solutions (using numerical methods) and finding out a few things about every solution (using what may be called qualitative techniques). The sort of thing we can expect to discover ranges from the simple determination of possible steady state operating conditions, through the stability of these steady states to small disturbances to the system (e.g. small changes in v or S_f), to (perhaps) a determination of which initial conditions (values of X and S at the start of operation or after a large change in, say v) lead to the designed steady state operation, and which do not.

In the steady state dX/dt and dS/dt are both zero and so we must find values of X and S which make the right-hand sides of (2.4) and (2.5) simultaneously zero. There are two possible solutions:

$$X = 0, S = S_f \tag{2.6}$$

and

$$X = Y[S_f - K_s/(\hat{\mu}\theta - 1)], S = K_s/(\hat{\mu}\theta - 1) \tag{2.7}$$

The first corresponds to washout of all microorganisms and the second is therefore the only desirable steady operating condition (for this model). An elementary first step is to see if this operating condition is physically realizable (since it only makes sense for positive values of S and X) and then we can try and find out whether it is stable, and hence whether it will occur in practice.

All the parameters are positive, so we require $\hat{\mu}\theta > 1$ to make S positive and $S_f > K_n/(\hat{\mu}\theta - 1)$ to make X positive. This puts an upper limit on the flow rate v for fixed V and S_f, namely

$$v < \hat{\mu} V S_f/(K_s + S_f) \tag{2.8}$$

Alternatively, we can think of v as fixed and look on this as a limitation on the concentration of substrate, S_f, which the reactor requires to sustain a population of microorganisms. Then we have

$$S_f > K_s v/(\hat{\mu} V - v) \tag{2.9}$$

This presentation is based on the view that the reactor (i.e. V), the kinetics (i.e. $\hat{\mu}$ and K_s) and the stoichometry (i.e. Y) are fixed and that v and S_f are the process variables which may alter or be altered. Clearly it is a trivial matter to rearrange the equations if we want to take a different view.

We can establish the local stability of the desired operating point by linearization about that point and use of standard results which give us, as necessary and sufficient conditions for stability to small disturbances,

$$-[K_s + (\hat{\mu}\theta - 1)^2 S_f]/[\hat{\mu}\theta^2 K_s] < 0 \tag{2.10}$$

and

$$[1/\theta Y][\hat{\mu}\theta - 1][S_f(\hat{\mu}\theta - 1) - K_s] Y/(\hat{\mu}\theta^2 K_s) > 0 \tag{2.11}$$

Both inequalities are clearly satisfied if $S_f(\hat{\mu}\theta - 1) > K_s$, the condition we have already required to make the operating point physically meaningful, so that this model predicts stability of the steady state for all values of the parameters which allow a steady state other than washout. Details of the techniques used here are outlined in the first section of the Appendix.

We may show in the same way that the other steady state (2.6) is unstable unless it is the only possible steady state, and then it is stable. It is important to realize that we have been using approximations to solution near the steady states in this analysis, so that conclusions only hold for small disturbances. Formally we have in fact considered small disturbances (from their steady state values) to S and X, but it is not too hard to extend the conclusions to small changes in any of the other parameters. (We think of the process as running in its steady state, then the change of, say, S_f moves the steady state. The values of X and S—the old steady state—are then effectively values slightly disturbed from the new steady state values).

Techniques for investigating stability to large disturbances are less systematic and (a personal view again) I think it is fair to suggest that the engineer who wants this sort of information should seek professional mathematical advice unless he is familiar with phase plane ideas or the name of Liapounov in one of its many transliterations. A brief look at the system we have been considering gave arguments which suggest strongly that the operating point is globally stable, i.e. stable to any disturbances (provided that they do not move the operating point out of the region of positive X and S, and provided that they are not so large as to invalidate any of the assumptions implicit in the model). A complete proof of this conjecture could be sought if it was thought worth having—either economically justifiable to the engineer or aesthetically justifiable to the mathematician.

Probably more interesting to both would be a consideration of improvements to the model, or of modifications to the process (which may be suggested by the foregoing analysis). For example, the idea of separating out the microorganisms in the stream leaving the reactor and recycling them to reduce the loss due to microorganisms washed out of the system could be obtained in this way, though like most good ideas in engineering it probably antedates any mathematical modelling of the system. Automatic control or regulation of the system to ensure that changes in flow rate or substrate concentration in the feed do not have an adverse (possibly catastrophic) effect on operations would again be the sort of application where mathematical modelling could contribute a great deal to the design process.

In concluding this case study we shall discuss briefly an improvement in the modelling of the kinetics proposed by Andrews (1971) to take account of the inhibition of growth of microorganisms at high substrate concentrations. We replace equation (2.3) by

$$\hat{\mu} = \hat{\mu}/[1 + (K_s/S) + (S/K_i)] \tag{2.12}$$

in which K_i is a measure of the substrate concentration at which inhibition becomes important. The limit of $K_i \to \infty$ will correspond to our earlier model with no inhibition. We now obtain three possible steady states given by (2.6).

$$S = S_1 = \tfrac{1}{2}K_i(\hat{\mu}\theta - 1) - \sqrt{[\tfrac{1}{4}K_i^2(\hat{\mu}\theta - 1)^2 - K_i K_s]}, X = Y(S_f - S_1) \tag{2.13}$$

and

$$S = S_2 = \tfrac{1}{2}K_i(\hat{\mu}\theta - 1) + \sqrt{[\tfrac{1}{4}K_i^2(\hat{\mu}\theta - 1)^2 - K_i K_s]}, X = Y(S_f - S_2) \tag{2.14}$$

The algebra involved in studying these in general becomes quite heavy, and it is doubtful if clear conclusions could be obtained. (But see section (2) of the Appendix.)

The reason for introducing this more complicated model is that we can use it to illustrate briefly the idea of approximation when we have a large or small parameter in our model. Here we can set out to gain a qualitative idea of how inhibition will affect our earlier conclusions by studying solutions for large values of K. One of the principal tools here is the binomial theorem which we use to rearrange equations (2.13) and (2.14) in the form

$$S_1 = [K_s/(\hat{\mu}\theta - 1)][1 + 2K_s/(K_i(\hat{\mu}\theta - 1)^2) + \text{terms in } 1/K_i^2, \text{ etc.}] \tag{2.15}$$

and

$$S_2 = [K_i(\hat{\mu}\theta - 1)][1 - K_s/(K_i(\hat{\mu}\theta - 1)^2) + \text{terms in } 1/K_i^2, \text{ etc.}] \quad (2.16)$$

Clearly in the limit of very large K_i (no inhibition), S_1 approaches the value of S given by equation (2.7), so our approximation is consistent, and the fact that S_2 tends to infinity with K_i tells us where the third steady state goes in this limiting case. We obtain an estimate of the effect of inhibition by taking into consideration the second term in each expression, neglecting the terms in $1/K_i^2$ because these will be much smaller than terms in $1/K_i$ if we make K_i large enough.

The practical question of how large K_i must be before the approximations we are considering are useful in giving numerical estimates for practical processes is not a question the mathematics can answer fully. The best answer for the engineer is one based on the closeness with which the approximate model can predict experimental results, since this takes account of all the imperfections of the model, and not just those we have deliberately introduced in the search for a mathematically tractable formulation. Further details of this investigation are given in the second section of the Appendix.

2.3 CASE STUDY 2—EFFLUENT DISPERSION IN A RIVER

We consider here a simple one-dimensional steady-state model of the concentration of organic material in a river. The three factors affecting the concentration are dispersion (molecular diffusion and mixing due to turbulent eddies), advection (material carried along with the mean river velocity) and chemical reaction (decomposition, which we assume to be first order). We denote the concentration (BOD) by c, which is a function of downstream distance x only, and obtain the linear second order ordinary differential equation,

$$d[E_x A(dc/dx)]/dx - Q dc/dx - kAc = 0 \quad (2.17)$$

Here A is the cross-sectional area of the river, which we allow to vary with x, and the other parameters, which for simplicity we take to be constant, are the dispersion coefficient in the x-direction, E_x, the total volumetric flow rate, Q, and the rate constant, k, for the first order decay of the organic material. The mean velocity of the river is $U = Q/A$, which must be a function of x since A varies. The net flow of organic material across any cross-section of the river, which we shall need later, is N (expressed as a mass flow rate) given by

$$N = Qc - AE_x dc/dx \quad (2.18)$$

There are two problems here which are not altogether trivial, the mathematical one of obtaining a solution of (2.17), to which we shall return later, and the more interesting step, which logically comes first, of completing the specification of the problem. First we consider the source of the organic material and suppose that there is an effluent discharge at $x = 0$ which adds M (mass per unit time) to the organic material in the river. Now note that we cannot use the differential equation (2.17) across the cross-section $x = 0$, since the basic mass balance used in deriving the equation equates the net flow into any stretch of the river to the rate of decay of the organic material within the stretch. There must therefore be a jump in N at $x = 0$, with M being the difference between N just below the discharge and N just above the discharge, or

$$M = \lim_{x \to 0+} [N(x)] - \lim_{x \to 0-} [N(x)] \qquad (2.19)$$

[the notation denotes limits as x tends to zero for x positive (downstream) and negative (upstream) respectively].

The system model is linear, so we need only consider the effect of one effluent discharge. (Linearity implies that we can add the effects of each discharge on its own to obtain the effect of a number of discharges at different points — a principle of superposition.) In passing it may be observed that most techniques for obtaining solutions for linear systems depend on the possibility of superposition, and it is for this reason that there are few standard techniques for nonlinear systems, even for very restricted classes of equation.

We have not yet completed our model, since we have a differential equation but no boundary conditions. In order to obtain a solution, we could specify c and dc/dx at $x = 0$, but there is no physical justification for this — the boundary conditions are part of the model and must be deduced from our picture of the physical system. (And what values would we assign to c and dc/dx anyway ?). The corresponding (slightly simpler) problem for a tubular reactor with longitudinal mixing has received extensive discussion in the chemical engineering literature (see, for example, Danckwerts, 1953; Pearson, 1959; Fan and Ahu, 1962; Aris, 1969; or Cooper and Jeffreys, 1971) which should convince us that the problem is non-trivial. Let us confine our attention to a definite stretch of the river between the effluent discharge at $x = 0$ and the river mouth at $x = L$. It is clear that conditions upstream of $x = 0$ will affect the concentration we wish to calculate, as will conditions downstream of $x = L$ (in the lake or sea). This tells us that we have a two-point boundary-value problem, but does not tell us precisely what the two boundary conditions, one at $x = 0$ and one at $x = L$, should be.

Essentially, these boundary conditions have to approximate the idea that the concentration and net mass flow must vary smoothly with x, i.e. that c and N are continuous functions of x, except where we have an effluent discharge into the river. The idea of approximation arises because the alternative is to model the complete river system and the whole lake or sea into which the river flows. The modelling of a particular stretch of river requires upstream and downstream conditions which simulate the requirement of matching conditions in this stretch with those upstream and downstream. One way of doing this is to consider a hypothetical lengthening of the stretch we are interested in and apply the conditions that if we go far enough all the organic material will have decayed, i.e. we require $c \to 0$ as $x \to -\infty$ and as $x \to +\infty$.

It may be better to look in more detail at the actual conditions upstream and downstream, and perhaps impose a condition like $dc/dx = 0$ at a weir or a dam (approximating these by solid boundaries into which there is no diffusion of organic material). An interesting model of the downstream condition might be obtained by considering a semi-infinite lake with the river as a point source of organic material. The model would involve two-dimensional steady diffusion and decay by first order kinetics but no advection, and we should impose the condition that $c \to 0$ as we get far enough away from the river mouth. This is somewhat speculative, and perhaps of mathematical rather than engineering interest. Here we shall consider a numerical example with two simpler boundary conditions. We suppose that the river is flowing reasonably fast so that there is effectively no organic material upstream of the effluent discharge. Then we have

$$M = \lim_{x \to 0+} [N(x)]$$

and for simplicity we write this

$$M = N(0) \tag{2.20}$$

since we are not considering $x < 0$ any more. Downstream we shall take

$$dc/dx = 0 \text{ at } x = L \tag{2.21}$$

following Danckwerts (1953) and other chemical engineering authors.

The key question for the engineer here is how much the solution is affected by the choice of boundary conditions (provided we do not make an unreasonable choice). There is an important general point here, relating to the robustness of mathematical models, or their sensitivity to inaccuracies. This really contains two ideas; we want a model which is not excessively sensitive to changes in numerical values of parameters we have to measure, so that the effect of unavoidable errors in measurement is not magnified by the model to produce predictions of no practical value. Also we want a model which produces results that are acceptable even though the model itself is imperfect, for example it does not matter if dc/dx is in fact non-zero at $x = L$ provided that our predictions for the function $c(x)$ or, say, the rate of discharge of organic material into the lake or sea $[N(L)]$ are reasonable.

For our numerical example we take a river whose area increases linearly (e.g. it has straight diverging banks and constant depth, or parallel banks and linearly increasing depth) and use the values

$$
\begin{aligned}
A &= 2000 + 0.3x \text{ m}^2 (x \text{ in metres}) \\
Q &= 50 \text{ m}^3 \text{ sec}^{-1} \\
k &= 10^{-6} \text{ sec}^{-1} \\
E_x &= 100 \text{ m}^2 \text{ sec}^{-1} \\
M &= 2000 \text{ g sec}^{-1} \\
L &= 10^4 \text{ m}
\end{aligned}
$$

We change variable to $z = 10^{-4}x$, and obtain the differential equation

$$(2 + 3z)d^2c/dz^2 - 2dc/dz - (2 + 3z)c = 0 \tag{2.22}$$

Equation (2.18) gives us

$$N = 50(c - 0.4\,dc/dz) \tag{2.23}$$

and from this we obtain the boundary conditions

$$c - 0.4\,dc/dz = 40 \text{ at } z = 0 \tag{2.24}$$

and

$$dc/dz = 0 \text{ at } z = 1 \tag{2.25}$$

We shall consider four methods of solution, two numerical and two 'traditional'.

For boundary-value problems like the one we have here there are two approaches to numerical solution—'shooting' approach which uses techniques developed for initial value problems and the approach specially designed for boundary value problems. When we set out to solve an initial value problem numerically, we start with values of the unknown function, c, and its first derivative for some value of the independent variable, z. We then use the (second-order) differential equation to obtain estimates of c and its derivative for a nearby value of z, and proceed step-by-step, usually increasing z by a fixed amount (h, the step size) each time. When we use this for a boundary value problem we have to guess a value at the start of the process and hope to hit the correct boundary value when z reaches the other end of the interval. Obviously we shall not guess correctly very often, but we can use the process iteratively, and in fact with a linear problem two iterations are all that are needed, followed by inverse linear interpolation.

In our numerical example we have used a fourth-order Runge–Kutta method and a step size of 0.1. We guess values -10 and -20 for dc/dz at $z = 0$ then use equation (2.24) to give the corresponding value of c at $z = 0$. The results of the numerical integration are tabulated below.

dc/dz (at $z = 0$)	-10	-20	-18.8506
c (at $z = 0$)	30	32	32.4598
dc/dz (at $z = 1$)	29.0977	-3.7788	0
c (at $z = 1$)	43.2710	19.7019	22.4113

The third column is obtained by linear interpolation, and actual computation with these values of c and dc/dz gives a value of -9×10^{-5} for dc/dz at $z = 1$. With exact arithmetic this should be zero, and this is supported by the fact that calculation to 12 significant figures gives -8×10^{-10} for this value. Halving the step size (as a partial check on the accuracy of the numerical method) makes a difference of at most 1 in the fourth decimal place, so we conclude that these results are reasonably accurate.

The alternative, boundary value method, consists of approximating the differential equation at a series of equally spaced values of $z(z = 0, h, 2h, \ldots$. $Nh = 1$) by a series of algebraic equations. These are obtained by using finite difference approximations to the derivatives of c in the differential equation. The boundary conditions are similarly approximated, and we end up with $N + 1$

equations in $N + 1$ unknown values of c. For a linear system these equations are linear, and the solution of simultaneous linear algebraic equations is reasonably straightforward (though care has to be taken when the number of equations becomes large). The results we obtain here for $N = 40$ are

$$
\begin{array}{lll}
\mathrm{d}c/\mathrm{d}z & (\text{at } z = 0) & -18\cdot6238 \\
c & (\text{at } z = 0) & 32\cdot5505 \\
\mathrm{d}c/\mathrm{d}z & (\text{at } z = 1) & 0 \\
c & (\text{at } z = 1) & 22\cdot6533
\end{array}
$$

(We choose $N = 40$ for our comparison because the fourth-order Runge–Kutta method uses the differential equation four times on each of the ten steps we have used). For this example the boundary value method shows up less well, but both give adequate answers. Some details of the use of these methods and a brief discussion of their relative merits in general is given in the Appendix.

An alternative to numerical solution which is applicable to a fairly wide class of differential equations with polynomial coefficients is solution in series. (We cannot, as with constant coefficient equations, find solutions in terms of exponentials or other elementary functions, and the best we can hope for is a solution expressed as a power series in the independent variable.) In this case we may obtain the general solution (with arbitrary constants P and Q)

$$
\begin{aligned}
C = P(1 &+ \tfrac{1}{2}z^2 + z^3/6 - z^4/24 + \ldots + p_k z^k + \ldots) \\
&+ Q(z + \tfrac{1}{2}z^2 + z^3/12 + z^4/8 + \ldots + q_k z^k + \ldots)
\end{aligned} \tag{2.26}
$$

where P_k is given by the recurrence relation

$$
2(k + 3)(k + 2)p_{k+3} = 3p_k + 2p_{k+1} - (3k + 1)(k + 2)p_{k+2} \tag{2.27}
$$

and q_k by the same relation with p replaced by q everywhere. Unfortunately we have been a little too hasty here, and have obtained a series solution which is only valid for z between $-\tfrac{2}{3}$ and $\tfrac{2}{3}$, which is of no use for our boundary value problem. (It is easy to verify that the series are divergent if z exceeds $\tfrac{2}{3}$.)

There is, however, a well-established body of mathematical theory to call on here, and this can tell us two things: first that we should have foreseen the non-convergence since the equation is singular (the coefficient of its highest derivative is zero) for $z = -\tfrac{2}{3}$, and second that we can obtain a series which will converge for a wider range of values of z if we obtain our series in powers of a variable which is zero where the equation is singular. We therefore change variable to $y = 2 + 3z$, giving the equation

$$
\mathrm{d}^2c/\mathrm{d}y^2 - (2/3y)\mathrm{d}c/\mathrm{d}y - c/9 = 0 \tag{2.28}
$$

and the series solution

$$
\begin{aligned}
c = A(1 &+ y^2/6 + y^4/504 + \ldots + a_{2r}y^{2r} + \ldots) \\
&+ By^{5/3}(1 + y^2/66 + y^4/13464 + \ldots + b_{2r}y^{2r} + \ldots)
\end{aligned} \tag{2.29}
$$

Here the coefficients are given by

$$a_{2r} = a_{2r-2}/[6r(6r-5)]$$
$$b_{2r} = b_{2r-2}/[6r(6r+5)]$$

(2.30)

and the series are convergent for all values of y, and in fact are quite rapidly convergent for values of y between 2 and 5 (z between 0 and 1).

This last point is of some importance, since there is a considerable difference between obtaining a formal solution as a convergent series and obtaining a solution we can use to compute numerical results. (With arithmetic carried out to a finite number of figures it can easily happen that a slowly convergent series gives inaccurate or even completely unreliable results.) Here we can get more than adequate accuracy with five terms in each series (and four terms give us results which agree to better than 0·1%), and the results are

$$dc/dz \text{ (at } z = 0) \quad -18·8501$$
$$c \quad \text{(at } z = 0) \quad 32·4599$$
$$dc/dz \text{ (at } z = 1) \quad -1·5 \times 10^{-9}$$
$$c \quad \text{(at } z = 1) \quad 22·4122.$$

(We have used the boundary conditions to evaluate the constants A and B.)

The fourth method we shall discuss here is essentially a trial and error approach—we may either look up a book like Kamke (1967) which has tabulated several hundred differential equations and their solutions or seek a change of variable which will bring the equation into a familiar form. In general this sort of approach is unlikely to be fruitful—we may not succeed in finding a solution, or if we do it may be so complicated that it is of no more help than a numerical solution, or we may encounter difficulties at a later stage, as in the example considered here. We observe that equation (2.28) is quite similar to Bessel's equation, and reference to Kamke gives us the solution

$$c = y^{5/6}[AI_{5/6}(y/3) + BI_{-5/6}(y/3)]$$

(2.31)

where $I_{5/6}$ and $I_{-5/6}$ are modified Bessel functions of order 5/6 and −5/6, respectively. We may look up these functions, to discover their properties, in textbooks like Sneddon (1961) or Whittaker and Watson (1927). Bessel functions are tabulated in many mathematical handbooks, for example Jahnke, Emde and Lösch (1960) or Abramowitz and Stegun (1964), but unfortunately these tables do not include Bessel functions of order 5/6, (nor, apparently, do any other tables in the libraries of the University of Newcastle upon Tyne) so the formula is of limited use.

The sort of advantage that solution in terms of known functions has (even if the acquaintance with the functions is slight) is that questions concerning the effect of boundary conditions of the type $c \to 0$ as $z \to \infty$ can be answered rather more easily than where we have numerical solutions. In this case that condition would require a solution proportional to $I_{5/6} - I_{-5/6}$, and if we had tables of this, the rest would be easy. As it is we might hope to obtain a series for this function from one of the textbooks.

We shall conclude our discussion of methods of solution of boundary value problems as we started, touching on the most important question of the formulation of the boundary conditions. The approaches outlined above may all be used in investigating the effect of choice of boundary condition on the solution, and if we have difficulty in doing this we might try simpler approximations – for example by considering a river of constant cross-section. If it is reasonable to suppose that the correct boundary conditions (or the best approximation to the real boundary conditions) will be the same whatever the shape of the river, then this approach would be the sensible one—simplicity is the most desirable attribute of any mathematical model (provided it is a reasonable representation of the real system).

2.4 CASE STUDY 3—THE TRICKLING FILTER

The aspect of this process I want to pick out for detailed consideration is the modelling of the intermittent dosing and its effect on the mass transfer or organic material to the bacterial film. We are now concerned with modelling the details of what goes on in the filter bed, and since we are emphasizing the time-dependent aspect of the process we must attempt to find the concentration of organic material as a function of position and time. This inevitably leads us to a partial differential equation for concentration which we want to solve for the time interval between dosings. One complication at a time is enough, so we take the simplest possible geometrical model—a plane film of liquid in contact with a plane bacterial film. We assume that the overall result of each dosing is to supply liquid with a uniform concentration of organic material (c_0, measured as BOD) to a zone at the top of the filter bed. This displaces the material there from the previous dosing to the next zone down, and so on.

What we want to find is the mean BOD, c_m, at the end of the dosing cycle. Because the model we shall use is linear, c_m will be a linear function of c_0, in other words the fractional reduction in BOD, $r(= c_m/c_0)$, is independent of c_0. Hence each zone in the filter will cause the same reduction, and if there are N zones (i.e. the mean retention time of the filter is NT, where T is the cycle time for dosing) the final concentration of organic material in the liquid leaving the filter will be $c_0^{r^N}$.

With our chosen simple geometry we consider the concentration to be a function of distance, x, from the bacterial film and time, t, from the start of the dosing cycle, and write $c(x,t)$ for this concentration. We assume that the liquid film is still, so that the only mechanism for transport of material to the bacterial film is molecular diffusion, and we may readily deduce the standard 'diffusion equation',

$$D\partial^2 c/\partial x^2 = \partial c/\partial t \tag{2.32}$$

where D is the molecular diffusion coefficient. This is a parabolic partial differential equation, so we know that we need one initial condition

$$c = c_0 \text{ at } t = 0 \text{ for all } x \text{ between } 0 \text{ and } L \tag{2.33}$$

(where L is the thickness of the film of liquid) and two boundary conditions. These are

$$\partial c/\partial x = 0 \text{ at } x = L \text{ for all } t > 0 \qquad (2.34)$$

(i.e. no mass transfer through the free surface) and

$$c = 0 \text{ at } x = 0 \text{ for all } t > 0 \qquad (2.35)$$

assuming for simplicity that the organic material is consumed by the microorganisms as soon as it reaches the surface of the bacterial film. We shall return to this particular simplifying assumption later.

We have now a standard problem in the solution of a partial differential equation, and the method of separation of variables gives the solution

$$c(xt) = \frac{4c_0}{\pi} \sum_{n=0}^{\infty} \frac{1}{(2n+1)} \sin\left\{\frac{(2n+1)\pi x}{2L}\right\} e^{-(2n+1)^2\pi^2 Dt/(4L^2)} \qquad (2.36)$$

Now we want the mean concentration at the end of the dosing cycle, which is given by

$$c_m = (1/L)\int_0^L c(x,T)\,dx$$

$$= c_0 \sum_{n=0}^{\infty} \frac{8}{(2n+1)^2\pi^2} e^{-(2n+1)^2\pi^2 DT/(4L^2)} \qquad (2.37)$$

Hence we may calculate r, and the mass transfer per unit area per dosing cycle is $(c_o - c_m)L$ for the top zone of the filter.

If we take typical numerical values of $D = 10^{-3}, L = 2, T = 300, c_o = 2 \times 10^{-4}$ (in mm, sec, mg units) we have

$$r = c_m/c_o = 0.69098$$

and hence

$$c_m = 1.38196 \times 10^{-4} \text{ mg mm}^{-3}$$

This only needs four terms of the infinite series. The number of terms needed, and the value of r obtained, depend only on the dimensionless number $DT^2/(4L^2)$. The smaller this is, the more terms are needed, and the closer r gets to 1 (no mass transfer). For example, if $D = 10^{-4}$, and L and T are unaltered, we need about twelve terms in the series and find $r = 0.90228$.

It would be possible to model in a similar way diffusion from a liquid film to a bacterial film on a spherical support, and it seems likely that a solution could be obtained in a similar way to the above. This is, perhaps, a better geometrical model of the situation in a real filter bed, and it would be of interest to see how the geometry affects the mass transfer. For the engineer the motivation here is the desire to get away with as simple a model as gives adequate results.

The other major simplifying assumption was that the rate controlling step in the overall mass transfer was the diffusion through the liquid film, and that the bacterial consumption of the organic material is virtually instantaneous. We can clearly model the process of diffusion within the bacterial film if the organic material survives there long enough for this to be significant—the model would have similarities to both our previous case studies, but would again be a partial differential equation. This time the solution would be less easy, unless we were able to use a linear approximation to the kinetics of the bacterial consumption of organic material. However, assuming we could obtain the solution, our troubles are not over, since the solution for the liquid film is now wrong. We must replace the condition $c = 0$ at $x = 0$ by a matching of the solutions for the two regions—the liquid between $x = 0$ and $x = L$ and the bacterial film between $x = 0$ and, say, $x = -b$. We should require that the concentration and the rate of mass transfer were both continuous (i.e. approached the same value as we approached $x = 0$ whether we were considering the solution for the liquid or the bacterial film). This problem for a continuous process is discussed by Atkinson in Coulson and Richardson (1971) and elsewhere.

Let us pursue this problem a little further to illustrate the alternative (numerical) approach to solving such equations. We should be forced to use such methods if the model were complicated by, say, a non-uniform diffusion coefficient (perhaps dependent on concentration or on temperature). One of the simplest methods for the diffusion equation is to calculate c for an array of equally spaced points ($x = 0$, h, $2h$,... up to $x = Nh = 2$) for times $t = k$, $2k$,... up to $t = Mk = 300$. We replace the partial derivatives by differences in x or t and obtain a set of algebraic equations. With the approximations

$$[c(x, t + k) - c(x, t)]/k \text{ for } \partial c/\partial t \tag{2.38}$$

$$[c(x + h, t) - 2c(x, t) + c(x - h, t)]/h^2 \text{ for } \partial^2 c/\partial x^2 \tag{2.39}$$

and for the boundary condition at $x = 2$,

$$[c(x, t) - c(x - h, t)]/h \text{ for } \partial c/\partial x \tag{2.40}$$

we obtain the typical equation

$$c_{n,m+1} = c_{n,m} + r(c_{n+1,m} - 2c_{n,m} + c_{n-1,m}) \tag{2.41}$$

where $r = Dk/h^2$ and $c_{n,m} = c(nh, mk)$, i.e. the value of c for $x = nh$, $t = mk$. We have $N - 1$ equations like this for each time step and need two extra equations in order to be able to calculate values of c at time $t = (m + 1)k$ from values at $t = mk$. We obtain these from the boundary conditions, which give

$$c_{0, m+1} = 0 \tag{2.42}$$

and

$$c_{N,m+1} = c_{N-1,m+1} \tag{2.43}$$

for all values of m. The initial condition gives us the values of c for $m = 0$ and this is all we need to start the calculation.

The following table presents typical results for step sizes $h = 0.2$ and $k = 10$, showing c at $t = 300$ as a function of x.

x	0·0	0·2	0·4	0·6	0·8	1·0
$c(x,300)$	0·0	0·000041	0·000079	0·000113	0·000140	0·000161
x		1·2	1·4	1·6	1·8	2·0
$c(x,300)$		0·000176	0·000186	0·000192	0·000195	0·000195

It only remains to integrate this numerically to give us the value of the mean concentration remaining, $C_m = 1·384 \times 10^{-4}$ mg mm^{-3}, in good agreement with the earlier solution.

2.5 DISCUSSION

The case studies presented here may be interesting in showing how problems in public health engineering look to a mathematician, but in addition they illustrate the variety of techniques that are available for initial value and boundary value problems for ordinary differential equations, and for partial differential equations. The Appendix contains some of the mathematical detail, together with references and a bibliography which, while it reflects my own personal taste, covers most mathematical aspects of these problems. The emphasis so far has been on methods that work, and raises questions of how far to pursue a particular mathematical line—questions on which the engineer and the mathematician may not always agree.

It would be wrong to conclude without a brief cautionary tale—consider the numerical solution of equation (2.32) again and notice what a large number of time steps we took. Let us try and cut down on the computing time for the problem by increasing the time step, since it seems that we are being unnecessarily accurate in approximating $\partial c/\partial t$ compared with $\partial^2 c/\partial x^2$. We choose a time step of 30, so that now we are taking an equal number of steps in each independent variable. After 300 seconds the numerical estimate of the concentration as a function of x is given in the following table.

x	0·0	0·2	0·4	0·6	0·8	1·0
$c(x,300)$	0·0	− 0·00096	0·00173	− 0·00166	0·00161	− 0·00081
x		1·2	1.4	1·6	1·8	2·0
$c(x,300)$		0·00069	− 0·00003	0·00026	0·00018	0·00018

We do not need to compare this with an accurate answer to deduce that all is not well, and indeed if we do compare with the answer obtained earlier we see that things have gone wrong out of all proportion with the relatively small change we have made in the computation.

This demonstrates the concept of the computer as a sewage plant operating in reverse—the old catch-phrase was 'garbage in, garbage out', which we seem to have extended to the less satisfying (and much less satisfactory) 'good data in, garbage out'. However, while there are in fact numerical methods which exhibit this phenomenon of instability under all circumstances, there are methods which are stable for all simple problems and which are more efficient than the simple method we are discussing here. Thus, provided the engineer is aware of the potential hazards when he decides to construct his own numerical method,

54

and provided that he will go and talk to a numerical analyst when he has a new and fiendishly complicated model to solve, he can place a good deal of reliance on computed results.

If I have talked about problems which are too elementary here, and if I have caused frustration by avoiding the full details of the mathematics, it is not because I wish to monopolize the hard problems. My reasons are partly didactic, since I have tried to use the case studies as vehicles for some general ideas which I consider to be important, and partly occasioned by the fact that I haven't solved any hard problems yet. Nor do I want to see methods guarded by a high priesthood of mathematicians, subscribing to the view put forward by R. E. Bellman at a recent colloquium that 'mathematicians should not solve problems too quickly in case they find themselves out of a job', for I am sure that there are enough significant and interesting problems to keep us all busy for the foreseeable future.

2.6 REFERENCES

Abramowitz, M., and Stegun, I. A. (1964). *Handbook of Mathematical Functions*, National Bureau of Standards, Washington.

Andrews, J. F. (1971). Kinetic Models of Biological Waste Treatment Processes, *Biotechnol. & Bioeng., Symp. No. 2*, 5.

Aris, R. (1969). *Elementary Chemical Reactor Analysis*, Prentice Hall, Englewood Cliffs, N. J.

Bellman, R. E. (1961). *Adaptive Control Processes: A Guided Tour*, Princeton University Press, Princeton, N. J., p. 4.

Cooper, A. R., and Jeffreys, G. V. (1971). *Chemical Kinetics and Reactor Design*, Oliver & Boyd, Edinburgh.

Coulson, J. M., and Richardson, J. F. (1971). *Chemical Engineering, Volume 3*, (J. F. Richardson and D. G. Peacock, Eds) Pergamon Press, Oxford. (Chapter 5—Biochemical Reaction Engineering by B. Atkinson).

Danckwerts, P. V. (1953). Continuous Flow Systems, Distribution of Residence Times, *Chem. Eng. Sci.*, **2** (1), 1.

Fan, L.-T., and Ahn, Y.-K. (1962). Critical Evaluation of Boundary Conditions for Tubular Flow Reactors, *Ind. Eng. Chem. Proc. Des. & Dev*, **1**, 190.

Jahnke, E., Emde, F., and Lösch, F. (1960). *Tables of Higher Functions*, 6th ed. McGraw-Hill, New York, (Teubner, Stuttgart).

Kamke, E. (1967). *Differentialgleichungen, Lösungmethoden und Lösungen. I: Gewöhnlichen Differentialgleichungen*, 8th ed. Geest & Portig, Leipzig.

Monod, J. (1949). The Growth of Bacterial Cultures, *Ann. Rev. Microbiol.*, **3**, 371.

Pearson, J. R. A. (1959). A Note on the 'Danckwerts' Boundary Conditions for Continuous Flow Reactors, *Chem. Eng. Sci.*, **10**, 281.

Sneddon, I. N. (1961). *Special Functions of Mathematical Physics and Chemistry*, Oliver & Boyd, Edinburgh.

Whittaker, E. T., and Watson, G. N. (1927). *A Course of Modern Analysis*, Cambridge University Press, Cambridge.

2.7 APPENDIX: SOME MATHEMATICAL DETAIL

(1) Stability of the steady state solution of (2.4) and (2.5)

We write

$$y = X - Y[S_f - K_s/(\hat{\mu}\theta - 1)] \tag{A.1}$$

$$z = S - K_s/(\hat{\mu}\theta - 1) \tag{A.2}$$

so that the desired operating point is at the origin in the (y, z) coordinates. Replacing X and S by y and z in equations (2.4) and (2.5), and then using the binomial theorem for the term

$$1/[K_s + S] = 1/[K_s/(\hat{\mu}\theta - 1)]$$
$$= [(\hat{\mu}\theta - 1)/(\hat{\mu}\theta K_s)] [1 + (\hat{\mu}\theta - 1)z/(\hat{\mu}\theta K_s)]^{-1}$$
$$= [(\hat{\mu}\theta - 1)/(\hat{\mu}\theta K_s)][1 - (\hat{\mu}\theta - 1)z/(\hat{\mu}\theta K_s) + ...]$$

we obtain

$$dy/dt = z(\hat{\mu}\theta - 1)[(S_f(\hat{\mu}\theta - 1) - K_s]Y/(\hat{\mu}\theta^2 K_s) + ... \tag{A.3}$$

$$dz/dt = -y/(\theta Y) - z[S_f(\hat{\mu}\theta - 1) + K_s]/(\hat{\mu}\theta^2 K_s) + G \tag{A.4}$$

where the omitted terms $(+...)$ are $0\,(y^2 + z^2)$, i.e. they go to zero at least as rapidly as $y^2 + z^2$ does as we approach the origin (the steady state solution).

Next we shall show that the linearized system (A.3) and (A.4) (obtained by neglecting the omitted terms) has solutions which all tend to zero as t increases, so that the steady state is stable for the linearized system. (We use the term asymptotically stable for this sort of stability, which is only one of many that it is possible to define.) We can quote theorems to back up (in this case) or contradict (when for example the linearized system is neutrally stable) the common-sense view the results we obtain for the linearized system will also hold for the nonlinear system, provided that the departures from the steady state are small enough. Asymptotic stability is a property which does carry over from the linearized to the nonlinear system—see any of the books in the ordinary differential equations section of the bibliography.

Of the various forms of condition for the asymptotic stability of the solution $y = 0$, $z = 0$ of a linear constant coefficient system of the form

$$\begin{aligned} dy/dt &= ay + bz \\ dz/dt &= cy + dz \end{aligned} \tag{A.5}$$

we choose the pair

$$a + d < 0$$

and

$$ad - bc > 0 \tag{A.6}$$

which are both necessary and sufficient. These give us conditions (2.10) and (2.11) in the main text.

For the 'washout' steady state we set

$$Z = S - S_f$$

and obtain the linearized equations

$$dX/dt = X[S_f(\hat{\mu}\theta - 1) - K_s]/[\theta(K_s + S_f)]$$

$$dZ/dt = X\hat{\mu}\theta S_f/[\theta Y(K_s + S_f)] - Z/\theta \tag{A.7}$$

Hence both of conditions (A.6) are satisfied if (and only if) the coefficient of X

in the first equation is negative, i.e.

$$S_f(\hat{\mu}\theta - 1) < K_s$$

which is the condition for there to be only the one realizable steady state. Thus if there is a possible steady state other than washout of the micro-organisms it is the only stable steady state.

If we wish to go further and try to find out how large a disturbance from the steady state the system can stand (and still return to the same steady state) we may, for this system (and for any second-order autonomous system, i.e. one in which t does not appear explicitly), sketch solution curves in the (S, x) plane (the 'phase-plane' for this system). This can give us a rough, qualitative, idea of how the solutions behave and probably some clues on how to prove formally that the steady state is stable to all disturbances (or to a restricted class of disturbances). The main alternative to this is Liapounov's directed method, which proceeds by discovering a function analogous to a total energy for the system and basing a proof on the necessary decrease of this function. Again the details (of finding the energy-like function and of establishing that it decreases along a solution curve) are not always obvious or straightforward, and the establishment of useful results is something of an art.

(2) The effect of inhibition in case study 1

We can simplify the algebra a little if we use (X_e, S_e) to denote either one of the steady state solutions (2.13) and (2.14). We put

$$x = (X - X_e)/Y; \; y = S - S_e \tag{A.8}$$

change the independent variable from t to $z = t/\theta$ (i.e. we use the retention time of the reactor as our time unit) and write m for $\hat{\mu}\theta$. We also make use of the equations

$$X_e = Y(S_f - S_e) \tag{A.9}$$

and

$$S_e^2/K_i - (m - 1)S_e + K_s = 0 \tag{A.10}$$

and obtain the linearized equations

$$\begin{aligned} dx/dz &= y[S_f - S_e][2K_s - (m - 1)S_e]/[mS_e^2] \\ dx/dz &= -x - y[S_e^2 + S_f(2K_s - (m - 1)S_e)]/[mS_e^2] \end{aligned} \tag{A.11}$$

Hence we have stability if and only if

$$[S_f - S_e][2K_s - (m - 1)S_e]/[mS_e^2] > 0 \tag{A.12}$$

and

$$[S_e^2 + S_f(2K_s - (m - 1)S_e)]/[mS_e^2] > 0 \tag{A.13}$$

We also require S_e and X_e to be positive, which places some restrictions on the values of m, S_f, K_s and K_i for which the two steady state solutions (2.13) and (2.14) are physically relevant; namely $m > 1$ (for the solutions of (A.10) to be positive, otherwise both are negative), $S_f > S_e$ (to keep X_e positive) and (to ensure that S_e is real)

$$4K_s/[(m-1)^2 K_i] < 1$$

Taking all these together, and recalling that all the parameters in the equations are positive, we obtain the necessary and sufficient set of conditions

$$2K_s > (m-1)S_e \tag{A.14}$$

$$S_f > S_e \tag{A.15}$$

$$m > 1 + 2\sqrt{(K_s/K_i)} \tag{A.16}$$

where S_e is a solution of equation (A.10).

When the main text was written it was thought that the only likely methods of making further progress in understanding the system were to put in numerical values for the constants or to use an approximation such as that for large K_i shown in the main text. This has been left since it serves as a very brief introduction to the important ideas of approximations of this type, which come under the general heading of asymptotic expansions or perturbation methods. The section in the bibliography on perturbation methods gives some advanced reading on this topic which is also referred to in Struble's book listed in the ordinary differential equations section.

However, a fresh look at the problem has produced the following result. Whenever (A.15) [for $S_e = S_1$, equation (2.13)] and (A.16) are both satisfied, the steady state solution (X_1, S_1) is stable and whenever (A.15) is also satisfied for $S_e = S_2$ (equation 2.14) the steady state solution (X_2, S_2) is unstable. The proof is obtained by putting the explicit solution of (A.10) in (A.14). We write the solution

$$S_e = [(m-1)K_i/2][1 \pm \sqrt{\{1 - 4K_s/[(m-1)^2 K_i]\}}] \tag{A.17}$$

and if we put a for $4K_s/[(m-1)^2 K_i]$ then (A.14) gives us

$$a > 1 \pm \sqrt{(1-a)}.$$

Now we must have $a < 1$, from (A.16), so this inequality can never be satisfied with the + sign, i.e. for $S_e = S_2$. For $S_e = S_1$ we require

$$a > 1 - \sqrt{(1-a)}$$

for stability, and this is always satisfied for a between 0 and 1. Thus, given (A.16), (A.14) is always satisfied for S_1 and never for S_2.

We summarize the conclusions (and return to the original notation) as follows: There is a stable steady state (corresponding to that for the uninhibited model of the process)

$$S = S_1 = (\hat{\mu}\theta - 1)K_i/2 - \sqrt{[(\hat{\mu}\theta - 1)^2/4 - K_s K_i]} \tag{2.13}$$

$$X = X_1 = Y(S_f - S_1)$$

58

provided that

$$S_f > S_1 \tag{A.15a}$$

and

$$\hat{\mu}\theta > 1 + 2\sqrt{(K_s/K_i)} \tag{A.18}$$

and there is an additional unstable steady state

$$S = S_2 = (\hat{\mu}\theta - 1)K_i/2 + \sqrt{[(\hat{\mu}\theta - 1)^2 K_i^2/4 - K_s K_i]}$$
$$X = X_2 = Y(S_f - S_2) \tag{2.14}$$

if, in addition, we have

$$S_f > S_2 \tag{A.15b}$$

Andrews (1971) has made similar deductions, apparently on heuristic grounds, and these are in broad agreement, though the inference (perhaps wrongly drawn from his discussion) that there might only be an unstable steady solution in some circumstances does not stand up to our analysis.

(3) Numerical solution of boundary-value problems

Details of initial-value methods, such as Runge–Kutta or predictor–corrector, may be found in any of the books listed in the sections of the bibliography on general numerical methods and on numerical methods for ordinary differential equations. Many computer 'packages' and subroutine libraries (see bibliography) contain a choice of such methods. The N.A.C. Subroutine Library is currently leading the way with some up-to-date and sophisticated methods. The basic steps in the solution of a second-order initial-value problem are usually

  (a) Rewrite the equation as a pair of first-order equations,
     e.g.

$$dc/dz = g$$
$$dg/dz = c + 2g/(2 + 3z) \tag{A.19}$$

  (b) Choose a numerical method and a step size, h, or (in more sophisticated routines) a desired accuracy.
  (c) Choose initial values for the dependent variables, here c and g, for the starting value of the independent variable ($z = 0$).
  (d) Calculate values of c and g step by step for values of z increasing from 0 to 1 in steps of h (or in steps of a suitable size to give the specified accuracy). Each step uses the chosen method, which in turn uses the differential equations (A.19) several times.

As an estimate of the amount of work required in obtaining a numerical solution we may use the number of times the differential equation is used in obtaining values of c and g at $z = 1$ (to an acceptable accuracy). This is not always the same as the number of steps taken multiplied by the order of the method—which is defined in terms of the behaviour of the error inherent in the method as h tends to zero. The fourth order Runge–Kutta method used has an error per step proportional to h^5 for small enough h and requires four uses of (A.19) per step, but has no estimate of the errors made, which some methods have.

Turning now to the application of this to our boundary-value problem, we denote the values of c and g at $z = 0$ and 1 by suffixes 0 and 1, respectively. We have to guess at values of g_0 and then use equation (2.24) to calculate c_0 and integrate numerically from $z = 0$ to $z = 1$ to obtain c_1 and g_1. Now we want g_1 to be zero by the correct choice of g_0, and we may see how to do this by looking on g_1 as a function of g_0. We write $g_1(g_0)$ for the value of g_1 obtained from a particular choice of g_0 and in effect wish to solve the equation

$$g_1(g_0) = 0 \qquad (A.20)$$

for g_0. We can easily show that for our linear homogeneous system the function g_1 is linear, and we can then solve (A.20) by drawing a straight line through two points on a graph of g_1 against g_0, i.e. by inverse linear interpolation. We write

$$g_1(g_0^{(1)}) = g_1^{(1)}$$
$$g_1(g_0^{(2)}) = g_1^{(2)}$$

for the two values of g_1 and obtain the formula

$$g_0^{(c)} = (g_1^{(1)}g_0^{(2)} - g_1^{(2)}g_0^{(1)}/(g_1^{(1)} - g_1^{(2)}) \qquad (A.21)$$

for the correct value of g_0.

In the case of a nonlinear problem we may proceed in the same way, but $g_0^{(c)}$ will not (unless by coincidence) be the correct value of g_0. Hopefully, it will be a better estimate than either $g_0^{(1)}$ or $g_0^{(2)}$, and we may calculate the corresponding $g_1^{(c)}$ and use this to replace one of our first two values in (A.21). This gives us an iterative process of successive approximation which we hope will converge to the correct answer. There are ways of speeding up the convergence of this process, based on standard methods of solving nonlinear algebraic equations of the form of (A.20).

The alternative boundary-value methods are slightly less convenient to use and are less commonly available as ready-to-use subroutines. In fact the Harwell Subroutine Library does have two routines, one for linear problems of the form

$$d^2y/dx^2 + f(x)dy/dx + g(x)y = r(x)$$

with boundary conditions of the form

$$a\,dy/dx + by = c$$

for two values of x, and a similar one for nonlinear problems which deals with equations of the same form, except that the functions f, g, and r may all depend on y and dy/dx as well as x. This latter routine works by repeated linearization, and should certainly work for mildly nonlinear problems.

The principal situation in which a boundary-value method may be chosen is where initial-value methods lead to instability. It is possible to have a problem where use of an initial-value method gives solutions which amplify excessively any departure from the true solution and so, because of the inherent errors due to rounding (in the computer arithmetic) and due to the approximations made in the use of the numerical method ('truncation errors'), numerical results are likely to be meaningless. It may sometimes be possible to get round this by starting the integration from the other end of the interval ($z = 1$ in our case), but some problems are unstable to integration in either direction. Boundary-value methods generally keep this sort of instability under control (except to the extent to which this is inherent in the problem as it is posed).

In our example the use of a boundary-value method is not necessary, and in fact the very simple method we have used is significantly less accurate for the same amount of work than the shooting method. However, it does serve to illustrate the approach, and for this reason we give few details. We choose a step size $h \,(= 1/N)$ and denote values of c at $z = 0, h, 2h,..., kh,..., Nh$ by by $c_0, c_1, c_2,..., c_N$. We use the approximations

$$(c_{k+1} - 2c_k + c_{k-1})/h^2 \text{ for } d^2c/dz^2 \tag{A.22}$$

$$(c_{k+1} - c_{k-1})/(2h) \text{ for } dc/dz \tag{A.23}$$

and, for the boundary condition only, the slightly less accurate

$$(c_k - c_{k-1})/h \text{ for } dc/dz \text{ at } z = 0 \text{ and } 1 \tag{A.24}$$

(where we take $k = 1$ for the $z = 0$ condition and $k = N$ for $z = 1$). Then we approximate the differential equation at each value of z from h to $(N - 1)h$ using these, which gives the typical equation (at $z = kh$)

$$[(2 + 3kh)/h^2 - 1/h]c_{k+1} - (2 + 3kh)(2/h^2 + 1)c_k$$
$$+ [(2 + 3kh)/h^2 + 1/h]c_{k-1} = 0 \tag{A.25}$$

We have now $N - 1$ equations in $N + 1$ unknown values of c, and shall obtain the other two form the two boundary conditions:

$$c_0 - (0\cdot 4/h)(c_1 - c_0) = 40 \tag{A.26}$$

from (2.24) and, from (2.25),

$$(c_N - c_{N-1})/h = 0 \tag{A.27}$$

The problem has now been reduced to that of solving a set of simultaneous linear algebraic equations, which presents no difficulty in principle. In practice we may have a very large number of these equations so some care has to be taken in the organization of the calculation, preferably by using expertly written subroutines for the solution of such systems. In this case (as for most problems arising from the approximation of differential equations by algebraic equations) the array of coefficients of the equations has a particularly simple structure and special routines which take advantage of this should be sought, as they will be more efficient.

The results quoted in the main text were obtained without this sort of care, but using double precision arithmetic on the IBM 360/67 in Newcastle (about 16 decimal digits as opposed to 7 for single precision). The same program was run using single precision, and the results differ by about 2 in the 4th decimal place. With single precision and $N = 1000$ we get a solution $C = 40$, $dc/dz = 0$ for all z, which should be warning enough of the dangers of too naive an approach. With double precision we can increase N beyond this value, and in fact we get accuracy similar to that of the initial value method for a value of about 8000 for N.

We may note finally that the solution of a nonlinear boundary-value problem by this sort of method raises problems an order of magnitude greater—since solving a large number of nonlinear algebraic equations is not at all straightforward.

(4) Series solution of differential equations

We consider equation (2.22) first, and assume a solution of the form

$$c = a_0 + a_1 z + a_2 z_2 + \ldots + a_k z_k \ldots \tag{A.28}$$

This has derivatives

$$dc/dz = a_1 + 2a_2 z + \ldots + k a_k z^{k-1} + \ldots$$

etc., and putting these series in (2.22) gives us

$$(2 + 3z)(2a_2 + 6a_3 z + \ldots + k(k-1)a_k z^{k-2} + \ldots)$$
$$- 2(a_1 + 2a_2 z + \ldots + k a_k z^{k-1} + \ldots)$$
$$- (2 + 3z)(a_0 + a_1 z + \ldots + a_k z^k + \ldots) = 0$$

If this is to be satisfied for all values of z in some interval, the coefficients of the powers of z must all be zero. We collect terms and get

$$[4a_2 - 2a_1 - 2a_0]$$
$$+ [12a_3 + 6a_2 - 4a_2 - 2a_1 - 3a_0]z$$
$$+ \ldots$$
$$+ [2(k+3)(k+2)a_{k+3} + 3(k+2)(k+1)a_{k+2}$$
$$- 2(k+2)a_{k+2} - 2a_{k+1} - 3a_k]z^{k+1}$$
$$+ \ldots \tag{A.29}$$

Hence we require

$$a_2 = \tfrac{1}{2}a_0 + \tfrac{1}{2}a_1$$

and, for $k = 0, 1, 2, \ldots$

$$2(k+3)(k+2)a_{k+3} = 3a_k + 2a_{k+1} - (k+2)(3k+1)a_{k+2}$$

which is equation (2.27). The two linearly independent solutions are obtained by setting $a_0 = P, a_1 = 0$ and $a_0 = 0, a_1 = Q$. Divergence for $|z| > \frac{2}{3}$ is established by the ratio test since as $k \to \infty$, $a_{k+3} \to \frac{3}{2} a_{k+2}$ and

$$\lim_{k \to \infty} \left(\left| \frac{a_{k+3} z^{k+3}}{a_{k+2} z^{k+2}} \right| \right) > 1 \text{ for } |z| > \frac{2}{3}$$

For equation (2.28) we use series of the form

$$c = y^m (b_0 + b_1 y + \ldots + b_k y^k + \ldots) \tag{A.30}$$

where m is chosen to make b_0 non-zero, and obtained by substitution in (2.28)

$$y^{m-2} \{ m(m-1)b_0 + (m+1)mb_1 y + \ldots + (m+k)(m+k-1)b_k y^k + \ldots \}$$
$$- (2/3y)y^{m-1} \{ mb_0 + (m+1)b_1 y + \ldots + (m+k)b_k y^k + \ldots \}$$
$$- (1/9)y^m \{ b_0 + b_1 y + \ldots + b_k y^k + \ldots \} = 0$$

which, on rearrangement, gives

$$\left\{ m(m-1) - \frac{2}{3}m \right\} b_0 y^{m-2}$$
$$+ \left\{ (m+1)m - \frac{2}{3}(m+1) \right\} b_1 y^{m-1}$$
$$+ \left\{ \left[(m+2)(m+1) - \frac{2}{3}(m+2) \right] b_2 - b_0/9 \right\} y^m$$
$$+ \ldots$$
$$+ \left\{ \left[(m+k)(m+k-1) - \frac{2}{3}(m+k) \right] b_k - b_{k-2}/9 \right\} y^{m+k-2}$$
$$+ \ldots \qquad = 0 \tag{A.31}$$

Now, since $b_0 \neq 0$, the first term here gives

$$m(m - 5/3) = 0 \tag{A.32}$$

(the 'indicial equation') and we have two solutions, one with $m = 0$ and one with $m = 5/3$. The second term can only be zero for these choices of m if b_1 is zero, and then clearly all odd coefficients are zero, and the even coefficients are given by

$$b_{2r} = b_{2r-2}/[9(m+2r)m + 2r - 5/3)]$$

with $m = 0$ and $5/3$ giving the two recurrence relations (2.30). The ratio test here proves convergence for all values of y.

(5) Solution of partial differential equations

Here we give some details of the method of separation of variables for the solution of the diffusion equation (2.32) with conditions (2.33) to (2.34). We seek a solution of the form

$$c(x, t) = X(x) T(t) \tag{A.33}$$

which satisfies the equation and the boundary conditions (but not necessarily the initial condition). Putting (A.33) in (2.32) gives, after rearrangement

$$X''(x)/X(x) = T'(t)/DT(t) \tag{A.34}$$

where we use dashes (') to denote differentiation with respect to the appropriate independent variable. Now the left-hand side of (A.34) is a function of x only, while the right-hand side is a function of t only. The only way in which (A.34) can be satisfied for all values of x between 0 and L and for all values of t between 0 and T is for each side of the equation to be constant. We call the constant value of each function the separation constant, and at present this is arbitrary. We try first a negative separation constant, $-k^2$, and obtain

$$X''(x) = -k^2 X(x); \; T'(t) = -k^2 DT(t)$$

with general solutions

$$X(x) = A \sin kx + B \cos kx; \; T(t) = e^{-k^2 Dt}$$

(Since the choice of X and T whose product is c is not unique, we may choose T to be 1 when $t = 0$.) Now we try to choose A and B to satisfy the boundary conditions

$$c(0, t) = X(0) T(t) = 0 \tag{A.35}$$

and

$$dc/dx|_{x=L} = X'(L) T(t) = 0 \tag{A.36}$$

We do not want $T(t) = 0$ for all t, since this gives the trivial solution $c(x, t) = 0$ always, and this can never satisfy the given initial condition. Then we deduce from (A.35) that $B = 0$, and (A.36) gives us either $A = 0$ which we do not want as again it gives the trivial solution or

$$k \cos kL = 0 \tag{A.37}$$

and hence $k = (2n + 1)\pi/(2L)$ for some integer n.

Had we chosen a positive or zero separation constant we should have discovered that the only possible solution was the trivial solution, and so we have, as the only possible solutions of the form (A.33)

$$c_n(x, t) = A_n \sin \{(2n + 1)\pi x/(2L)\} e^{-(2n+1)^2\pi^2 Dt/(4L^2)}$$

for any integer n. These solutions satisfy the partial differential equation and the boundary conditions, and because these are linear and homogeneous any linear combination of functions c_n also does so. We therefore seek a solution which is a linear combination of these solutions and which in addition satisfies the initial condition. We require

$$c(x,0) = \sum_{n=0}^{\infty} c_n(x,0) = c_0$$

and this gives us a Fourier series

$$c_0 = \sum_{n=0}^{\infty} A_n \sin\{(2n+1)\pi x/(2L)\}$$

in the interval $0 < x < L$.

Standard results for definite integrals of trigonometric functions give us

$$A_n \int_0^L \sin^2\{(2n+1)\pi x/(2L)\}\,dx = \int_0^L c_0 \sin\{(2n+1)\pi x/(2L)\}\,dx \qquad (A.39)$$

and so

$$A_n = (2c_0/L)\int_0^L \sin(2n+1)\pi x/(2L)\,dx$$
$$= 4c_0/\{(2n+1)\pi\} \qquad (A.40)$$

and this completes the solution.

We could have found this solution by Laplace transform methods also, and the two techniques are described in the books listed in the sections on partial differential equations and on engineering mathematics in the bibliography. It is possible to extend the technique to deal with non-homogeneous problems for linear constant coefficient partial differential equations with a little extra work. If things get much more complicated than that it is likely that numerical methods, such as that described in the main text, will have to be used.

2.8 BIBLIOGRAPHY

Mathematical modelling

T. W. F. Russell and M. M. Denn (1972). *Introduction to Chemical Engineering Analysis*, Wiley, New York.

Engineering mathematics

E. Kreyszig (1972). *Advanced Engineering Mathematics*, (3rd ed.). Wiley, New York.
G. Stephenson (1961). *Mathematical Methods for Science Students*, Longmans, London.

Ordinary differential equations

W. E. Boyce and R. C. Di Prima (1969). *Elementary Differential Equations and Boundary Value Problems*, 2nd ed. Wiley, New York.
M. W. Hirsch and S. Smale (1974). *Differential Equations, Dynamical Systems, and Linear Algebra*, Academic Press, New York.

D. A. Sanchez (1968). *Ordinary Differential Equations and Stability Theory: an Introduction,* W. H. Freeman & Co., San Francisco.
G. Sansone and R. Conti (1964). *Non-Linear Differential Equations.* Pergamon, Oxford.
R. A. Struble (1962). *Nonlinear Differential Equations.* McGraw-Hill, New York.

Partial differential equations

G. F. D. Duff and D. Naylor (1966). *Differential Equations of Applied Mathematics,* Wiley, New York.
M. G. Smith (1967). *Introduction to the Theory of Partial Differential Equations,* Van Nostrand, Princeton, N. J.
G. Stephenson (1970). *An Introduction to Partial Differential Equations for Science Students,* 2nd ed. Longmans, London.

Perturbation methods

J. D. Cole (1968). *Perturbation Methods in Applied Mathematics,* Blaisdell, Waltham, Mass.
M. Van Dyke (1964). *Perturbation Methods in Fluid Mechanics,* Academic Press, New York.

Numerical methods (general)

W. S. Dorn and D. D. McCracken (1972). *Numerical Methods with Fortran IV Case Studies,* Wiley, New York.
L. Fox and D. F. Mayers (1968). *Computing Methods for Scientists and Engineers,* Oxford University Press, Oxford.
M. V. Wilkes (1966). *A Short Introduction to Numerical Analysis,* Cambridge University Press, Cambridge.

Numerical methods (differential equations)

L. Fox (Ed.) (1962). *Numerical Solution of Ordinary and Partial Differential Equations,* Pergamon, Oxford.
P. Henrici (1962). *Discrete Variable Methods in Ordinary Differential Equations,* Wiley, New York.
J. D. Lambert (1973). *Computational Methods in Ordinary Differential Equations,* Wiley, New York.
A. R. Mitchell (1969). *Computational Methods in Partial Differential Equations,* Wiley, New York.
G. D. Smith (1965). *Numerical Solution of Partial Differential Equations.* Oxford University Press, Oxford.

Computer Programs

A. E. R. E. Harwell (1971). *The Harwell Subroutine Library: a Catalogue of Subroutines,* HMSO, London (Ref: AERE-R. 6912).
IBM (1970). *System/360 Scientific Subroutine Package,* IBM, New York, (Manual ref: GH20-0205).
IBM (1967). *System 360 Continuous System Modelling Program,* IBM, New York, (Manual ref: H20-0367).
N. A. G. (1974). *N. A. G. Mini Manual,* Numerical Algorithms Group, Oxford.

3

Statistical Techniques in the Field of Water Pollution Control

N. M. D. GREEN

3.1 INTRODUCTION

In writing a short paper on the application of statistical techniques in water pollution control it is not possible to cover every aspect of a rapidly expanding field. For simplicity, it is convenient also to subdivide the subject into three sections that is:

(1) Design of experiments for point estimation of parameters
(2) Descriptive statistics;
(3) Statistical synthesis + simulation for the purposes of prediction.

The rationale behind this subdivision can best be illustrated by an example (see Figure 3.1).

This example examines the statistical techniques required in:

(1) The survey of a polluted river system
(2) A proposal for a model of the system
(3) Subsequent prediction of system behaviour under different operating policies.

Many factors influence the design of the sampling programme used for the survey. One important consideration is the accuracy with which measurements of the instantaneous or time averaged value of a parameter have to be made. Measurement error in the survey results are reproduced throughout any work based on the survey.

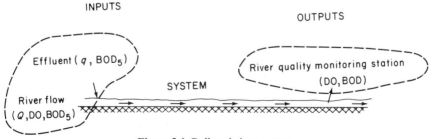

Figure 3.1 Polluted river system

For example, it is a waste of time constructing a sophisticated model of a system and using this model for prediction, if the survey results used to prove the model are subject to large sampling error. Sampling error and similar problems will be discussed in the first section.

Having obtained a series of values of each different parameter from the survey, it is useful to describe their principal characteristics in a recognized format which facilitates comparison.

Most variables which describe natural processes illustrate one component which is dependent on time and another which is dependent upon other variables. Generally the interaction between the variable of interest and others is too complex to be described mathematically. The relationship between the parameter and time can be estimated by the method of harmonic analysis and this component can be removed to leave a residual series. The residual series is usually stochastic in nature. That is each point value of this series will illustrate a random deviation caused by the interaction of a large number of influences. The series as a total, however, will have certain statistical properties that should be constant.

By analysing the total series for each parameter in this way, it is possible to express its characteristics in a statistical 'shorthand' which has most of the advantages of a mathematical relationship.

The methods of analysis used in descriptive statistics together with some examples in application are discussed in Section 3.2.

Finally, one of the most useful applications of statistical techniques is in the field of modelling and simulation of a system for predicting the effects of future loading and control policies.

In the example in Figure 3.1 it would be possible to apply a steady state model to predict DO levels at the river quality station, using the mean values of each input parameter. The DO output would be near to the expected mean of the system. However it would *not* be possible to estimate the probability of occurrence of low DO values and relate these back to decisions on operating the system.

However, if a working model of the system which can accept synthetic statistical input parameters is constructed, the performance of the system can be tested by simulating the system over a long period of real time and sampling the output. Future operating policies can be tested out by changing the inputs or modifying the model of the system.

The above example illustrates just one application of statistical methods for analysing system performance. The final section of this paper will elaborate on these techniques and will point out some of the advantages and pitfalls inherent in the methodology, and the analysis of the output.

3.2 DESIGN OF EXPERIMENTS FOR POINT ESTIMATION OF PARAMETERS

A single measurement of a parameter by sampling and chemical analysis, or by a direct recording, represents the true value of the parameter plus an error term, which reflects the sum of individual deficiencies in the measuring technique.

Each error term may be regarded as a sum of individual random events and should be distributed according to a normal distribution (of central limit theorem, Mood & Graybill (1963)). Certain deficiencies in the apparatus, such as incorrect calibration, will give a definite bias in the *precision* of each measurement. The overall *accuracy* of measurement can be defined by replicate observations on the same sample.

An example using replicate measurements of turbidity using two different measuring techniques will serve to illustrate the difference between precision and accuracy. See Figure 3.2.

Precision can be defined as the difference between the true value and the mean value from measurement. Usually it is not possible to measure precision but the effect of departures in precision can be limited by using a standardized procedure of measurement for both analysis and subsequent control in a particular study. Accuracy can be assessed by the spread of the results about the mean observation, i.e. the sample variance (s^2):

$$s^2 = \frac{\sum (x - \bar{x})^2}{n - 1}$$

By carrying out replicate observations to establish $\bar{x}$ and s, it is possible to place an *approximate* confidence interval on the accuracy of any individual measurement of turbidity. For example: First replace x by the standardizing transform:

$$y = \frac{x - \bar{x}}{s},$$

then

$$f(y) = \frac{1\,e^{-\frac{1}{2}y^2}}{\sqrt{2\pi}}$$

i.e. normally distributed with zero mean and unit variance.

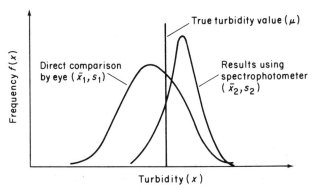

Figure 3.2 Frequency distributions for replicate observations of turbidity using two different methods of measurement

Since only 5% of the area under the normal frequency curve lies beyond the value of y equal to -1.96 then:

$$P(-1.96 \leqslant y < 1.96) = 0.95$$

i.e. 95% confidence intervals on y are ± 1.96
therefore giving 95% confidence intervals on x as $\bar{x} \pm 1.96 \times s$.

This example gives only an approximate result since the errors in estimating $\bar{x}$ and s have been ignored. This error increases with small sample sizes.

The above technique can be used in devising sampling programmes. For example, if the accuracy in measuring turbidity in a survey needs to be within ± 3 units, and twenty replicate measurements on water samples taken at the same time and place gave $\bar{x} = 30$ and $s = 2$, then a single measurement would with 95% confidence lie in the range of $\bar{x} \pm 1.96 \times 2$ units. This is outside the desired accuracy.

If the mean of two sample measurements is taken then the variance of the means will be given by $s^2/2$ and 95% confidence intervals on the means of two samples will be $\bar{x} \pm 1.96 \times 2/\sqrt{2}$, which is within the limits of accuracy required.

In setting up an experiment it may be necessary to compare different experimental procedures to test if there is any significant difference in the results. The method of making this comparison depends on the available number of results for each method. The reader is referred to Mood and Graybill (1963) Chapter 11 and following for further information on this problem.

A further class of problems arise when a parameter cannot be measured directly.

This problem has been discussed in detail by Berthouex and Hunter (1971 a and b). An example taken from their papers will illustrate some of the main features of experimental design.

It will be assumed that the two coefficients L_a and k_1 are to be estimated in the first order BOD equation:

$$y = L_a (1 - e^{-k_1 t})$$

where $\qquad\qquad\qquad y = $ BOD measurements at time t.

The quantity most often used to estimate how well the calculated and observed results agree is the sum of the squares S

$$S = \sum_{i=1}^{n} \{y_{obs,i} - L_a(1 - e^{-k_1 t_i})\}^2$$

where n is the number of observations.

The best estimates of L_a and k_1 are those that minimize the sum of squares S.

By using a computer it is relatively simple to evaluate S over a range of L_a and k_1 values and determine the minimum sum of squares S^* and therefore the best estimates of L_a and k_1. It is essential to place some confidence in the accuracy of the estimates. This can be carried out by mapping a confidence region around the estimates such that the sum of the square error terms within

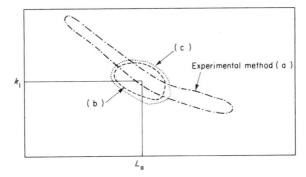

Figure 3.3 Approximate 95% confidence regions for different experimental methods

the region all fall below a critical sum, S_c:

$$S_c = S^* + s^2 p F\alpha(p, n - p)$$

where s^2 is an independent estimate of experimental variance given approximately by:

$$\frac{S^*}{n - p}$$

where p is the number of parameters indicated (2 in this case), and $F\alpha(p, n - p)$ is the $(1 - \alpha)$ 100% point of the F distribution with p and $n - p$ degrees of freedom.
Berthouex and Hunter (1971 a and b) give examples of the 95% confidence interval ($\alpha = 0.05$) for three different experimental methods:

(a) 30 consecutive observations of y up to $t = 5$ days
(b) 50 consecutive observations of y up to $t = 20$ days
(c) 12 observations of y, 6 on the 4th day and 6 on the 20th day.

The results are summarized graphically in Figure 3.3.
It is obvious from this example that the first method gives a very inaccurate determination, whereas the third method produces comparable accuracy with the second with much less experimental work.
The design of statistical experiments which produce small well conditioned confidence regions with the minimum experimental effort is discussed in detail by Berthouex and Hunter (1971 a and b).

3.3 DESCRIPTIVE STATISTICS

If two series of daily effluent BOD data are available from two different treatment works then comparisons can be made using certain statistical parameters of each record.
Green (1972) has shown that BOD data from two treatment works in this

72

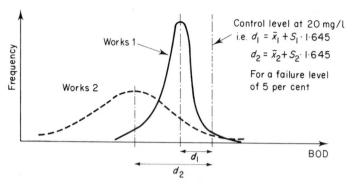

Figure 3.4 Two frequency distributions of BOD output from separate
treatment works

country can be described by the normal frequency distribution, and two typical
distributions are shown in Figure 3.4.

It is obvious that a control level say at 20 mg/l will be violated at some time,
since the tails of the frequency distribution asymtope to zero at a BOD of
infinity. A probabilistic specification of the control (say BOD 20 mg/l for
95% of the time) can be met in different ways. An analogy can be made between
the degree of treatment and the mean BOD output and the degree of control
at the works and the output variance. Both works in Figure 3.4 are meeting
the same probabilistic standards in a different way. Statistical control of either
treatment plant could be based on daily sampling and sequential testing. The
above example just illustrates the use of statistics in describing naturally varying
data. However, the mean, variance and frequency distribution of the variable
are insufficient to describe the dependence of the series in time.

(1) Time series analysis

The simplest way to illustrate all the properties contained in a series of data is
by example. Separate series of air and water temperatures taken at a river
quality station at the same time each day are shown in Figure 3.5.

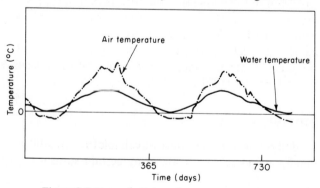

Figure 3.5 Record of air and water temperatures

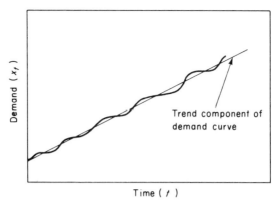

Figure 3.6 Demand curve for domestic water

The causes in the variation of both air and water temperatures can in part be attributed to the rotation of the earth on its own axis (diurnal variation) and around the sun (seasonal component). The variation can be represented by a series of harmonics superimposed on the mean value:

$$x_t = \bar{x} + \sum_{i=1}^{m} A_i \sin(i\omega t) + \sum_{i=1}^{m} B_i \cos(i\omega t) + x_{r,t}$$

where $x_{r,t}$ = the residual value of x_t

A_i and B_i = coefficients of ith harmonic and t is the time from the origin.

The series may also contain long term changes which may be *cyclic* or form a *trend*. These changes may be due to natural or artificial variations in climate or due to artificial changes in control. In the case of water temperature, the construction of a large power station discharging cooling water would change the characteristics of the series. The available length of record is not usually long enough to identify cyclic or trend components due to climatic changes. Trends due to artificial variations can be detected and must be removed before any analysis is undertaken for harmonics. A simple example will serve to illustrate. See Figure 3.6.

Providing the record length is long compared with the annual harmonic the trend can be removed by linear regression analysis of the actual record, i.e.

$$x_t = a + b\,T_t + e_t$$

where a and b are the intercept and slope of the regression line of x_t upon T_t, the temperature, e_t is an error term.

The harmonic analysis can then be based on the transformed series: $y_t = e_t$.

If the record length is small, then the raw data will have to be transformed by a moving average process before regression analysis.

(2) Harmonic analysis

The best estimates of the coefficients A_i and B_i, in a harmonic analysis are given by:

$$A_i = \frac{2}{N} \sum_{t=1}^{N} x_t \sin(i\omega t)$$

$$B_i = \frac{2}{N} \sum_{t=1}^{N} x_t \cos(i\omega t)$$

The harmonic can be replaced by a single cosine term:

$$C_i \cos(i\omega t - \phi_i)$$

$$\text{where } C_i = \sqrt{A_i^2 + B_i^2}$$

$$\phi_i = \arctan(A_i/B_i)$$

The variance accounted for by the ith harmonic is equal to $C_i^2/2$ except when $i = N/2$ then it is C_i^2.

In most cases it is only worthwhile fitting those harmonics that could be expected to be present. For example dealing with similar data, Kothanderaman (1971) fitted the first ten harmonics, but found that the first harmonic, the annual harmonic, explained 95% of the total variance.

It is possible to examine the importance of the various harmonics by two alternative methods: spectral analysis and autocorrelation analysis. Quimpo (1968) studied the advantages of each method for analysing hydrological data and concluded that the comparison was marginal, further since auto-correlation requires less computation and can be used to analyse the $x_{r,t}$ series it will be used in this paper.

A correlogram is a graphical plot of the serial correlation coefficient for increasing lags. The serial correlation coefficient for lag k is defined as:

$$r_k = \frac{\dfrac{1}{N-k} \sum\limits_{t=1}^{N-k} x_t \cdot x_{t+k} - \left(\dfrac{1}{N-k}\right)^2 \sum\limits_{t=1}^{2N-k} x_t \cdot \sum\limits_{t=1}^{N-k} x_{t+k}}{S^2}$$

The correlogram of the raw data for water temperature is shown in Figure 3.7. The presence of the annual harmonic is pronounced.

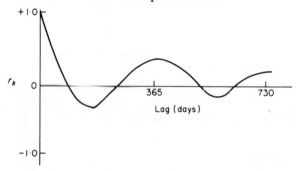

Figure 3.7 Correlogram for raw water temperature data

It is now necessary to define the two essential components of the series. The *deterministic* component is that component which is time dependent, i.e. the part of any parameter that can be predicted knowing only time. The *stochastic* component is that part of the parameter left after removal of the deterministic component—it may be random, dependent on previous values of the parameter, or on other parameters. In the example chosen

$$x_t = \bar{x} + \underbrace{\sum_{i=1}^{N/2} C_i \cos(i\omega t - \phi_i)}_{\substack{\text{deterministic} \\ \text{components}}} \underbrace{- x_{r,t}}_{\substack{\text{stochastic} \\ \text{components}}}$$

(3) Analysis of stochastic component

The residual $x_{r,t}$ series can be examined by removing the deterministic component from the original series. A correlogram for the residual series should show no significant component for the higher lags. Anderson (1942) described a method for setting confidence intervals on the serial correlation coefficient. The correlogram for the residuals is likely to show significant values for low lags demonstrating persistence in the data, i.e. a high temperature on one day is likely to be followed by high temperatures on the following day. In stochastic analysis the object is to fit a model to the persistence so that its effect can be removed to leave a completely random series.

There are several models available. The best model for hydrological data (Kottegoda, 1970) and therefore probably for data in the water pollution field is a Markov model of first or second order.

Typical correlograms for these models are shown in Figure 3.8.

Generally a first order Markov model is adequate and can be represented

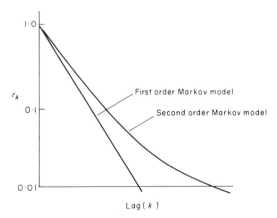

Figure 3.8 Log r_k against k for 1st and 2nd order Markov models

statistically as:

$$x_{r,t} = \rho_1 x_{r,t-1} + e_t \sqrt{1 - \rho_1^2}$$

where ρ_1 is the lag 1 serial correlation coefficient estimated directly from r_1 of the $x_{r,t}$ series, and e_t is a normally distributed random variable with zero mean and standard deviation $(\sqrt{s^2})$ equal to the standard deviation of the $x_{r,t}$ series.

The choice of a Markov model of any particular order can be based on the minimization of the observed variance of the e_t series, taking into account the additional effort involved for fitting higher order models.

After a Markov model has been chosen it is possible to calculate the e_t series and check their structure by correlogram analysis. There should not be any significant structure if the e_t are a random series, see Figure 3.9.

The probability distribution of the e_t series should be checked to ensure that they can be described by a normal distribution. A Chi-squared test (Williams, 1950) or the more sensitive Kolmorgoroff–Smirnov test (Massey, 1951) can be used for this purpose.

The analysis is now complete and the salient characteristics of the water temperature data can be expressed by

$$x_t = \bar{x} + C_i \cos\left(\frac{2nt}{365} - \phi_1\right) + \rho_1 x_{r,t-1} + e_t \sqrt{1 - \rho_1^2}$$

Two assumptions have been made in the analysis of the data, i.e.

(1) The data is stationary
(2) That the x_t and e_t series are normally distributed

Stationarity means the non-variance, except by chance fluctuations of the statistical properties such as the mean, standard deviation, and covariance, when compared for different sections of a series of data. For a complete analysis of a series, stationarity at the highest possible level is a necessary criterion.

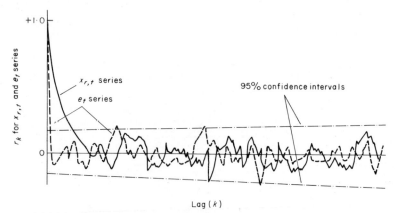

Figure 3.9 Correlogram for $x_{r,t}$ and e_t series

A simple example of non-stationarity is with river flow data where both the mean and standard deviation are different for winter and summer flows. Non-stationary data will lead to bias in the estimation of the statistical parameters of the time series, e.g. changes in variance with time will lead to errors in estimating the coefficients in the harmonic analysis.

On a similar basis departures from normality of the e_t series will result in bias of the serial correlation coefficient.

One method of ensuring stationarity and the normality of the e_t series is to transform the original data. The choice of a 'transform' can be based on the distribution of the original series. For example, monthly streamflow data which are highly skewed may be rendered stationary by taking the logarithms of the original series. Gamma and beta transformers may also be used.

3.4 STATISTICAL SYNTHESIS AND SIMULATION FOR THE PURPOSES OF PREDICTION

One of the most useful applications of statistics is in the field of model building for prediction. This approach is illustrated diagrammatically in Figure 3.10a and 3.10b.

The model can take the form of a mathematical function that attempts to describe the essentials of the system response, or may be a 'black box' relationship such as a multiple regression equation, which is based on past measurements of input and output.

Using a model of the system it is possible to simulate the system by routing the input through the model to obtain output that would be characteristic of the system. The inputs are usually time series which can be described using

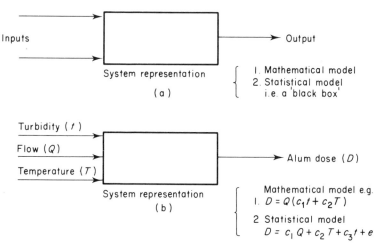

Figure 3.10 (a) Idealized case
(b) Simple model representation linking Alum dose to input variables in an upflow clarifier (fictitious)

the analysis in the previous section, also the resulting output will be a time series. Certain statistical characteristics of the output can be relevant to different design purposes. For instance, in the example given in Figure 10b, the mean and standard deviation of the dose would give an indication of running costs; extreme value of dose would give storage requirements. Unfortunately, the estimates of extreme events from the short records usually available are unstable. In this case the records can be extended to an adequate length by synthesis.

(1) Synthesis of time series

In the example of section 3.3 it would be possible to generate a synthetic trace of water temperatures by generating a new set of e_t terms and then reversing the analysis. The synthesized record should have the same statistical properties as the original series and generation of an adequate record would not be possible.

An alternative method of generating water temperatures would be to use a cross correlation model. For instance, if a long record of air temperatures was available, these could be used to extend the record of water temperatures, providing the cross-correlation coefficient was high. A frequent mistake made by many engineers is to cross correlate the raw series. However, this practice introduces a spurious correlation, since the data points are not independent. To avoid this error both series have to be reduced to random sets, i.e. the e_t series before correlation. The model for generating water temperature would then be:

$$x_t = \bar{x} + c_1 \cos(\omega t - \phi_1) + \rho_1 x_{r,t-1} + (1 - \rho_1^2)^{1/2} \{be_t^1 + f_t(1 - \rho_c^2)^{1/2}\}$$

where e_t' are the residuals of air temperature

ρ_c is the cross correlation coefficient between the e_t and e_t' series
b is the slope of the regression line of e_t on e_t'

and f_t is a random series with zero mean and standard deviation equal to the standard deviation of the e_t series

Generation of the water temperatures can be made by choosing zero value for the first x_r term and using the e_t residuals, remembering also that the x_r term for the next value to be generated comprises the term:

$$\rho_1 x_{r,t-1} + (1 - \rho_1^2)^{1/2} \{be_t^1 + f_t(1 - \rho_c^2)^{1/2}\}$$

from the previously generated result.

In the example in Figure 3.10b it is apparent that cross correlation coefficients of the residual series in the inputs should be preserved if synthetic inputs are to be generated.

Prediction of future conditions can be made by changing the input parameters. For instance an increase in demand can be accounted for by changing the definition of the flow input and rerunning the system.

(2) System sensitivity

It is highly desirable to know the sensitivity of the system to a particular parameter. First, as a guide to the collection of further data, and secondly to assist further model improvements. For example, by changing the inputs it may be found that the Alum dose in the system in Figure 3.10b is sensitive to turbidity but not to temperature. It would be useful therefore to increase the accuracy of future work for measuring turbidity. There would also be little point in improving the statistical definition of temperature in the model.

Although the example used in this section, Figure 10b, is fictitious it does demonstrate the potential of statistical synthesis and simulation. Applications in the water pollution control field are described by various authors, i.e.

Kothanderaman (1971)	— dealing with water temperature
Thomann (1970)	— waste treatment plant characteristics
Berthouex and Polkowski (1970)	— growth of demand and design capacities
Montgomery and Lynn (1964)	— strategies for operating sewage treatment works
Green (1972)	— DO model for polluted rivers

There have also been many applications in the operation of water resource systems.

Finally it is worthwhile looking at multiple regression models for estimating one parameter from others. For instance, in Figure 10b a statistical model relating Alum dose to other parameters is proposed, i.e.

$$D = c_1 Q + c_2 T + c_3 t + e$$

dependent parameter   independent parameter

In many engineering problems where such a relationship is required, the engineer is tempted to carry out a multiple regression analysis adding parameters he thinks may be important (possibly rationalized by dimensional groupings) until a respectable value for the multiple regression coefficient emerges. In doing this, he frequently violates basic statistical requirements for the independent parameters, i.e. that they are error free and independent. In the above equation all the independent parameters are likely to be correlated and instability in the coefficients will be present unless principal component regression or some other technique is used. The error term (e) will also contain all the effects of measurement error in the other parameters.

Problems of this nature are investigated by Wallis (1965) who compares several methods of analysis that take into account correlations between the independent parameters.

3.5 CONCLUSION

A variety of statistical techniques have been presented, which the author feels have widespread application in a field where exact mathematical relationships cannot be expected. In particular the use of simulation using statistically generated data will prove increasingly useful as systems for water pollution control become more sophisticated and integrated into general plans for water use.

3.6 REFERENCES

Anderson, R. L. (1942). 'Distribution of the serial correlation coefficient', *Annals of Math.Stat.*, **13**, 1–13.

Berthouex, P. M., and Hunter, W. G. (1971a), 'Statistical experimental design—BOD tests; *Am.Soc.Civ.Eng.*, *Jour.San.Div.*, **97** (SA4), 393–407.

Berthouex, P. M., and Hunter, W. G. (1971b). 'Problems associated with planning BOD experiments! *Am.Soc.Civ.Eng.*, *Jour.San.Div.*, **97** (SA3), 333–344.

Berthouex, P. M., and Polkowski, L. B. (1970). 'Design capacities to accommodate forecast uncertainties', *Am.Soc.Civ.Eng.*, *Jour.San.Div.*, **96** (SA5), 1183–1210.

Green, N. M. D. (1972). '*A stochastic model for a polluted river*', unpublished Ph.D. thesis, University of Birmingham, England.

Kothanderaman, V. (1971). 'Analysis of water temperature variations in large river, *Am.Soc.Civ.Eng.*, *Jour.San.Div.*, **97** (SA1), 19–31.

Kottegoda, N. T. (1970). 'Statistical method of river flow synthesis for water resources assessment', *Proc.Inst.Civ.Eng.*, Supplement 18.

Massey, F. J. (1951). 'The Kolmogorow-Smirnov test for goodness of fit', *Jour.Am.Stat. Assoc.*, **46**, 68–79.

Montgomery, M. M., and Lynn, W. R. (1964). 'Analysis of sewage treatment by simulation', *Am.Soc.Civ.Eng.*, *Jour.San.Div.*, **90** (SA1), 73–79.

Mood, A. M., and Graybill, F. A. (1963). *Introduction to the theory of statistics*, McGraw-Hill.

Quimpo, R. G. (1968). 'Autocorrelation and spectral analysis in hydrology', *Am.Soc.Civ. Eng.,Jour.HY.Div.*, **94** (HY2), 363–373.

Thomann, R. V. (1970). 'Variability of Waste Treatment Plant Performance', *Am. Soc. Civ. Eng.*, *Jour.San.Div.*, **96** (SA3), 819–837.

Wallis, J. R. (1965). 'Multivariate Statistical Methods in Hydrology—a comparison using data of known functional relationship'. *Wat.Res.Res.*, **1**, (4), 447–461.

Williams, C. A. (1950). 'On the choice of the number and width of classes for the Chi-square test of goodness of fit'. *Jou.Am.Stat.Assoc.*, **45**, 77–86.

4

Optimization and its Application to a Unit Process Design Problem

J. KNAPTON

4.1 INTRODUCTION

Although many reviews of optimization methods have been undertaken (see References), most are written from a mathematical viewpoint and few seek to help the engineer in solving real problems. The methods presented here comprise those which either have been used or show promise of being used to solve engineering problems. They are presented in a simple manner which should ensure their applicability. As examples of typical applications, two methods are applied to a design problem.

4.2 CLASSIFICATION

Although optimization is a developing discipline, the methods available fall into a generally accepted classification. The three main groups are linear mathematical programming, nonlinear mathematical programming and non-linear iterative methods.

Of the linear programming methods, the linear simplex method (Chung, 1963; Dantzig, 1963) is the most widely used. However, it can be used for solving only continuous variable problems. If other variables exist, cutting plane (Kelly, 1960), zero-one (Greenberg, 1970) or branch and bound methods (Lawler and Wood, 1966) are employed.

Nonlinear mathematical programming methods are currently receiving much attention from research workers. With all the methods, except dynamic programming, the objective function and constraints must be written explicitly. The third group, nonlinear iterative methods, comprises those methods in which an algorithm is used to search for the optimum. Neither constraints nor objective functions need be written explicitly.

A classification of nonlinear iterative methods is presented by Spang (1962) who divides them into sequential and non-sequential ones. Whereas sequential methods use information at a given design point to determine the position of the next trial point, non-sequential methods choose new designs randomly. This latter group is less efficient than the former and is, therefore, disregarded henceforth. Spang (1962) goes on to subdivide the sequential group into three,

82

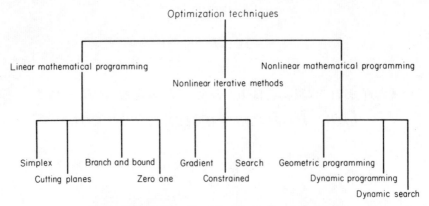

Figure 4.1 Classification of optimization techniques

i.e. those methods which use the cost function's gradient, those which solve constrained problems and those which reduce the design space. Only the first two are relevant. Spang's classification is now extended to form the one shown in Figure 4.1. Optimization methods are now described according to the classification.

4.3 LINEAR MATHEMATICAL PROGRAMMING

The simplex method is described first since the other methods, i.e. cutting planes, zero-one, and branch and bound, all require an initial solution provided by the simplex method.

(1) Simplex method

An excellent account of the method is given by Metzger (1967). If the problem is

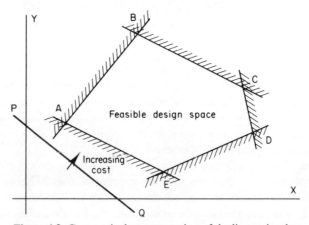

Figure 4.2 Geometrical representation of the linear simplex method

continuous and both the cost function and the constraints can be written as linear inequalities, then the method offers an efficient solution.

The method can be understood by considering the geometrical representation of the two independent variable cases as shown in Figure 4.2. Feasible design space is defined by the polygon ABCDE, the sides of which are the constraints. Thus, any point within the polygon is feasible but the optimum is where the cost line PQ takes its maximum value. Figure 4.2 shows that this always occurs at one of the polygon vertices. Thus the simplex method examines each vertex until it finds the one that cannot be improved upon. The simplex algorithm to accomplish this was first developed by Dantzig (1963) and has been programmed frequently (IBM, 1968).

(2) Cutting plane methods (Kelly, 1960)

When the solution to the linear programming problem is to be integer in some or all of the independent variables, the solution provided by the simplex method is generally infeasible. Cutting plane methods take the simplex method's continuous solution and amend it to form an integer solution by generating extra constraints. Considering the previous example, if the independent variable Y must take an integer value, then horizontal lines, known as cutting planes, define feasible zones within the continuous polygon. Typical cutting planes are shown in Figure 4.3, and their equations are given by Toakley (1968). The method ensures that the vertices of the simplex occur at integer values, thereby forcing an integer solution. The addition of extra constraints reduces the size of the feasible polygon so that generally the optimum cost of an integer design is greater than the corresponding continuous one. The continuous solution is, therefore, termed a lower bound of the integer problem.

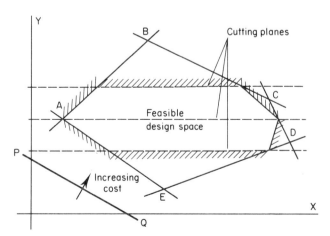

Figure 4.3 Reduction of feasible design space by cutting planes

(3) Branch and bound methods (Greenberg, 1970 and Lawler and Wood, 1966)

Whereas cutting plane methods use the simplex method to solve a problem which has been further constrained, branch and bound methods solve the continuous problem and develop the integer solution from it. The method finds that integer solution which is nearest the continuous one. From the continuous solution, two new problems are generated, each of which is a descendant of the original. In the first the constraint

$$x_n \leqslant i$$

is added and in the second,

$$x_n \geqslant j$$

is added where

x_n = a continuous variable, obtained by simplex
i = nearest integer below x_n
j = nearest integer above x_n

repeated for each independent variable until the solution is sufficiently integer. Each step uses the better of the previous descendants as its original so that each new solution has one more integer variable.

(4) Zero-one programming (Plane and McMillan, 1971)

The method most commonly used to solve zero-one problems is implicit enumeration in which all solutions are considered but the vast majority are evaluated implicitly. The method first sets all the independent variables to zero. The constraints are examined, one is selected, and some of the variables are set to 1 to satisfy that constraint in such a way that as little cost as possible is incurred. With these variables set to 1 and the rest at 0, the remaining constraints are examined. If the setting of the variables to 1 has not made the satisfying of the remaining constraints impossible, one or more of the variables is next set to zero in the partial solution. Once a partial solution has been obtained, all the solutions for which the variables are opposite to those forming that partial solution can be rejected and are said to be implicitly evaluated. The problem, now smaller, is approached afresh and more variables are appended to the partial solution. This is continued until all the variables are included in the partial solution. Balas (1965) suggests an amendment in which only one variable is set from 0 to 1 in a given step, that variable being the one incurring least cost. Its effect on each constraint is examined and providing all the constraints can still be satisfied, it remains at 1, otherwise it reverts to zero. This is repeated for each variable until a complete solution is provided.

4.4 NONLINEAR ITERATIVE METHODS

One of the early workers in the nonlinear iterative methods field was Box (1957, 1951, 1954) who pioneered gradient methods. Gradient methods can be sub-divided into those which require derivatives and those which do not (Crockett and Chernoff, 1955).

The nonlinear iterative division comprises both these groups plus techniques for handling constraints, all of which forms Spang's (1962) sequential group.

(1) Iterative methods requiring derivatives

The steepest descent method (Crockett and Chernoff, 1955, Fiacco and McCormick, 1963, Kowalik and Osborne, 1968) requires an initial design from which small steps are taken through design space until the optimum is attained. If the original design is x_0, then the first step is represented by the equation

$$x_1 = x_0 + h\mathbf{d}$$

where x_1 = new design
h = step length
$\mathbf{d}$ = direction vector
A value for h is chosen so that the distance between successive designs is acceptable, while the vector $\mathbf{d}$ is chosen to ensure that the step is downhill in 'the direction of steepest descent'.
The ith term of $\mathbf{d}$ is given by (Powell, 1964):

$$d_i = \frac{\dfrac{\delta(\text{cost})}{\delta x_i}}{\sum_{j=1}^{n} \dfrac{\delta(\text{cost})}{\delta x_j}}$$

where n = number of independent variables.

Thus, the projection of the step length on each variable's axis is proportional to that variable's partial derivative. Therefore, as interactions proceed, the path of successive designs follows the direction of steepest descent.

Algorithms have been written in which the step length varies (Box, Davies, and Swann, 1969) so that a shorter step length can be used near the optimum.

If derivatives cannot be calculated analytically, they may be found by considering finite differences (Zanwgill, 1967). Thus, if two near designs have been evaluated at design points x_n and x_{n+1}, the partial derivative for the jth variable is given by:

$$\frac{\delta(\text{cost})}{\delta x_j} = \frac{\text{cost}(x_{n+1}) - \text{cost}(x_n)}{h_j} + O(h_j^2)$$

where h_j = step length projected onto x_j axis.

The last term represents the error in assuming linearity. Therefore, h_j must be small enough for h_j^2 to be negligible.

A disadvantage of steepest descent is that the search direction coincides with the steepest descent direction only at the design point. This problem is overcome in Newton's method (Box, Davies, and Swann, 1969) which offers a one step solution for quadratic functions and an economical solution for functions approximating a quadratic. Consider the second order Taylor expansion of the cost function $f(x_1, x_2, \ldots x_n)$ about the optimum point $\mathbf{x}^*$. Then:

$$f(\mathbf{x}) = f(\mathbf{x}^* + \mathbf{h})$$

where $\mathbf{x}$ = a design point nearer the optimum

      $\mathbf{h}$ = vector of components of line joining $\mathbf{x}^*$ to $\mathbf{x}$

      $f(\mathbf{x})$ = cost of $\mathbf{x}$

Taylor's expansion gives:

$$f(\mathbf{x}) = f(\mathbf{x}^*) + \sum_{j=1}^{n} h_j \frac{\delta f}{\delta x_j}\bigg|_{\mathbf{x}^*} + \tfrac{1}{2} \sum_{j=1}^{n} \sum_{k=1}^{n} h_j h_k \frac{\delta^2 f}{\delta x_j \delta x_k}\bigg|_{\mathbf{x}^*}$$

where h_j = component in j^{th} direction of line joining $\mathbf{x}^*$

      n = number of independent variables

The above function defines the cost at design point $\mathbf{x}$ in terms of the optimum cost and the properties of the cost function. If the function is differentiated, the following equations can be written for each independent variable x_1.

$$\frac{\delta f}{\delta x_1} = \frac{\delta f}{\delta x_1}\bigg|_{\mathbf{x}^*} + \sum_{j=1}^{n} h_j \frac{\delta}{\delta x_j \delta x_1}\bigg|_{\mathbf{x}^*}$$

By definition, the slope at the optimum is zero. Therefore, the above equation can be written in matrix form:

$$\mathbf{g} = \mathbf{h} \cdot \mathbf{s}$$

where $\mathbf{g}$ = gradient vector at $\mathbf{x}$

      $\mathbf{s}$ = matrix of second order differentials at the optimum

Now, at any design point $\mathbf{x}$, if first and second order partial derivatives can be evaluated, the above set of simultaneous linear equations can be solved to yield $\mathbf{h}$ which is a vector of design changes. If the cost function is quadratic, $\mathbf{s}$ is constant and its value at $\mathbf{x}$ will be the same as at $\mathbf{x}^*$. In this case, the optimum will be found in one step. For non-quadratic surfaces, an iterative procedure must be used, as with steepest descent.

It might appear that this method is of limited use since it offers no improvement on the steepest descent method for all but quadratic functions. However, near the optimum many functions approximate to quadratics. Therefore, this method is used extensively (Fletcher and Reeves, 1964; Fletcher and Powell, 1963; Rosenbrook, 1960; Wild and Reightler, 1967) to complete the search undertaken by other methods.

(2) Iterative methods requiring no derivatives

In 1957, Box (1957) made an important contribution to the development of

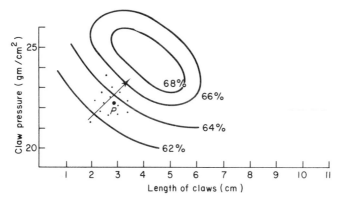

Figure 4.4 Natural evolution example of Box (1957)

gradient methods when he presented the concept of evolutionary operation. Box (1957) used the following example to illustrate this method:

Natural evolution is caused by natural selection and genetic variability due to mutation. Taking the example of a lobster, the objective function can be thought of as the percentage surviving long enough to reproduce, while the two independent variables are claw length and pressure exerted between the claws. The problem can be seen geometrically in Figure 4.4 which shows the contours of the objective function. Point P represents a lobster and the dots around it are its offspring. Since those in the direction of the arrow have a greater chance of survival, after a few generations the point P will creep nearer the optimum.

Evolutionary operation developed into the non-linear simplex method (Fletcher, 1970a; Nelder and Mead, 1965; Spendley, Hext, and Himsworth, 1962) which replaces the scatter of points by a simplex. A design is evaluated at each corner of the simplex and the most expensive is found. This design is replaced by a new one which is found by applying a set of rules. This step is continually repeated in such a way that the simplex creeps through design space until the optimum is attained. The method at first proposed (Spendley,

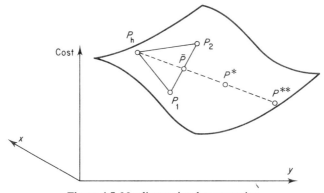

Figure 4.5 Nonlinear simplex operation

Hext, and Himsworth, 1962) can be understood by reference to the two independent variable case in Figure 4.5. If there are n independent variables, then $n + 1$ designs are generated to form a regular simplex. Thus in Figure 4.5, the simplex is P_h, the most expensive design, P_1, the least expensive design and P_2 the intermediate one. The design to replace P_h is found as follows. Let P be the centroid of all designs excluding $\bar{P}_h$, then the jth variable of P, i.e. x_j^c, is given by:

$$x_j^c = \frac{\sum\limits_{i=1}^{n+1} [x_j^i - x_j^h]}{n}$$

where n = number of independent variables

x_j^i = value of jth variable of design i

x_j^h = value of jth variable of worst design

The point P_h is reflected through $\bar{P}$ to obtain the point P^*. The jth variable x_j^d at P^* is given by:

$$x_j^d = x_j^c + \alpha(x_j^c - x_j^h)$$

where α = reflexion constant.

The point P^* is evaluated and depending upon its value, one of four alternative steps is taken.

(a) If P* becomes the best design

This indicates a successful move and suggests that more improvement will be gained if the line P_hP^* is extrapolated to form point P^{**}. The jth term of P^{**}, i.e. x_j^e, is given by:

$$x_j^e = x_j^c + \gamma(x_j^d - x_j^c)$$

where γ = expansion coefficient.

P^{**} is evaluated and if it is a better design than P^*, it is accepted as the new design, otherwise P^* is taken instead. The new simplex is thus either P_1, P_2 P^* or P_1, P_2, P^{**}.

(b) If P* is worse than every point except the worst

A new point, P^{**}, is generated by interpolating between $\bar{P}$ and P^*. The jth term of P^{**}, i.e. x_j^e, is given by:

$$x_j^e = x_j^c + \beta(x_j^h - x_j^c)$$

where β = contraction coefficient.

The contracted design is evaluated and the better of P^* and P^{**} is taken to replace P_h so that the new simplex becomes either P_1, P_2, P^* or P_1, P_2, P^{**}.

(c) *If P* is worse than some of the vertices*

In this case, there is no evidence to indicate that either an expansion or a contraction will yield a better design. Therefore, *P** is taken to be the new vertex.

(d) *If P* is worse than every point*

The simplex is reduced in size since it is probably near the optimum. The conventional approach is to halve the simplex side length, retaining only the current best point. This shrinkage is accomplished for the ith point by the equation

$$P_i = 0.5(P_1 + P_i)$$

The aim of each of the four alternatives is to generate a new simplex nearer the optimum than the current one. Thus, the simplex creeps towards the optimum by continual rejection of the worst design point. Many techniques have been proposed to improve the method's efficiency. In the original version, the coefficients α, β, and γ were 1, 2, and 0.5 respectively. Davidon (1959) proposed a method in which the coefficients could be varied dynamically. Nelder and Mead (1965) altered the strategy for dealing with unsuccessful points. Instead of a contraction, they advocated reflecting the next worse point. However, it would be wrong to state that one technique is better than another since the success of the method depends primarily on the cost surface, as is reported by Box (1966). An example is presented later illustrating this method.

(3) Constraint processing for iterative methods

Introduction

During the description of the iterative methods, no mention has been made of constraints. There exist several techniques for converting constrained non-linear problems into unconstrained ones. It is usual to use one of these methods followed by an unconstrained optimization method to solve constrained problems. The three constraint elimination methods to be outlined are Lagrangian multipliers (Wilde and Reightler, 1967), penalty functions (Marcal and Gellatly, 1968; Fox, 1971; Fletcher, 1971) and that due to Paviani (1969).

Lagrangian Multipliers

This method is particularly applicable to problems having equality constraints although Dorn (1960) has extended the method to include inequalities. Consider a problem subject to the S equality constraints:

$$e_j(x_1 \ldots x_n) = 0 \text{ for } j = 1, 2 \ldots S$$

where n = number of independent variables.

If the objective function is $f(\mathbf{x})$, then it can also be written:

$$f(\mathbf{x}) + \sum_{j=1}^{s} [\lambda_j e_j(\mathbf{x})]$$

where λ_j is the jth Lagrangian multiplier.

The above function is termed the Lagrangian function. One of the properties of this function is that one of its turning points is the optimum design of the constrained problem. Thus, if the Lagrangian function is partially differentiated with respect to each independent variable and the result is set to zero, a set of equations is generated which, when combined with the equality constraint equations, form an unconstrained optimization problem. Therefore, the problem becomes, solve:

$$e_j(x)_1 \ldots x_n) = 0 \text{ for } j = 1, 2, \ldots s$$

$$\frac{\delta f}{\delta x_1} + \sum_{j=1}^{s} \gamma^j \frac{\delta e_j}{\delta x_1} = 0 \text{ for } i = 1, 2, \ldots n$$

This set of nonlinear equations can be solved by one of the methods described previously.

The extension to include inequality constraints requires the use of slack variables as described by Metzger (1967).

Penalty functions

The concept of penalty function methods is simple. With exterior penalty functions, a sum is added to the cost of a design which violates constraints. Exterior penalty functions are illustrated by Marcal and Gallattey (1968) and Moe (1972) in which the penalty transforms a slope into a valley as shown in Figure 4.6. The effect is to generate a surface upon which an unconstrained

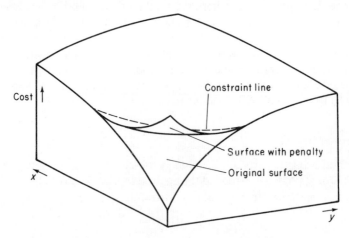

Figure 4.6 Effect of exterior penalty function

method will succeed. Instead of finding designs near the bottom of the slope, the unconstrained method is deflected into the artificial valley.

Interior penalty function techniques were proposed by Carroll (1961) whose created response surface method replaces a constrained problem by a series of unconstrained ones. The method requires that all constraints be expressed in the form $R_i \geqslant 0$ where R_i represents the ith constraint. If C represents the cost, then a new cost function, E, is defined as:

$$E = C + r \sum_{i=1}^{m} \frac{W_i}{R_i}$$

where m = number of constraints

r = constant

W_i = weight applied to the ith constraint

The constant, r, controls the effect of the penalty function. If r is zero, the effect is nil and the surfaces E and C coincide. The method consists of repeatedly optimizing on the created response surface E. During the first optimization r is set to 1 and is incrementally reduced for each successive optimization. The optimum point found with $r = 1$ is used as the starting point for the next optimization, with r at say 0·5. When r has become small, say 0·01, the difference between C and E is sufficiently small for the optimum to be taken as the true optimum. The effect of this method can be seen on Figure 7 in which the optimum cost is required for the function $C = x$ subject to $x \geqslant X_1$. It can be seen that as r is reduced, the created surface approximates to the true one. Successive designs obtained by Carroll (1961) are shown on Figure 6. Carroll (1961) suggests using the method of steepest descent, although there is no reason why any nonlinear iterative method should not be used. Interior penalty function methods have the advantage of not permitting infeasible points. Their chief disadvantages are that many function evaluations are required and that the optimum can never be the true optimum exactly.

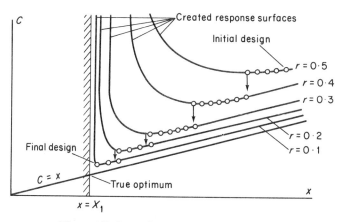

Figure 4.7 Created response surface technique

Paviani's Method

Paviani (1969) suggests a constraint processing method in which constraints are allowed to be violated by a variable amount. He suggests using the nonlinear simplex method together with a sub-optimization technique which ensures that constraint violation is tolerably small. Initially, a large violation is permitted, equal to twice the simplex side length. As each new design is generated, it is checked for feasibility and if it is unsafe, it is adjusted until it falls within the tolerance criterion. As the simplex progresses, the violation tolerance is reduced so that when the optimum is found, the tolerance is negligible and the design is acceptable.

The sub-optimization problem of ensuring that each trial design is within the tolerance criterion consists of minimizing the amount of violation. The minimum value is zero and the variables are adjusted until the violation is reduced to that value. Thus, the problem is reduced to a series of unconstrained ones on which an iterative technique can be used.

Summary

The three constraint processing techniques all turn the constrained problem into either one or several unconstrained problems. Fletcher (1970b) describes a method which solves constrained problems directly but this is of little relevance. Other methods (Box, Davies, and Swann, 1969; Fletcher, 1969) search along the constraint, a technique termed 'riding the constraint', while others oscillate across the constraint, i.e. 'hemstitching'. These methods require many function evaluations and use complicated algorithms. For these reasons, they are omitted from this survey.

4.5 NONLINEAR PROGRAMMING

(1) Introduction

Of the many mathematical programming techniques, as reviewed by Hadley (1964) and Abadie (1967), the three which have been used most successfully in engineering are dynamic and geometric programming and dynamic search. These three methods are now discussed.

(2) Dynamic programming

Dynamic programming is based upon what has become known as Bellman's Principle (Bellman, 1957; Bellman and Dreyfus, 1962). In the context of this paper, this can be paraphrased into: 'If a series of decisions forms an optimal design, then any given decisions affect only those decisions which follow.'

Nemhouser (1967) explains the method in terms of multi-stage decision theory, while Douty (1972) applies the principle to a structural design problem.

This method can be applied to sequential problems only, and is best described by example, see Section 4.6 (7).

Dynamic programming offers advantages over other mathematical programming methods. Continuous, discrete, integer and zero-one variables can be handled directly. Unimodality of the cost surface is not essential and constraints are handled in the design of the elements forming the design chain (White, 1969). The two disadvantages are firstly that many function evaluations are required and secondly the problem must be sequential.

(3) Geometric programming

In contrast to dynamic programming, geometric programming offers a one step solution to a problem whose cost function constraints can be formulated algebraically. The method has been exploited by Templeman (1970), and is summarized by Duffin, Peterson, and Zener (1965) in the following way.

Firstly the cost must be written as the sum of several components, i.e.

$$g = U_1 + U_2 + \ldots + U_n + \ldots$$

where g = total cost
U_n = cost of nth component

Secondly each cost component must be written as the product of a set of m design variables (t), i.e.

$$U_n = C_n \cdot t_1^{an_1} \cdot t_2^{an_2} \ldots t_m^{an_m}$$

If each constant C_n and all the variables t are positive, then the cost function, g, is termed a posynomial and geometric programming can be applied.

The method is based on the principle that the arithmetic mean of a set of positive numbers is at least as great as the geometric mean (Jacobs, 1967), i.e.

$$\delta_1 U_1 + \delta_2 U_2 + \ldots + \delta_n U_n \geqslant U_1^{\delta_1} \cdot U_2^{\delta_2} \ldots U_n^{\delta_n}$$

where $\delta_1 + \delta_2 + \ldots + \delta_n = 1$
Replacing each $\delta_i U_i$ by U_i, the above inequality can be written:

$$U_1 + U_2 + \ldots + U_n \geqslant \frac{U_1^{\delta_1}}{\delta_1} \cdot \frac{U_2^{\delta_2}}{\delta_2} \ldots \frac{U_n^{\delta_n}}{\delta_n}$$

The left-hand side is now in the same form as the cost function. Therefore, one can now write:

$$g \geqslant \frac{C_1^{\delta_1}}{\delta_1} \cdot \frac{C_2^{\delta_2}}{\delta_2} \ldots \frac{C_n^{\delta_n}}{\delta_n} \cdot t_2^{d_1} \cdot t_2^{d_2} \ldots t_m^{d_m}$$

where $d_j = \sum_{i=1}^{n} \delta_i a_{ij}$ for $j = 1, 2, \ldots, m$

If the δ_i values are chosen to make each d_j zero, all the variables t are eliminated from the cost function, which can be rewritten:

$$g \geqslant \left(\frac{C_1}{\delta_1}\right)^{\delta_1} \cdot \left(\frac{C_2}{\delta_2}\right)^{\delta_2} \ldots \left(\frac{C_n}{\delta_n}\right)^{\delta_n}$$

If the cost g is the optimum, then the inequality can be replaced by an equality so that the above equation gives the optimum cost. In order that the above equation will be valid, the following conditions have been assumed:

$$\sum_{i=1}^{n} d_i = 1 \qquad \text{(the 'normality' condition)}$$

$$\sum_{j=1}^{m} \sum_{i=1}^{n} \delta_i a_{ij} = 0 \qquad \text{(the 'orthogonal' condition)}$$

The above relationships represent a set of simultaneous linear equations which can be solved for the δs. These are substituted into the cost equation to give the optimum cost.

It has been assumed that there are sufficient equations to obtain all the weights δ. If this is not the case, it is necessary to obtain some of the weights in terms of the others. Values must then be assigned to the unknown weights and a search initiated to find the optimum set. The number of unknown weights is termed the number of 'degrees of difficulty' (Templeman, 1972) of the problem.

If constraints are present, they are handled in a similar manner to the objective function. This is illustrated by the following example. Minimize:

$$g = \frac{40}{t_1 . t_2 . t_3} + 40 . t_2 . t_3$$

Subject to:

$$\frac{t_1 . t_3}{2} + \frac{t_1 . t_2}{4} \leqslant 1$$

Rewriting the cost function:

$$g = 40 . t_1^{-1} . t_2^{-1} . t_3^{-1} + 40 . t_2 . t_3$$

The normality condition is:

$$\delta_1 + \delta_2 = 1$$

The condition that all the d_i's are zero is called the orthogonal condition. Each variable has an orthogonal condition equation, which is written below. In these equations, the weights δ_1 and δ_2 refer to the cost function terms while δ_3 and δ_4 refer to the constraint terms. The constants in the equations are the powers to which the corresponding variables are raised in the cost and constraint functions. Thus, the orthogonal condition gives:

$$
\begin{aligned}
\text{for } t_1 \quad & -\delta_1 + \delta_3 + \delta_4 = 0 \\
\text{for } t_2 \quad & -\delta_1 + \delta_2 + \delta_4 = 0 \\
\text{for } t_3 \quad & -\delta_1 + \delta_2 + \delta_3 = 0
\end{aligned}
$$

Solving the orthogonal and normality equations gives the weights δ. These are substituted into the rewritten cost function to give the optimum cost. In order to obtain the design variables, the physical meaning of the weights must be considered. Each object function δ_i is given by:

$$\delta_i = \frac{\text{cost of } i^{\text{th}} \text{ cost function term}}{\text{optimum cost}}$$

Similarly, for each constraint δ_i:

$$\delta_i = \frac{\text{cost of } i^{\text{th}} \text{ term in constraint}}{1/(\text{sum of weights for this constraint})}$$

Thus, for each term in the object function, one can write:

$$\text{cost term} = \delta_i \text{ (optimum cost)}$$

Similarly, for each constraint term one can write:

$$\text{constraint term} = \frac{\delta_i}{\sum (\text{all } \delta \text{s for the constraint})}$$

The above two relationships yield a set of equations which can be solved to yield the optimum set of independent variables.

It can be seen that geometric programming offers an efficient solution to problems having continuous variables, providing the cost and constraint equations can be written algebraically.

(4) Dynamic search

The final method to be considered is dynamic search which has been applied by Majid and Elliott (1971). The method is one of the more sophisticated mathematical programming methods for dealing with continuous variable problems. Consider a problem in n independent variables whose optimum combination lies on an algebraically defined nonlinear constraint. The method takes two of the variables and obtains the optimum solution with these as variables and the other variables held constant. Having found this sub-optimum, the method replaces one of the pair of variables by another one and finds the optimum combination of the new pair. This is repeated until all the variables have been considered and the procedure is terminated.

If the two variables in a given step are U and V, the problem is shown geometrically in Figure 4.8. The lines Z represent constant cost and the constraints are shown hatched on their infeasible side. An initial feasible solution, A, is chosen with U and V equal. The solution is improved by moving perpendicular to the constant cost line until a constraint is reached at B. At the point, the slope of the constraint $\delta U/\delta V_{\text{constraint}}$ and the slope of the objective function $\delta U/\delta V_{\text{objective}}$ are found and a new slope, which is the mean of these, is calculated. Small steps are taken in this direction until a constraint is reached at C. The slope of the cost function is compared with that of the constraint at C. If they

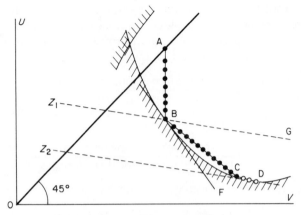

Figure 4.8 Dynamic search technique

are sufficiently near, C is taken as the optimum, otherwise the mean slope is calculated and a further step is taken. This procedure is repeated until a constraint point is found at which the cost function's slope is almost equal to that of the constraint. In Figure 4.8 this occurred at D, where the cost is Z_2. The method offers a very efficient solution for a problem with algebraically defined constraints and a linear cost function. Thus, the method is applicable to a very narrow range of problems. Nonetheless, it illustrates the nature of the more sophisticated nonlinear programming methods (Rosen, 1960; Rosen, 1961).

4.6 APPLICATION OF OPTIMIZATION METHODS

(1) Introduction

Two methods are applied to the optimum cost design of a simple sewage system. The methods to be illustrated are:
(i) Nonlinear Simplex Method
(ii) Dynamic programming

(2) The problem

The plant is as shown in Figure 4.9. It comprises a primary clarifier, an aeration tank, and a secondary clarifier. The only requirement of the plant is that it reduces the BOD from its initial value to 20 ppm. It is assumed that the opera-

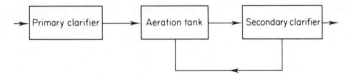

Figure 4.9 Model sewage plant

tion of the three units is governed by the equations below (Malina, 1972). In both methods, it is assumed that the initial BOD and solids are both 400 ppm. The influent flowrate is 1 mgd^{-1} at a temperature of 17°C.

(3) Clarifier assumption

$$P = (\cdot00043\,S + 0\cdot51)\,(1 - e^{-0\cdot7T})$$
$$B = 0\cdot860^2 - 0\cdot029P$$

where P = proportion of solids removed
S = amount of solids (ppm)
B = proportion of BOD removed
T = retention time (h)

$$\left(\text{Note: } T = \frac{\text{volume of clarifier}}{\text{sewage flowrate}}\right)$$

(4) Aeration tank assumptions

$$B_2 = 9\cdot3 + (B_1 - 9\cdot3)e^{-CTm/2000}$$

where B_1 = initial BOD (ppm)
B_2 = final BOD (ppm)
T = retention time (h)
m = mixed liquor concentration = 3000 ppm
C = temperature coefficient
$= 0\cdot0073t^2 + 0\cdot0827t + 0\cdot7162$
where t = temp. (°C)

(5) Design problem formulation

The design should ensure that the effluent has a BOD not greater than 20 ppm. The BOD is reduced to the figure by the primary clarifier, the aeration tank, and the secondary clarifier. In order that each of these three units can be designed, it is necessary to specify intermediate BOD values, i.e. those between the primary clarifier and the aeration tank and those between the aeration tank and the secondary clarifier. Thus, the two independent variables are the two intermediate BOD values since once they have been specified, the volumes and retention times can be determined by the equations in the assumptions.

Therefore, the optimization problem consists of selecting that combination of intermediate BOD values which provides the least cost design. In this context, cost is considered to be initial capital cost and it is assumed that this is a function of the volume. It is further assumed that the cost of each unit is independent of the other units so that cost can be written:

$$\text{cost} = f(V_1) + f(V_2) = f(V_3)$$

where V_1 = volume of primary clarifier
V_2 = volume of aeration tank
V_3 = volume of secondary clarifier

Functions have been chosen which ensure a unimodal cost surface, so allowing a gradient method to operate on the surface.

(6) Nonlinear simplex solution

Two simplex runs are shown superimposed on a contour map of the cost response surface in Figure 4.10. The simplex strategy is as follows. A reflexion is made wherever the worst point yields a design which is better than the current worst. If the reflexion is worse than the current worst, the simplex is shrunk. Shrinkage is accomplished by retaining the current best point and halving the simplex side lengths. The program to accomplish this is shown in the Appendix.

In both simplex examples (see Figure 4.10), two shrinkage steps were made before the optimum was attained. In the first example, the shrinkages occurred as consecutive moves while in the second, two distinct shrinkage steps are taken. Progress was terminated when the simplex side length became less than 1. It can be seen that both examples travel towards the same area and that initial progress is perpendicular to the cost contours. When the elongated valley

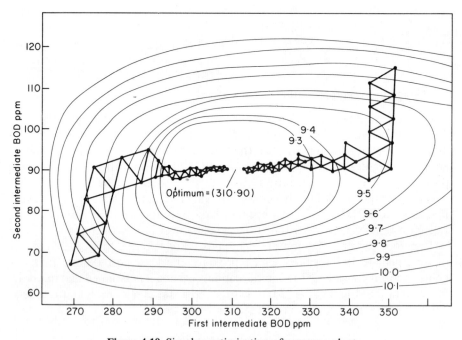

Figure 4.10 Simplex optimization of a sewage plant

Table 4.1

| Design No. | Volume of | | | | |
	Unit 1	Unit 2	Unit 3	Total	Cost
1	379·7	70·6	119·7	570	10·244
2	326·5	64·7	139.0	530	9·510
3	353·8	59·9	151·1	564	9·476
4	316·8	59·8	155·2	531	9·299
5	307·6	55·5	172·3	535	9·281
6	286·6	59·8	159·4	506	9·187
7	260·9	59·8	163·6	484	9·121
8	255·1	59·8	164·7	480	9·110
9	251·0	60·9	161·5	473	9·106
10	249·4	59·8	165·8	474	9·101
11	245·4	60·8	162·6	469	9·096
12	243·9	59·8	166·8	470	9·094
13	240·1	60·8	163·6	464	9·088
14	234·8	60·8	164·7	460	9·082

Table 4.2

| Design No. | Volume of | | | | |
	Unit 1	Unit 2	Unit 3	Total	Cost
1	123·5	55·1	226·9	405	10·034
2	135·1	58·4	202·9	396	9·464
3	123·5	60·7	195·4	380	9·362
4	135·1	61·4	188·2	385	9·254
5	123·5	63·9	181·3	369	9·216
6	135·1	64·7	174·5	374	9·161
7	147·0	65·5	167·9	380	9·130
8	153·1	64·2	171·2	388	9·119
9	159·4	64·6	167·9	392	9·107
10	165·8	63·6	171·2	400	9·099
11	172·3	63·8	167·9	404	9·089
12	175·6	63·1	169·6	408	9·085
13	179·0	63·3	167·9	410	9·081
14	185·9	63·7	164·7	414	9·078

bottom is reached, the direction changes and progress is along the valley. The volumes and costs for the two examples are shown in Tables 4.1 and 4.2.

(7) Dynamic programming solution

Consider the problem solved by the nonlinear simplex method (i.e. reducing the BOD from 400 ppm to 20 ppm at least cost). Let the problem be further simplified by allowing only three possible values for the first intermediate BOD

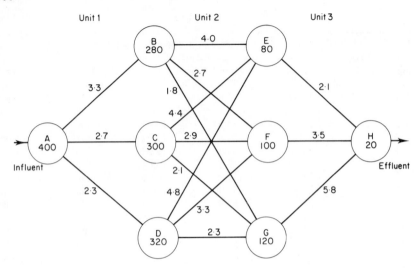

Figure 4.11 Dynamic programming problem formulation

(280 ppm, 300 ppm, and 320 ppm) and three possible values for the second intermediate BOD (i.e. 80 ppm, 100 ppm, and 120 ppm). Thus, there are nine alternative possible designs, the cheapest of which is sought. Figure 4.11 shows the problem in network form. The influent enters at A and the effluent leaves at H. Thus, any path connecting A to H represents a feasible design. The number on a path represents the cost of using that path. Thus, if the primary clarifier reduced the BOD from 400 ppm to 280 ppm, its cost is 3·3 units. The total cost of a typical design is obtained by summing all the costs on links forming that design.

Dynamic programming operates as follows, it being assumed that the unit costs have already been obtained. Once E, F or G have been reached, the remainder of the design is fixed. Once B is reached, there is a choice of subsequent routes (i.e. via E, F or G to H). From B, the total cost via E is 6·1 units, if B is used, the remainder of the optimal route is $B \rightarrow E \rightarrow H$, since the other two alternatives are more expensive. A similar argument applies to C and D such that if C is reached, the remaining optimal route is $C \rightarrow F \rightarrow H$ and if D is reached the route is $D \rightarrow F \rightarrow H$. Consider next point A. From A, it is possible to go through B, C or D. Consider the route to B. If this link is made, the rest of the route has been previously determined ($A \rightarrow B \rightarrow E \rightarrow H$). Similarly for C and D. Thus, if the costs of the three links leaving A are added to the running totals at B, C and D, it becomes obvious that the optimal route is $A \rightarrow D \rightarrow F \rightarrow H$.

Thus, dynamic programming operates by considering the design process as a sequential operation. Therefore, the method allows the optimal design to be obtained without each feasible design being calculated. Its chief drawback is that it can operate only on 'sequential' type problems. Its chief advantage is that it does not require a unimodal cost surface (c.f. steepest descent, simplex).

4.7 REFERENCES

Abadie, J. (1967). *Non-linear Programming*, North Holland Publishing Co., Amsterdam.

Aris, R. (1963). *Dynamic Programming*, Socony Mobil, New York, N. Y.

Aris, R. (1964). *Discrete Dynamic Programming*, Blaisdell Inc., Washington D. C.

Balas, E. (1965). 'An Additive Algorithm for solving Linear Programs with Zero-One Variables', *Operations Research*, **13**, 517.

Bellman, R. E. (1957). *Dynamic Programming*, Princeton U. P., Princeton, N. J.

Bellman, R. E., and Dreyfus, S. E. (1962). *Applied Dynamic Programming*, Princeton U. P., Princeton, N. J.

Box, G. E. P. (1957). 'Evolutionary Operation: a method for increasing Industrial Productivity', *Applied Statistics*, **6**, 81.

Box, G. E. P. (1951). 'On the Experimental Attainment of Optimum Conditions', *Journal of the Royal Statistical Society*, **13**, 1.

Box, G. E. P. (1954). 'The Exploration and Exploitation of Response Surfaces', *Biometrics*, **10**, 16.

Box, M. J., Davies, D., and Swann, K. H. (1969). *Non-linear Optimization Techniques*, Oliver and Boyd, Edinburgh.

Box, M. J. (1966). 'A comparison of Several Current Optimization Methods', *The Computer Journal*, **9**, 67.

Box, M. J. (1965). 'A new method of Constrained Optimization and a comparison with other methods', *The Computer Journal*, **8**, 42.

Brooks, S. H. (1959). 'A comparison of maximum seeking methods', *Journal of the Operations Research Society of America*, **7**, 430.

Carroll, C. W. (1961). 'The created response surface technique for optimizing non-linear restrained systems', *Operations Research*, **9**, 169.

Chung, A. M. (1963). *Linear Programming*, Merrill Books Inc., New York, N. Y.

Cooper, L., and Steinberg, D. (1970). *Introduction to methods of optimization*, W. B. Saunders, London.

Crockett, J. B., and Chernoff, H. (1955). 'Gradient methods of maximisation', *Pacific Journal of Mathematics*, **5**, 33.

Dantzig, G. B. (1963). *Linear Programming and Extensions*, Princeton U. P., Princeton, N. J.

Davidon, W. C. (1959). 'Variable metric method for minimisation', *Argonne National Laboratories Bulletin ANL-5990*.

Davies, D. (1970). 'Some practical methods of optimization', *Integer and Nonlinear Programming*, North Holland Publishing Co., Amsterdam.

Dorn, W. S. (1960). 'On Lagrange Multipliers and Inequalities', *Journal of the Operations Research Society*, **9**, 95.

Douty, R. T. (1972). 'Structural optimization with Dynamic Programming utility tables', *International Symposium on Computer Aided Design*, Warwick.

Duffin, R. J., Peterson, E. L., and Zener, C. (1965). *Geometric Programming*, John Wiley Inc., New York, N. Y.

Fiacco, A. V., and McCormick, G. P. (1963). 'Programming under nonlinear constraints by unconstrained minimisation', *RAC-TP-96*, Research Analysis Corporation, Maryland.

Fletcher, R., and Reeves, C. M. (1964). 'Function minimisation by conjugate gradients', *The Computer Journal*, **7**, 149.

Fletcher, R. (1971). 'Methods for the solution of optimization problems', *Symposium on Computer-Aided Engineering*, University of Waterloo, Canada.

Fletcher, R. (1969). *Optimization* Academic Press, London.

Fletcher, R. and Powell (1963). 'A rapidly convergent descent method for minimisation', *The Computer Journal*, **6**, 163.

Fletcher, R. (1970a). 'The calculation of feasible points for linearly constrained optimization problems', *UKAEA Report AERE R. 6354, HMSO*.

Fletcher, R. (1970b). 'A new approach to variable metric algorithms', *The Computer Journal*, **13**, 317.

Fox, R. L. (1971). 'Unconstrained minimization approaches to constrained problems', *Structural Design applications of Mathematical Programming Techniques*, AGARD 149.

Greenberg, G. (1970). *Integer Programming*, Academic Press, London.

Hadley, G. (1964). *Nonlinear and Dynamic Programming*, Addison Wesley Inc., Mass.

Hadley, G. (1962). *Linear Programming*, Addison Wesley Inc., Mass.

IBM (1968). *The Scientific Subroutine Package*, IBM Corporation.

Jacobs, D. L. R. (1967). *An introduction to dynamic programming*, Chapman and Hall, London.

Kelly, J. E. (1960). 'The cutting plane method for solving convex problems', *S. I. A. M. Journal*, **8**, 703.

Kowalik, J., and Osborne, M. R. (1968). 'Methods for unconstrained optimization problems', *Elsevier*, 82.

Künzi, H. P., Tzschach, H. G., and Zehnden, C. A. (1968). *Numerical methods of Mathematical Optimization*, Academic Press, London.

Lawler, E. L., and Wood, D. F. (1966). 'Branch and Bound Methods: a survey', *Operations Research*, **14**, 699.

Majid, K. I. and Elliott, D. W. C. (1971). 'Dynamic Search Optimization Method', *The Structural Engineer*, **49**, 179.

Malina, J. F. (1972). *Design Guides for Biological Wastewater Treatment Processes*, University of Texas, Austin, Texas.

Marcal, P. V., and Gellatly, R. A. (1968). 'Applications of the created response surface technique to structural optimization', *2nd Conf. on Matrix Methods in Structural Mechanics*.

Metzger, R. W. (1967). *Elementary Mathematical Programming*, John Wiley, London.

Moe, J. (1972). 'Penalty function methods', *International Symposium on Computers in Optimization of Structural Design*, Swansea.

Nelder, J. A., and Mead, R. (1965). 'A simplex method for function minimisation', *The Computer Journal*, **7**, 308.

Nemhauser, G. L. (1967). *Introduction to Dynamic Programming*, John Wiley Inc., New York, N. Y.

Paviani, D. A. (1969). 'Constrained nonlinear optimization by heuristic programming', *Operations Research*, **17**, 117.

Plane, D. R., and McMillan, C. (1971). *Discrete Optimization*, Prentice-Hall, N. J.

Powell, M. J. D. (1964). 'An efficient method for finding the minimum of a function of several variables without calculating derivatives', *The Computer Journal*, **7**, 155.

Rosen, J. B. (1960). 'The gradient projection method for nonlinear programming. Part 1, linear constraints', *S. I. A. M. Journal*, **8**, 181.

Rosen, J. B. (1961). 'The gradient projection method for nonlinear programming. Part 2, nonlinear constraints', *S. I. A. M. Journal*, **9**, 514.

Rosenbrook, H. H. (1960). 'An automatic method for finding the greatest or least value of a function', *The Computer Journal*, **3**, 175.

Spang, H. A. (1962). 'A review of minimisation techniques for nonlinear functions', *S. I. A. M. Journal*, **4**, 343.

Spendley, W., Hext, G. R., and Himsworth, F. R. (1962). 'Sequential Application of Simplex Designs in Optimization and Evolutionary Operation', *Technometrics*, **4**, 441.

Templeman, A. B. (1970). 'Structural design for minimum cost using the method of geometric programming', *Proc. Inst. Civ. Engrs.*, **46**, 459.

Templeman, A. B. (1972). 'Geometric programming with examples of the optimum design of floor and roof system', *International Symposium of Computer Aided Structural Design, Warwick*.

Toakley, A. R. (1968). 'Optimum design using available sections', *Proc. Amsoc.*, **94**, (ST5), 1219.
Wilde, D. J., and Reightler, C. S. (1967). *Foundations of Optimization*, Prentice Hall, N. J.
White, D. J. (1969). *Dynamic Programming*, Oliver and Boyd, Edinburgh.
Zangwill, W. I. (1967). 'Minimising a function without calculating derivatives', *The Computer Journal*, **10**, 293.

4.8 APPENDIX: SIMPLEX METHOD FORTRAN PROGRAM

```
1          DIMENSION  A(21), POINTS(8), VALUE(4)
2          DATA  A/100.,17.,400.,400./
3   C
4   C  INPUT COORDINATES OF THE INITIAL SIMPLEX
5   C
6          READ(5,100) (POINTS(I), I = 1,6)
7   100    FORMAT(6F5.0)
8   1      WRITE(6,101) (POINTS(I), I = 1,6)
9   101    FORMAT(////,'THE  SIMPLEX  IS', 2F8.0,/,15X,2F8.0,/,15X,
            2F8.0
10         &)
11  C
12  C  EVALUATE EACH DESIGN IN THE SIMPLEX
13  C
14         DO 2 I = 1, 3
15         A(5) = POINTS(1*2-1)
16         A(6) = POINTS(1*2)
17         CALL  EVAL(A)
18  2      VALUE(I) = A(21)
19  C
20  C  DETERMINE THE BEST & WORST DESIGNS IN THE
        SIMPLEX
21  C
22         IWER = 1
23         IBES = 1
24         IF(VALUE(3).GT.VALUE(2)) GO TO 4
25         IF(VALUE(2).GT.VALUE(1)) IWER = 2
26         GO TO 6
27  4      IF(VALUE(3).GT.VALUE(1)) IWER = 3
28  6      IF(VALUE(3).LT.VALUE(2)) GO TO 7
29         IF(VALUE(2).LT VALUE(1)) IBFS = 2
30         GO TO 8
31  7      IF(VALUE(3).LT.VALUE(1)) IBFS = 3
32  8      P = 0
33         Q = 0
34  C
```

```
35  C  OBTAIN THE CENTROID OF ALL DESIGNS EXCEPT THE
           WORST
36  C
37         DO 9 I = 1,3
38         IF(I.EO.IWER) GO TO 9
39         P = P + 0.5*POINTS(I*2-1)
40         Q = Q + 0.5*POINTS(I*2)
41  9      CONTINUE
42  C
43  C  PERFORM A REFLEXION STEP
44  C
45         POINTS(7) = 2*P-POINTS(IWER*2-1)
46         POINTS(8) = 2*Q-POINTS(IWER*2)
47         A(5) = POINTS(7)
48         A(6) = POINTS(8)
49         CALL EVAL(A)
50         VALUE(4) = A(21)
51  C
52  C  JUMP TO LABEL 11 IF THE REFLEXION WAS A SUCCESS
53  C
54         IF(VALUE(4).LT.VALUE(IWER)) GO TO 11
55  C
56  C  SHRINK THE SIMPLEX ABOUT THE CURRENT BEST DESIGN
57  C
58         DO 10 I = 1,3
59         IF(I.EQ.IBES) GO TO 10
60         POINTS(I*2-1) = 0.5*(POINTS(I*2-1) + POINTS(IBES*2-1))
61         POINTS(I*2) = 0.5*(POINTS(I*2) + POINTS(IBES*2))
62  10     CONTINUE
63  C
64  C  STEP COMPLETE
65  C
66         GO to 1
67  C
68  C  REPLACE THE CURRENT WORST DESIGN BY THE
           REFLEXION
69  C
70  11     POINTS(IWER*2-1) = POINTS(7)
71         POINTS(IWER*2) = POINTS(8)
72         GO TO 1
73  C
74  C  STEP COMPLETE
75  C
76         END
```

II

Application to Polluted Environments

5

Reservoir Algal Productivity

J. A. STEEL

5.1 INTRODUCTION

It is only fairly recently that attempts to use mathematical modelling techniques in ecological investigation have become widespread. Undoubtedly much of the impetus has been derived from the now relatively easier access to electronic computers. This has led to some problems where the ecological insight and/or data has not matched the numerical analytical technique. Previously it might be true to assert that much modelling investigation was devoted to evolving functional descriptions which, whilst retaining descriptive veracity, were nonetheless analytically tractable. The great advances in numerical techniques which have recently occurred very much emphasize our basic need for a much greater capability for real description of the ecosystem dynamics.

It is clear that our ultimate goal must be to be able to specify the selected species biomass at any point in space and time; a goal still very much only a distant future vision. That being so, we must accept some curtailment in our immediate objectives in modelling. This may for instance mean attempting only short-term predictive accuracy, or a quantitative but not qualitative (or vice versa) assessment of a situation. It is therefore necessary to choose a model type which best accords with our present ecological knowledge and the objectives of the study.

Levins (1966) has identified, in this limited objective approach, three main types of models to which I would add a fourth:

Attributes	*Lacking*
(1) Precision	Realism and generality
(2) Precision and realism	Generality
(3) Generality and precision	Realism
(4) Generality and realism	Precision

The final objective would be:
(5) Generality, Precision and Realism

The grouping tends from site specific, highly quantitative artificial description through to a universal, fundamentally correct qualitative specification. Included in this grouping will be models which range from the extreme complexity of numerous simultaneous partial differential equations to mathematically relatively trivial factoral designs. This contribution is very much a member of the latter type.

I shall define my objective as a model which allows some explanation of the processes upon which organic production rests, attempting at the same time to throw out quantitative statements in order that any emergent hypotheses may be tested. Such testing is clearly necessary for establishing credibility. If this may be achieved then some general statements about production may be possible. These may prove useful in assessing the magnitude of some of the algal problems which may occur in a reservoir, the rapidity with which they may arise, and if there is any means by which they may be ameliorated.

The increasingly available capability for massive environmental intervention (Ridley, Cooley, and Steel, 1966; Symons, Irwin, Robinson, and Robeck, 1967; Steel, 1972a) ideally also requires some means of fundamentally interpreting and predicting response to operational decision. This would then allow an objective attempt to formulate a reservoir management strategy.

5.2 MODEL STRUCTURE

The model is based upon an attempt to describe the change of algal biomass as the resultant of a number of energy fluxes. Let the water body be deep and contain non-limiting nutrient concentrations. In vectorial notation the rate of change is:

$$\frac{\partial n}{\partial t} = \frac{1}{\rho}.\nabla^2(An) - (\nabla n).\mathbf{v} + n(P - R) - Hhn \text{ mg } a\,\text{m}^{-3}\text{d}^{-1}$$

= Turbulent diffusion exchanges through the volume − translation through the volume (regarding sedimentation as an apparent velocity of the volume) + net photosynthesis − grazing loss.

Where: n = Algal concentration   mg chlorophyll a m^{-3}
A = Turbulent diffusion coefficients
ρ = Density
$\mathbf{v}$ = Velocity vector
P = Photosynthetic rate per unit concentration
R = Respiration rate per unit concentration
H = Herbivore concentration
h = Grazing coefficient per unit herbivore concentration

Considerable simplifying conditions are required to render this expression amenable to easy manipulation. The first such assumption will be that the horizontal diffusion coefficients are sufficiently large, and the lateral extent of the water mass sufficiently small, to maintain near homogeneity at any particular depth. Constancy of all other coefficients other than the biological parameters will also be assumed as will any lateral transport being effective in only one direction. Then:

$$\frac{\partial n}{\partial t} = \frac{A_z}{\rho}\frac{\partial^2 n}{\partial z^2} - v\frac{\partial n}{\partial z} + n(P-R) - Hhn - q(n - n_R)\ \text{mg}\,a\,\text{m}^{-3}\,\text{d}^{-1}$$

Where: v = Sedimentation rate
q = Fractional volume change per day (Q/V)
n = River concentration

If we further restrict consideration to a non-grazed alga ($h = 0$) and a condition of no significant throughput, or a river concentration which is a reasonable approximation to that in the reservoir, then:

$$\frac{\partial n}{\partial t} = \frac{A}{\rho}\cdot\frac{\partial^2 n}{\partial z^2} - v\frac{\partial n}{\partial z} + n(P - R) \qquad\qquad \text{mg}\,a\,\text{m}^{-3}\,\text{d}^{-1}$$

This equation may be soluble with suitable boundary conditions; a classic paper illustrating analytic solution of the steady state condition is that of Riley, Stommel, and Bumpus (1949). Consider, however, the conditions in which

$$\frac{\partial^1 n}{\partial z^2} = \frac{\partial n}{\partial z} = 0 \qquad \text{i.e. with '}n\text{' homogeneously distributed with respect}$$

to depth, then:

$$\frac{\partial n}{\partial t} = n(P - R) \qquad\qquad\qquad \text{mg}\,a\,\text{m}^{-3}\,\text{d}^{-1}$$

For constant 'P' and 'R' this would clearly simply integrate into an exponential expression. It is the case, however, that 'P' is depth and time dependent and so there would be unequal growth at different depths, immediately invalidating the initial assumption of vertical homogeneity. Some sort of average value for 'P' and 'R' must be found.

Such consideration has led to a scheme of modelling which regards the energy fluxes per unit area rather than per unit volume, so overcoming the depth dependency of 'P'. The net photosynthesis is taken as the difference between the carbon or oxygen ($\equiv$ energy) captured by the column population and that released by its respiration. Sverdrup, Johnson, and Fleming (1942), Sverdrup (1953), Talling (1957a) and Vollenweider (1965) are all modelling procedures which aim to specify production and/or respiration per unit area.

Laboratory and field study of the light relationships of phytoplankton frequently show that as the light intensity increases from zero the photosynthetic rate increases linearly with it, until at some intensity the photosynthetic rate begins to saturate and become constant at the so-called light saturated photo-synthetic rate, 'P_{max}'. At high intensities reversible inhibition may occur although at extreme radiation levels irreversible damage may take place. 'P_{max}' is due basically to a biological reaction rate limitation and hence has a marked temperature coefficient, somewhat more than doubling for a 10°C increase in temperature.

Ignoring the high intensity inhibition, this behaviour is describable by a simple equation, Smith's equation:

$$P = \frac{aI}{\sqrt{1 + (aI)^2}} \cdot P_{max} \qquad\qquad \text{mg C (mg } a)^{-1} h^{-1}$$

If some recognition of inhibition is required then Vollenweider's functions may be used, the simplest of which is:

$$P = \frac{bI}{1 + (bI)^2} \cdot P_{opt} \qquad\qquad \text{mg C (mg } a)^{-1} h^{-1}$$

It is a simple matter to show that for these two functions to approximate as closely as possible before the inhibition carries the values derived from the second function away from those of the first, that:

$$P_{opt} = P_{max}/2 \text{ and } b = a/2$$

In order to be dimensionally correct, Talling points out that 'a' (or 'b') must be a reciprocal intensity and that when $aI = 1$, $P = 0.7 P_{max}$. He therefore defined an intensity, $I_k = 1/a$, as indicating a 'knee' in the characteristic at the onset of saturation. This definition allows:

$$P/P_{max} = (I/I_k)/\sqrt{1 + (I/I_k)^2}$$

This expression is non-dimensional and is frequently a convenient means of generalizing the light relationships of algae (Talling, 1957b; Steel, 1972b).

In the field, the radiation of the sun and sky falls upon the water surface and, after losing perhaps 10% by reflection, travels downwards through the water mass, being exponentially reduced in general accord with the Beer–Lambert relationship:

$$I_z = I_0 e^{-\varepsilon z}$$

Where: Z = depth
$\qquad \varepsilon$ = vertical attenuation coefficient

The radiation is normally measured at a number of wavelengths, for ε is spectrally sensitive. These various measurements may then be combined to provide some estimate of the effective energy flux. When parallel measurements of photosynthesis are performed it is possible, by restricting consideration to those intensities less than I_k to find some measure of agreement between the energy gradient and the photosynthetic gradient. It is convenient to measure the light at its most penetrating wavelength, which for many temperate inland waters will lie in the waveband 500–600 nm; 530 nm being very frequently used. I have found $\sigma\varepsilon_{530} \simeq \varepsilon_{PhAR}$ where $\sigma = 1.25$, although Talling suggests 1.33 for this coefficient. (PhAR—photosynthetically active radiation, approximately 0.46 total radiation.)
Then: $I_z, PhAR \simeq I_0, PhAR\ e^{-\sigma\varepsilon_{min} z}$

Which allows us to write:

$$nP_z = \frac{nP_{max}I_z/I_k}{\sqrt{1+(I_z/I_k)^2}} \qquad \text{mg C m}^{-3}\text{h}^{-1}$$

Then total amount of photosynthesis within the mixed water column (depth $= Z_m$) will be the integral of this expression with respect to depth.

$$n\int_0^{Z_m} P_z\,dZ = nP_{max}\int_0^{Z_m} \frac{(I_0 e^{-\sigma\varepsilon_{min}Z}/I_k)}{\sqrt{1+(I_0 e^{-\sigma\varepsilon_{min}Z}/I_k)^2}}\cdot dZ \qquad \text{mg C m}^{-2}\text{h}^{-1}$$

Let $\quad i = I_0 e^{-\sigma\varepsilon_{min}Z}/I_k, \quad$ then $\quad dZ = -\frac{1}{\sigma\varepsilon_{min}}\cdot\frac{di}{i}$

$$n\int_0^{Z_m} P\,dZ = -\frac{nP_{max}}{\sigma\varepsilon_{min}}\int_{I_0/I_k}^{I_0 e^{-\sigma\varepsilon_{min}Z_m}/I_k} \frac{i}{\sqrt{1+i^2}}\cdot\frac{di}{i}$$

$$= \frac{nP_{max}}{\sigma\varepsilon_{min}}\int_{I_0 e^{-\sigma\varepsilon_{min}Z}/I_k}^{I_0/I_k} \frac{di}{\sqrt{1+i^2}}$$

$$= \frac{nP_{max}}{\sigma\varepsilon_{min}}\{\sinh^{-1}(I_0/I_k) - \sinh^{-1}(I_0 e^{-\sigma\varepsilon_{min}Z_m})\}$$
$$\text{mg C m}^{-2}\text{h}^{-1}$$

Similar manipulation for the 'inhibition' function results in:

$$n\int_Z^0 P\,dZ = \frac{2nP_{opt}}{\sigma\varepsilon_{min}}\{\tan^{-1}(I_0/2I_k) - \tan^{-1}(I_0 e^{-\sigma\varepsilon_{min}Z_m}/2I_k)\} \quad \text{mg C m}^{-2}\text{h}^{-1}$$

This applies only for constant surface radiation, but in reality that intensity will vary during the day. This variation may be approximated by:

$$I_0 = I_{max}/2\left\{1 - \cos\frac{\tau.2\Pi}{\Delta}\right\}$$

Where: $\quad \Delta = $ Light daylength
$\quad\quad\quad \tau = $ Time within Δ

Incorporating this function into the column photosynthesis integral produces some integration difficulty and linear approximation to the light course will be used to allow easier solution. The daily gross photosynthesis will then be, approximately (Steel, 1972a):

$$G = n\int_{\tau=0}^{\tau=\Delta}\int P\,dZ\,dt = \frac{nP_{max}\Delta}{\sigma\varepsilon_{min}}0\cdot6$$

$$\times\left\{1\cdot333\sinh^{-1}\phi - \frac{1}{\phi}[\sqrt{(1+\phi^2)}-1]\right\}_{\phi=I_{0,max}/I_k e^{-\sigma\varepsilon_{min}Z}}^{\phi=I_{0,max}/I_k} \qquad \text{mg C m}^{-2}\text{d}^{-1}$$

Similarly, for 'inhibition':

$$G_2 = n \int_{\tau=0}^{\tau=\Delta} \int_Z^0 P dZ dt = \frac{nP_{max}\Delta}{\sigma \varepsilon_{min}} 1 \cdot 2$$

$$\times \left\{ 1 \cdot 333 \tan^{-1}(0 \cdot 5 \phi) - \frac{1}{\phi} \ln[1 + (0 \cdot 5 \phi)^2] \right\}_{\phi = I_{0,max}/I_k e^{-\sigma \varepsilon_{min} Z}}^{\phi = I_{0,max}/I_k} \qquad mg\, C\, m^{-2}\, d^{-1}$$

It is taken that no photosynthesis occurs in the dark, thus these integrals are the daily photosynthetic gain to the column. It is also assumed that the photosynthetic products do not produce an increase in algae during the light portion of the day.

The vertical attenuation coefficient may be regarded as an additional function, and be apportioned, for present purposes, between algal and non-algal material, thus:

$$\varepsilon_{min} = \varepsilon_q + \varepsilon_s . n$$

It is understood that the light wavelength used is that of the most penetrating component. 'ε_q' is the apparent vertical attenuation coefficient associated with all non-algal particulate material, the dissolved colour and the water itself. 'ε_s', is a 'self-shading' coefficient relating the increase in 'ε_{min}' to that of 'n'. 'ε_s' seems to have values which lie between 0·008 and 0·022 (Steel, unpublished). The smaller values are frequently associated with larger genera. The few reported values lie within this range (Talling, 1960; Ganf, 1970; Steel, 1972a).

The daily gross photosynthesis then becomes:

$$G_1 = \frac{n . P_{max} \Delta . 0 \cdot 6}{\sigma (\varepsilon_q + \varepsilon_s n)} \left\{ 1 \cdot 333 \sinh^{-1} \phi - \frac{1}{\phi} [\sqrt{(1 + \phi^2)} - 1] \right\}_{\phi = I_{0,max} e^{-\sigma \varepsilon_{min} Z}/I_k}^{\phi = I_{0,max}/I_k}$$

$$mg\, Cm^{-2} d^{-1}$$

and similarily for G_2.

Let:
$$G_1 = \frac{nP_{max}\Delta}{\sigma \varepsilon_{min}} 0 \cdot 6\, G(\phi, \sigma \varepsilon_{min} Z)$$

and:
$$G_2 = \frac{nP_{max}\Delta}{\sigma \varepsilon_{min}} 1 \cdot 2\, F(\phi, \sigma \varepsilon_{min} Z)$$

Let $B = nZ$ mg a m^{-2} (For convenience, let $B \equiv$ 'Biomass'; mg a m^{-2} and $n \equiv$ 'concentration'; mg a m^{-3})

$$G_1 = \frac{BP_{max}\Delta}{\eta} 0 \cdot 6\, G(\phi, \eta)$$

$$G_2 = \frac{BP_{max}\Delta}{\eta} 1 \cdot 2\, F(\phi, \eta)$$

Where:
$$\eta = \sigma Z \varepsilon_{min}$$

Generally:
$$G = \frac{BP_{max}\Delta}{\eta} \cdot L(\phi,\eta) \qquad \text{mg C m}^{-2}\text{ d}^{-1}$$

Figure 5.1 and 5.2 illustrate the values of these functions for various values of ϕ and η. It may be seen that $0.6\,G\,(\phi,\eta)$ is, naturally, always somewhat greater than $1.2\,F(\phi,\eta)$ and so will be used predominantly as it will always tend to give slightly more pessimistic predictions. It may also be seen that for values of $\phi = 20\text{--}25$ and $\eta > 4\text{--}5$, $0.6\,G(\phi,\eta)$ has a relatively constant value close to 2.5. This value of ϕ is reasonable for good growth conditions. In this radiation is judged as relative to 'I_k' and not by its absolute value.

For water columns which contain the complete photosynthetic profile, it

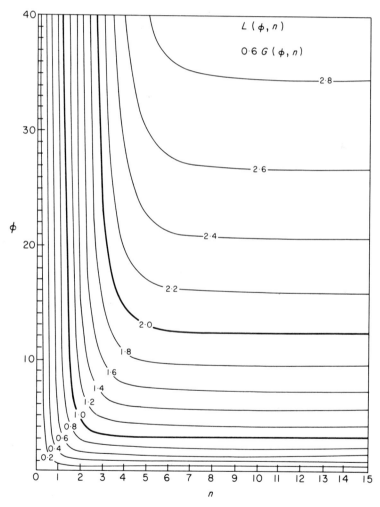

Figure 5.1 Values of the light function $0.6G\,(\phi,\eta)$, determined by the function arguments

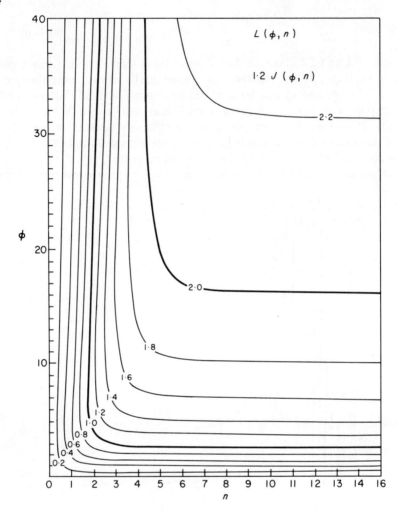

Figure 5.2 Values of the light function $1.2F(\phi, \eta)$, determined by the function arguments

is possible to define a depth, 'Z_{eu}', the euphotic depth, which is that depth which contains the profile. It has frequently been characterized by the level at which superficial illumination has been reduced to 1%.

$$I_{Z_{eu}} = 0.01\, I_o$$

$$0.01 = e^{-\sigma \varepsilon_{min} Z_{eu}}$$

$$Z_{eu}\varepsilon_{min} = -\frac{1}{\sigma}\ln 0.01$$
$$= 3.7 \text{ for } \sigma = 1.25$$

This figure is that derived by Talling (1965).

Thus if $Z\varepsilon_{min} \geqslant 3.7$, the column is always optically deep. Further increase in depth in such a column can add no further to the photosynthetic capacity and this is reflected in the essentially constant values of $L(\phi, \eta)$.

The integral 'G' may now be considered as having been derived from an uniformly distributed population all photosynthesizing at some average rate $\bar{P}$.

A similar process of reasoning must be applied to the respiration within the column. Conveniently, respiration will be taken as independent of short-term variation in light intensity and so will be taken as constant throughout. As such, the various integrals are particularly trivial:

Let

$$R_1 = n\int_o^{24}\int_o^{Z_m} R\,dZdt = nRZ_m24 \qquad \text{mg C m}^{-2}\,\text{d}^{-1}$$

Then:

$$c\frac{dB}{dt} = \frac{BP_{max}\Delta L(\phi, \eta)}{\eta} - BR24 \qquad \text{mg }a\,\text{m}^{-2}\,\text{d}^{-1}$$

where 'c' is a conversion factor for carbon: chlorophyll.

$$\text{As: } \frac{dB}{dt} = \frac{d(nZ)}{dt} = \frac{Zdn}{dt} \text{ for } \frac{d^2n}{dZ^2} = \frac{dn}{dZ} = \frac{dZ}{dt} = 0$$

this expression is equivalent to the originally required equation. It is convenient, however, to retain the relation with respect to area rather than to volume and the derived equation will therefore be used.

5.3 GROSS PHOTOSYNTHESIS AND MAXIMUM BIOMASS

As derived:

$$G = \frac{BP_{max}\Delta L(\phi, \eta)}{\eta}$$

$$n = \sigma Z(\varepsilon_q + \varepsilon_s n) = \sigma(Z\varepsilon_q + \varepsilon_s B)$$

$$G = \frac{1}{\sigma\varepsilon_s}(\eta - \sigma Z\varepsilon_q).P_{max}\Delta\frac{L(\sigma, \eta)}{\eta}$$

$$\frac{dG}{d\eta} = \frac{\Delta P_{max}}{\sigma\varepsilon_s}\left\{L\frac{(\phi, \eta)}{\eta} + (\eta - \sigma Z\varepsilon_q)D_\eta\frac{L(\sigma, \eta)}{\eta}\right\}$$

$$= \frac{\Delta P_{max}}{\sigma\varepsilon_s}\left\{\frac{L(\phi, \eta)}{\eta} + (\eta - \sigma Z\varepsilon_q)\left[\eta\frac{L'_\eta(\sigma, \eta) - L(\sigma, \eta)}{\eta^2}\right]\right\}$$

$$= \frac{\Delta P_{max}}{\sigma\varepsilon_s}\left\{L_n^1(\phi, \eta) - \sigma Z\varepsilon_q\frac{L'_\eta(\phi, \eta)}{\eta} + \sigma Z\varepsilon_q\frac{L(\phi, \eta)}{\eta^2}\right\}$$

At an extreme $\dfrac{dG}{d\eta} = 0$ at least, so

$$L'_\eta(\phi,\eta)(\eta - \sigma Z\varepsilon_q) + \sigma Z\varepsilon_q \frac{L(\sigma,\eta)}{\eta} = 0$$

now $\eta > \sigma Z\varepsilon_q \, L(\sigma,\eta) > 0; \; 0 < \eta < \infty, \sigma > 0$

Thus $\dfrac{dG}{d\eta} = 0$ could only arise if any value of $L'_\eta(\sigma,\eta) < 0$. It may be shown that:

$$0{\cdot}6\, G'_\eta(\sigma,\eta) = 0{\cdot}6\left\{\frac{1}{3}\frac{\psi}{\sqrt{1+\psi^2}} + \left[\frac{\sqrt{(1+\psi^2)}-1}{\psi}\right]\right\}, \text{ where } \psi = \sigma e^{-\eta}$$

This is clearly positive for any reasonable value of $\eta, \phi > 0$. The same is true for $1{\cdot}2\, F'_\eta \, 1\eta(\sigma,\eta)$.

'G' is therefore a monotonic function of η which implies, for constant '$Z\varepsilon_q$', that the greatest gross photosynthesis must be associated with the greatest biomass.

We have:

$$c\frac{dB}{dt} = \frac{BP_{max}\Delta.L(\sigma,\eta)}{\eta} - BR24 \qquad\qquad \text{mg } a\,\text{m}^{-2}\,\text{d}^{-1}$$

Any greatest value of 'B' would have to at least occur when this expression was zero, the trivial case of $B = 0$ is, of course, excluded.

$$0 = \frac{B_{max}P_{max}\Delta L(\phi,\eta_{max})}{\eta_{max}} - B_{max}R24$$

$$0 = \Delta P_{max}.B_{max}\left\{\frac{L(\phi,\eta_{max})}{\eta_{max}} - \frac{24r}{\Delta}\right\}$$

Where: $r = \dfrac{R}{P_{max}}$, the 'relative respiration'.

Now $\dfrac{\Delta}{R}$ may frequently be reasonably constant, so, simply:

$$\frac{L(\phi,\eta_{max})}{\eta_{max}} - \frac{24r}{\Delta} = 0$$

It is clearly possible to derive n_{max} for various values of ϕ and $\dfrac{24r}{\Delta}$ and then derive B_{max} from n_{max} by the relation:

$$n_{max} = \sigma Z(\varepsilon_q + \varepsilon_s n_{max})$$

$$\varepsilon_s B_{max} = \left\{\frac{n_{max}}{\sigma} - Z\varepsilon_q\right\}$$

Now
$$n_{max} = \frac{\Delta}{24r}L(\phi, \eta_{max})$$

$$\therefore \qquad \varepsilon_s B_{max} = \left\{ \frac{\Delta}{24r}\frac{L(\phi, \eta_{max})}{\sigma} - Z\varepsilon_q \right\}$$

Figure 5.3 is a set of solutions for this equation for various values of Δ/r and $Z\varepsilon_q$. For very low biomasses, if any growth is to occur, $\frac{dB}{dt} > 0$ and therefore:

$$\frac{L(\phi, \eta_{min})}{\eta_{min}} > \frac{24r}{\Delta}$$

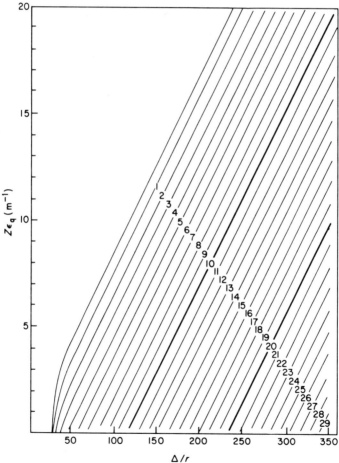

Figure 5.3 Values of $\varepsilon_s B_{max}$ derived from the ratio of daylength to relative respiration (Δ/r) in mixed waters of different extinction depth ($Z\varepsilon_q$)

Now $\eta_{min} \simeq \sigma Z \varepsilon_q$ and depending upon this value various differing requirements for growth may be stated.

1. $Z \varepsilon_q < 3$.

Then $\eta_{min} \simeq 3 \cdot 75$ and from Figure 5.1 it may be seen that for higher values of ϕ, $L(\phi,\eta) \simeq 0 \cdot 7 \eta$

Thus:

$$\frac{0 \cdot 7 \eta_{min}}{\eta_{min}} > \frac{24r}{\Delta}$$

or:

$$\frac{\Delta}{r} > 34$$

In temperate regions $\Delta_{min} \simeq 8$ hours and normally r_{max} is perhaps not more than $0 \cdot 15$, so $(\Delta/r)_{min} \simeq 53$ and such situations would always be photosynthetically suitable for growth in more than minimal radiation. In these situations there would probably be rapid response to light increases from winter to summer and less requirement for a marked ability to acclimate to low light.

2. $Z \varepsilon_q > 4$.

Then $\eta_{min} > 5$ and $L(\phi,\eta_{min}) \simeq 2 \cdot 5$
For growth:

$$\frac{2 \cdot 5}{\sigma Z \varepsilon_q} > \frac{24r}{\Delta}$$

$$\frac{\Delta}{r} > 12 Z \varepsilon_q$$

In these situations there will thus not only exist the need for adequate, acceptable radiation intensity, but also its existence for longer than minimum values. For marked success a high physiological efficiency as judged by the relative respiration may also be required.

It is possible that such relationships are important in allowing Spring outbursts of diatoms. These algae seem to frequently have low relative respiration after over-wintering, values of $0 \cdot 04$ occurring quite often (Steel, 1972a). It may be that the large cell dimensions of many Spring diatoms are of some consequence in this respect. For $\Delta = 10\text{--}12$, these values of 'r' cause $\Delta/r = 250\text{--}300$ which would allow some growth in a reservoir with $Z \varepsilon_q = 20\text{--}25$. For algae having $r = 0 \cdot 1$, which is often observed, reservoirs of such a character would completely preclude their growth at that time.

The delimited area of Figure 5.3 is the boundary for possible growth. The curves suggest some linear, inverse relation between B_{max} and $Z \varepsilon_q$ in a mixed column of water. This follows from $L(\phi,\eta_{max}) \simeq 2 \cdot 5$ for $\eta_{max} > 4$; $\phi > 20$.

Then:

$$\varepsilon_s B_{max} = \left\{ \frac{\Delta}{24r} \cdot \frac{2\cdot5}{\sigma} - Z\varepsilon_q \right\}$$

$$\therefore B_{max} = \frac{1}{\varepsilon_r} \left\{ \frac{\Delta}{12r} - Z\varepsilon_q \right\} \qquad \text{mg } a \text{ m}^{-2}$$

In this relation $Z\varepsilon_q$ acts as a combined variable and indicates that judgement as to the potential production of a column of water may only be made if something of its optical quality is also known. Thus equal limitation on maximum potential biomass will occur with $Z = 5$, $\varepsilon_q = 2\cdot0$ and with $Z = 20$, $\varepsilon_q = 0\cdot5$. Optical density can offset shallowness; depth offset extreme clarity in production dynamics. In view of the importance of this product, I shall term it the 'extinction depth' although it is dimensionless.

The figure indicates that for very low values of the extinction depth there is still some potential maximum biomass. As long as these low values are caused by low values of 'ε_q', then this is probably not unreasonable. When, however, shallow depths are the cause some difficulty arises for the implication is near infinite algal concentration. In such circumstances it is clear that, in the absence of sub-surface inhibition response, other factors must begin to operate in order to limit the concentrations to more intuitively acceptable levels.

Earlier consideration showed that $\Delta/r > 12Z\varepsilon_q$ was a required condition for growth, thus for non-growth:

$$\frac{12rZ\varepsilon_q}{\Delta} \geqslant 1$$

This equation represents a particular example of the condition, called 'column compensation' by Talling (1957a), in which the column energy income is exactly balanced by its expenditure. This relation is particular in the sense that this limiting balance is achieved with very low biomass and so would characterize a non-productive situation.

Let $\omega = 12rZ\varepsilon_q/\Delta$, then for growth, $\omega < 1$, and for limitation, $\omega \geqslant 1$. For convenience let $\Delta/r = 240$, then $Z\varepsilon_q \geqslant 20$ will impose considerable limitation on potential production.

$$B_{max} = \frac{1}{\varepsilon_s} \left\{ \frac{\Delta}{12r} - Z\varepsilon_q \right\} \qquad \text{mg } a \text{ m}^{-2}$$

$$B_{max} = \frac{\Delta}{12r\varepsilon_s} \left\{ 1 - \omega \right\}$$

Then for $\Delta/r = 300$,

$$B_{max} = \frac{25}{\varepsilon_s} \left\{ 1 < 0\cdot04\, Z\varepsilon_s \right\} \qquad \text{mg } a \text{ m}^{-2}$$

If $\varepsilon_s = 0\cdot01$ and $Z\varepsilon_q \sim 4$ are taken as minimal values:

$$B_{max} = 2500.0\cdot84$$
$$\simeq 2100 \qquad \text{mg } a \text{ m}^{-2}$$

120

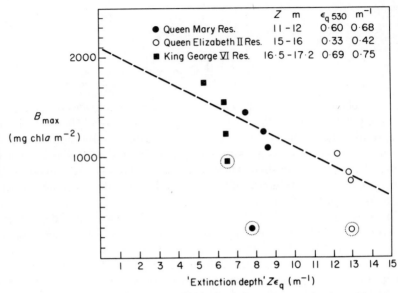

Figure 5.4 Observed maximum areal chlorophyll *a* biomass in three Thames Water reservoirs characterized by their particular extinction depths, during 1968–1971

For general conformity to the original stipulations therefore, the maximum amount of algae as characterized by chlorophyll *a* in a mixed column is about $2\,\mathrm{g}\,a\,\mathrm{m}^{-2}$ (see, however, Talling and others, 1973).

Figure 5.4 illustrates the maximum biomasses of three Metropolitan Water Board reservoirs for 1968–1971, plotted against the relevant values of the extinction depth. The data (excluding those values for which some alternative tenable hypothesis for limitation exists, shown circled by dots) suggests an approximate relation with $\Delta/r \simeq 150$–200. 'r' here is a complex which includes some effects of throughput sedimentation and any grazing. 'ε_s' was close to 0·01 for all the populations.

For small algae with 'ε_s' closer to 0·02 the maximum potential biomass will of course be closer to $1\,\mathrm{g}\,a\,\mathrm{m}^{-2}$.

It was indicated earlier that the maximum gross photosynthesis would occur when the biomass was maximal, thus:

$$G_{max} = \frac{B_{max} \cdot P_{max} \cdot \Delta \cdot L(\phi, \eta_{max})}{\eta_{max}} \qquad \mathrm{mg\,C\,m^{-2}\,d^{-1}}$$

$$= \frac{\Delta}{12 r \varepsilon_s}\left\{1 - \omega\right\} P_{max} \Delta \left\{\frac{L(\phi, \eta_{max})}{\eta_{max}}\right\}$$

$$= \frac{\Delta}{12 r \varepsilon_s}\left\{1 - \omega\right\} \Delta P_{max} \frac{24 r}{\Delta}$$

$$= \frac{2 P_{max} \Delta}{\varepsilon_s}\left\{1 - \omega\right\} \qquad \mathrm{mg\,C\,m^{-2}\,d^{-1}}$$

This suggests that the maximum potential daily photosynthetic rate will be:

$$G_{max} \simeq \frac{2P_{max}\Delta}{\varepsilon_s} \text{ as } \omega \to 0$$

'Δ', of course, is variable through the year. P_{max} is also variable due to its temperature dependency. It is possible that 'ε_s' may also be a time variable in that the cell sizes of greatest 'ε_s' quite frequently occur in warmer water.

I have found $P_{max} = 3\cdot1e^{0\cdot09\theta}$ mg $0(\text{mg}\,a)^{-1}\text{h}^{-1}$, where $\theta \equiv °C$. If $= 4(3 - \cos t.\Pi/183)$ is taken to describe the day-length through the year, and $\theta = 9(4/3 - \cos t - 30/183.\Pi)$ as a temperate water temperature regime,

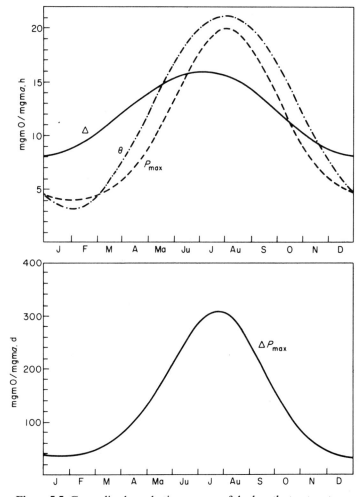

Figure 5.5 Generalized yearly time course of daylength, temperature and temperature-determined light-saturated photosynthetic rate, and the resultant product of daylength and light-saturated photosynthetic rate

then a theoretical $(\Delta P_{\max})$ yearly time course may be constructed and this with the previous relations is illustrated in Figure 5.5. It may be seen that the maximum $(\Delta P_{\max})$ is approximately $280-300$ mg 0 $(\text{mg } a)^{-1} \text{d}^{-1}$. In a tropical situation the constant day length of 12 hours may offset to some extent the higher temperatures and so maintain a value not far, if at all, in excess of the temperate situation. In temperate areas the maximum values of $(\Delta P_{\max})$ occur, naturally, just after mid-year. If $\varepsilon_s = 0.015-0.02$ is associated with this maximum then:

$$G_{\max} \simeq \frac{2 \times 300}{\varepsilon_s} \qquad \qquad \text{mg } 0 \text{ m}^{-2} \text{d}^{-1}$$

$$\simeq 30-40 \qquad \qquad \text{g } 0 \text{ m}^{-2} \text{d}^{-1}$$

$$\simeq 10-13 \qquad \qquad \text{g C m}^{-2} \text{d}^{-1}$$

This will constitute some estimate of the maximum attainable column photosynthesis. Any larger cell size capable of surviving in warm water may increase this capacity somewhat.

As $G_{\max} = 2P_{\max}\Delta/\varepsilon_s\{1 - 12rZ\varepsilon_q/\Delta\}$, it is clear that the gross photosynthesis is also linearly reduced by increasing extinction depth for any situations of similar conditions. For $\phi > 20$; $\Delta/r = 300$; $Z\varepsilon_q = 4$:

$$G_{\max} = \frac{2P_{\max}\Delta}{\varepsilon_s} \cdot 0.84$$

$$\simeq 8.5-11 \qquad \qquad \text{g C m}^{-2} \text{d}^{-1}$$

$$\text{Simply: } G_{\max} \simeq 12(1 - 0.04 Z\varepsilon_q) \qquad \qquad \text{g C m}^{-2} \text{d}^{-1}$$

$$\text{Now:} \quad G = BP_{\max}\Delta\frac{L(\phi,\eta)}{\eta}$$

$$\frac{G}{B} = \frac{P_{\max}\Delta.2.5}{\sigma Z \varepsilon_{\min}} \qquad \qquad \text{mg C (mg } a)^{-1} \text{d}^{-1}$$

$$= \frac{2P_{\max}\Delta}{Z\varepsilon_q + \varepsilon_s B}$$

Hence the relative amount of production per unit of population decreases with increasing biomass, until at the maximum:

$$\frac{G_{\max}}{B_{\max}} = \frac{2P_{\max}\Delta}{\varepsilon_s} \cdot \frac{12r\varepsilon_s}{}$$

$$= 24rP_{\max}$$

$$= 24R \qquad \qquad \text{mg C (mg } a)^{-1} \text{d}^{-1}$$

The final relative gross production is therefore dictated by the daily respiration rate.

5.4 NET PHOTOSYNTHESIS

Whilst the orders of the maximum column energy capture and biomasses are of interest, the possibility of their being achieved must also be considered. This possibility will be some reflection of the net photosynthetic rates which the conditions will allow. It may be, for instance, that the environmental situation only permits a net photosynthetic rate such that months of near constant, favourable conditions are required to produce the maximum inferred biomasses. Whilst this constancy may be a quality of a tropical situation, it is likely to occur only infrequently, if at all, in a temperate area.

As the net column photosynthesis is the difference between its gross photosynthesis and respiration:

$$N = \frac{nP_{max}\Delta L(\phi,\eta)}{(\varepsilon_q + \varepsilon_s n)} - n\,RZ\,24 \qquad\qquad mg\,C\,m^{-2}\,d^{-1}$$

$$= BP_{max}\Delta\frac{L(\phi,\eta)}{\eta} - B24R$$

$$= BP_{max}\Delta\left\{\frac{L(\sigma,\eta)}{\eta} - \frac{24r}{\Delta}\right\}$$

Then the population which causes the maximum net photosynthetic rate in any given conditions, will be characterized by a value of η such that $dN/d\eta = 0$ at least.

Now:

$$N = \frac{P_{max}\Delta}{\sigma\varepsilon_s}\left\{\eta - \sigma Z\varepsilon_q\right\}\left\{\frac{L(\phi,\eta)}{\eta} - \frac{24r}{\Delta}\right\}$$

$$\frac{dN}{d\eta} = \frac{P_{max}\Delta}{\sigma\varepsilon_s}\left[(\eta - \sigma Z\varepsilon_q)D_\eta\left\{\frac{L(\phi,\eta)}{\eta} - \frac{24r}{\Delta}\right\} + \left\{\frac{L(\phi,\eta)}{\eta} - \frac{24r}{\Delta}\right\}\right]$$

$$= \frac{P_{max}\Delta}{\sigma\varepsilon_s}\left[(\eta - \sigma Z\varepsilon_q)\left\{\eta\frac{L'_\eta(\phi,\eta) - L(\phi,\eta)}{\eta^2}\right\} + \left\{\frac{L(\phi,\eta)}{\eta} - \frac{24r}{\Delta}\right\}\right]$$

$$0 = (\eta_{opt} - \sigma Z\varepsilon_q)\left\{\eta_{opt}\frac{L'_\eta(\phi,\eta_{opt}) - L(\phi,\eta_{opt})}{\eta_{opt}^2}\right\} + \frac{L(\phi,\eta_{opt})}{\eta_{opt}} - \frac{24r}{\Delta}$$

$$0 = L'_\eta(\phi,\eta_{opt})\left\{1 - \frac{\sigma Z\varepsilon_q}{\eta_{opt}}\right\} + \sigma Z\varepsilon_q\frac{L(\phi,\eta_{opt})}{\eta_{opt}^2} - \frac{24r}{\Delta}$$

Figure 5.6 is derived from the values of η_{opt} which form this equality. It appears that for values of $Z\varepsilon_q > 4$ some reasonably simple relationship might exist between $\varepsilon_s B_{opt}$ and the extinction depth. Slight rearrangement gives:

$$\sigma\frac{L'_\eta(\phi,\eta_{opt})}{\eta_{opt}}\cdot\varepsilon_s B_{opt} + \sigma Z\varepsilon_q\frac{L(\phi,\eta_{opt})}{\eta_{opt}^2} - \frac{24r}{\Delta} = 0$$

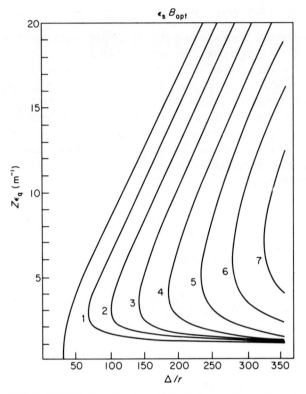

Figure 5.6 Values of $\varepsilon_s B_{opt}$ derived from the ratio of daylength to relative respiration (Δ/r) in mixed waters of different extinction depth $(Z\varepsilon_q)$

It is possible to show that for $(\varepsilon_s B_{opt})_{max} = 7, \eta_{opt} \simeq 6,$

$$\sigma \frac{L'_\eta(\phi, \eta_{opt})}{\eta_{opt}} \cdot \varepsilon_s B_{opt} < 0.7\, \sigma Z\varepsilon_q \frac{L(\phi, \eta_{opt})}{\eta_{opt}^2}$$

and vanishes proportionately rapidly with any further increase in η_{opt}.
Writing then, as an approximation; $Z\varepsilon_q > 5, \phi > 20$

$$\sigma Z\varepsilon_q \frac{L(\phi, \eta_{opt})}{\eta_{opt}^2} - \frac{24r}{\Delta} = 0$$

$$\eta_{opt} = \sqrt{\left(\frac{Z\varepsilon_q \Delta}{24r} \cdot \sigma 2.5\right)}$$

$$\therefore \varepsilon_s B_{opt} = \frac{1}{\sigma}\sqrt{\left(\frac{Z\varepsilon_q \Delta}{24r} \cdot \sigma 2.5\right)} - Z\varepsilon_q$$

$$= \frac{\Delta}{12r}\left\{ \sqrt{\left(\frac{12rZ\varepsilon_q}{\Delta}\right)} - \frac{12rZ\varepsilon_q}{\Delta} \right\}$$

$$= \frac{\Delta}{12r} \left\{ \sqrt{\omega} - \omega \right\}$$

$$\text{and } N_{\max} = B_{\text{opt}} P_{\max} \Delta \left\{ \frac{L(\phi \eta_{\text{opt}})}{\eta_{\text{opt}}} - \frac{24r}{\Delta} \right\}$$

$$= \frac{P_{\max} \Delta}{\sigma \varepsilon_s} \cdot \eta_{\text{opt}} \left\{ 1 - \frac{\sigma Z \varepsilon_q}{\eta_{\text{opt}}} \right\} \cdot \frac{24r \, \eta_{\text{opt}}}{\Delta \, \sigma Z \varepsilon_q} \left\{ 1 - \frac{\sigma Z \varepsilon_q}{\eta_{\text{opt}}} \right\}$$

$$= 2 \frac{P_{\max} \Delta}{\varepsilon_s} \left\{ 1 - \frac{\sigma Z \varepsilon_q}{\eta_{\text{opt}}} \right\}^2$$

$$= 2 \frac{P_{\max}}{\varepsilon_s} (1 - \sqrt{\omega})^2$$

For the optimal conditions used previously

$$N_{\max} = \frac{2 P_{\max} \Delta}{\varepsilon_s} \cdot 0 \cdot 36$$

$$\simeq 3 \cdot 6 - 4 \cdot 7 \qquad\qquad\qquad\qquad \text{g C m}^{-2} \text{d}^{-1}$$

5.5 ASSIMILATION 'EFFICIENCY'

This efficiency is indicated by the fractional conversion of gross into net photosynthesis.

$$\Gamma = \frac{N}{G}$$

$$= \frac{G - R}{G}$$

$$\Gamma = 1 - \frac{R}{G}$$

$$= \frac{24r}{\Delta} \cdot \frac{\eta}{L(\phi, \eta)}$$

$$\Gamma_{\max} = 1 - \left(\frac{24r}{\Delta} \right)_{\min} \left(\frac{\eta}{L(\phi, \eta)} \right)_{\min}$$

It is possible to show that for $\phi > 20$, $(\eta/L(\phi, \eta))_{\min} \simeq 4/3$, which value occurs as $\eta \to 0$

$$\Gamma_{\max} = 1 - \frac{24r}{\Delta} \cdot \frac{4}{3}$$

$$= 1 - \frac{32r}{\Delta}$$

$$= 0 \cdot 85$$

where:
$$Z\varepsilon_q > 3, \eta > 4\frac{\eta}{L(\phi,\eta)} \simeq \frac{Z\varepsilon_{min}}{2}$$

$$\Gamma = 1 - \frac{12rZ\varepsilon_q}{\Delta}$$

Now $(Z\varepsilon_{min}) = Z\varepsilon_q$

$$\Gamma_{max} = 1 - \frac{12rZ\varepsilon_q}{\Delta}$$

$$= 1 - \omega$$

With $\omega = 0.16$ as before

$$\Gamma_{max} = 0.84$$

Γ_{max} seems fairly constant for $0 < Z\varepsilon_q < 4$, but decreases linearly with any further increase in the extinction depth. It seems that the most 'efficient' photosynthetic population is the minimal population. It is to be noted that the maximum net photosynthesis is not achieved by the most efficient population.

In summary, $Z\varepsilon_q > 3, \phi > 20, Z > 2$:

$$\omega = \frac{12rZ\varepsilon_q}{\Delta} = \frac{12RZ\varepsilon_q}{P_{max}\Delta}$$

$$G_{max} = \frac{2P_{max}\Delta}{\varepsilon_s} \cdot (1 - \omega) \qquad \text{mg C m}^{-2}\,\text{d}^{-1}$$

$$N_{max} = \frac{2P_{max}\Delta}{\varepsilon_s}(1 - \sqrt{\omega})^2 \qquad \text{mg C m}^{-2}\,\text{d}^{-1}$$

$$B_{max} = \frac{2P_{max}\Delta}{\varepsilon_s} \cdot \frac{1}{24R}(1 - \omega) \qquad \text{mg } a\,\text{m}^{-2}$$

$$B_{opt} = \frac{2P_{max}\Delta}{\varepsilon_s} \cdot \frac{1}{24R}(\sqrt{\omega} - \omega) \qquad \text{mg } a\,\text{m}^{-2}$$

$$\Gamma_{max} = 1 - \omega$$

$$\frac{N_{max}}{G_{max}} = \frac{(1 - \sqrt{\omega})}{(1 + \sqrt{\omega})}$$

$$\frac{G_{max}}{B_{max}} = 24R \qquad \text{mg C (mg } a)\,\text{d}^{-1}$$

5.6 BIOMASS CHANGES AND NUTRIENTS

The rate of change of biomass in the water mass will be due to the net photosynthesis less all the other losses to sedimentation, outflow and grazing. B_{max} and N_{max} therefore represent the maximum possible yield and rate of production of organic material when those other losses are negligible. For convenience it is useful to have both B_{max} and N_{max} in the same units apart from the time dimension, then for cells which have a carbon to chlorophyll ratio of 30:1, B_{max} becomes:

$$B_{max} = \frac{2P_{max}}{\varepsilon_s} \cdot \frac{1 \cdot 25}{R}(1 - \omega) \qquad\qquad \text{mgC}\,\text{m}^{-2}$$

It is apparent from the formulae for B_{max} and N_{max} that as ω approaches 1, not only does the maximum possible yield decrease but the maximum growth rates are also lessened. As indicated earlier, larger values of ω will also probably mean that the onset of growth will be further displaced in time from conditions unfavourable to growth.

For depths greater than one or two metres, the algal concentration, $n = B/Z$, will be very markedly affected by any decrease in B_{max} if that decrease is caused by an increase in $Z\varepsilon_q$ due to a greater depth of circulation. Some indication of this interaction may be gained by solution with respect to time of the equation representing net photosynthesis. Some solutions for the given conditions are illustrated in Figure 5.7. It is obviously of considerable benefit in management terms to not only have lower potential crops to deal with, but have more time to organize against them. Where multiple source availability exists, the 'phasing'

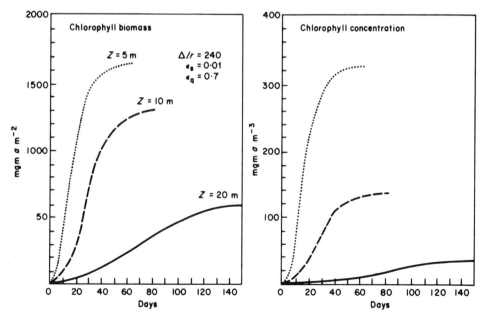

Figure 5.7 Calculated rates of change and yields of chlorophyll a concentrations and biomasses in waters differing only in mixed depth

may also prove of considerable benefit. Whilst the growth rates for the 10 m basin compare to those observed in reservoirs of the Thames Water Authority of similar depths at times when the conditions are near to those chosen for solution, the slow growth attainable in a population circulation in 20 m of water with $\varepsilon_q = 0.7$ would probably prevent even the indicated low maximum population. As much algal growth seems opportunistic, it seems clear that larger extinction depths can severely limit the number of occasions within the year when such opportunities may be exploited.

The yield and maximum rate calculations contain some information on the nutrient requirements of any algal production. If the minimum nutrient requirement of the algae are specified relative to the carbon content, then the maximum carbon biomass may be used to assess whether there is any possibility of nutrient limitation on yield. Let the nutrient concentration dissolved in the water before the growth starts be v_0 mg m^{-3}, then if $Zv_0 \gtrsim CB_{max}$, no rate or yield limitation of any great consequence may occur. If this initial dissolved nutrient content is such that $CB_{max} < Zv_0 < CB_{max}$ then, at least initially, no great rate limitation will occur but the ultimate yield will be lessened. If $Zv_0 \ll CB_{opt}$ severe nutrient limitation must occur.

The previous observations relate to a situation in which there is no throughput during the growing season. If the reservoir is in circulation, then the input nutrient must also be considered. In a mixed reservoir with an extinction depth of 7–8, calculation suggests that, allowing for some throughput loss and a raw-water source similar to the river Thames:

$$B_{max} = 50 - 55 \qquad\qquad \text{gC m}^{-2}$$

$$N_{max} = 2.5 \qquad\qquad \text{gC m}^{-2}\text{d}^{-1}$$

If, for nitrogen and phosphorus, $N:C = 1:10$ and $P:C = 1:100$, these figures represent, ultimately, 5gNm^{-2} and 0.5gPm^{-2}; and a maximum daily requirement 250mgNm^{-2} and 25mgPm^{-2}. In a reservoir 12 m deep, these figures represent 0.4 mgNl^{-1} and 0.04mgPl^{-1}. As the initial levels before growth are of the order of $5 - 8$ mgNl^{-1} and $1.0 - 1.5$mgPl^{-1}, it is clear that there is little chance of nitrogen or phosphorus ever imposing a limit on either maximum growth rate or yield.

For reservoirs fed from a source such as the river Thames, the mean daily input loadings are about 1.5mgNm^{-2}d^{-1} for nitrogen and about 360 mg Pm^{-2}d^{-1}. These are clearly so far in excess of the maximum possible daily requirement that the dissolved nitrogen and phosphorus in the reservoir should show virtually no response to the largest crops, as is indeed the case.

The figures suggest that it might be useful to be able to make some calculation as to the river nutrient status with respect to algal production. If near instantaneous mixing is assumed, then a simple nutrient model may be set-up for the flow-through condition:

$$\frac{dv}{dt} = -qv + q[\text{Riv}]Z - CN \qquad\qquad \text{mg m}^{-2}\text{ d}^{-1}$$

Where: v = Nutrient mass per unit area
          $[Riv]$ = River concentration of considered nutrient

Assume, at the very worst, that $N = N_{max}$ is constant from the onset of the growth of the biomass to its maximum, and that the river concentration is also constant, then,

$$v_t = v_0 e^{-qt} + \left([Riv]Z - \frac{CN_{max}}{q}\right)\left(1 - e^{-qt}\right) \qquad \text{mg m}^{-2}$$

Clearly, if $[Riv] > CN_{max}/Z_q$ then no nutrient limitation is possible. That is if:

$$[Riv] > CN_{max} \cdot \frac{V}{Q} \cdot \frac{1}{Z}$$

$$[Riv] > C \cdot \frac{2P_{max}\Delta}{\varepsilon_s} \cdot \frac{1}{q} \cdot \frac{(1 - \sqrt{\omega})^2}{Z}$$

In such circumstances, the daily input of the nutrient will always be sufficient to supply the greatest potential requirement. The curves illustrated in Figure 5.8 are the values of $f(Z) = (1 - \sqrt{\omega})^2/Z$ and Z for various values of ε_q, Δ, and r. They suggest that nutrient limitation is much more likely to occur in a shallow mixed depth. This need not, of course, imply that the biomasses of the deeper basins will be any greater; the reverse, if anything, being more probable.

If the maximum value for $2P_{max}\Delta/\varepsilon_s$ is taken as $10 \text{gCm}^{-2}\text{d}^{-1}$, then if $[Riv] \geqslant 10 \cdot c/q . f(Z)$ no limitation may occur, taking 'c' = 0·1 for nitrogen and $c = 0·01$ for phosphorus. The curves may be used to assess whether any possibility of nutrient limitation exists with a known raw-water quality, or what that quality must be for such a possibility to at least exist. For example, for a reservoir 17 m deep, $\varepsilon_q = 0·7$ and 2% continuous throughput, $f(Z) \simeq 0·006$ in favourable growth conditions. No limitation may exist if:

$$[N]_{Riv} \leqslant 1·50.0·006 \qquad \text{gNm}^{-3}$$
$$\geqslant 0·3 \qquad \text{gNm}^{-3}$$
$$\geqslant 0·3 \qquad \text{mgNl}^{-1}$$

It does not follow that if $[Riv] < c N_{max}/Z.q$ nutrient limitation will necessarily occur. This is because in temperate situations there will normally be a period of little, if any, nutrient uptake. This period will be defined by $\omega \geqslant 1$ and during it the input river water will replenish the reservoir nutrient content, assuming the river to have an higher concentration of the nutrient than the reservoir. The reservoir water will increase in its nutrient concentration in a logarithmic manner and when the growing season begins ($\omega < 1$) the nutrients available from the input may be supplemented as required from the internal nutrient store. The possibility of internal cell stores of nutrients may also allow active growth in conditions of very low extra-cellular nutrient concentrations.

The various requirements, capacities and time relationships make all of these interactions less simply calculable, but with slightly greater effort than

130

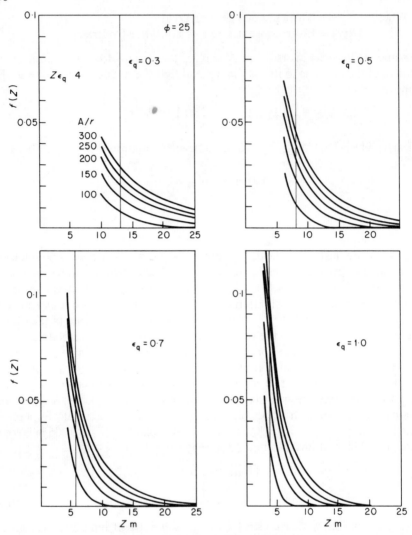

Figure 5.8 Values of the nutrient function $f(Z)$, for waters of differing optical quality and mixed depth

has been applied here, some considerable insights may be achieved. It would seem desirable to attempt to relate nutrients to requirements, so far as they are presently known and understood, rather than to suggest 'thresholds' which whilst valid in some areas may be completely meaningless in others.

5.7 ALGAL CONCENTRATIONS

So far, consideration has been restricted essentially to considering biomass, whereas it is frequently the concentration of algae that is of greatest concern.

For $Z > 2$, $Z\varepsilon_q > 3-4$;

$$N_{max} = \frac{B_{max}}{Z} \qquad \text{mg g m}^{-3}$$

In very shallow water, however, it is unlikely that this relation can hold. As just indicated, it is probable that a greater nutrient imposition on production will be felt in these circumstances, however in eutrophic situations the potential chlorophyll concentration may be as much as $2 - 3\,\text{g a m}^{-3}$. It seems a possibility, however, that before these concentrations are reached, a population density dependent physiological response will have taken place. This response may include an increase in respiration per unit population and a decrease in the

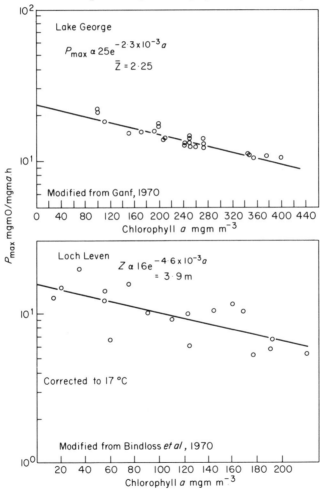

Figure 5.9 Depression in light-saturated photosynthetic rates related to the chlorophyll a concentrations of the photosynthetic populations. (After Ganf (1970) and Bindloss et al. (1970))

light saturated photosynthetic rate per unit population (Ganf, 1970; Bindloss, and colleagues 1970). Figure 5.9 shows reworked data from these two sources which illustrate the depression in P_{max} with increasing algal concentration as evidenced by chlorophyll a. It seems that the factors causing this response are still obscure, but it is possible that very high oxygen tension may be of some consequence (Ganf, 1970). It would seem that diffusion limitations on nutrient availability in dense algal concentrations and differing generic response may also be possible causes. Whatever the case may be shown to be, the results are inevitably serious for potential net photosynthesis. Although the net photosynthesis will be much reduced, any resulting product will, of course, be distributed in limited volumes. It is possible that for algae which are grazed, the confinement within a relatively small space of both herbivorous zooplankton and the food may be very disadvantageous for the latter.

Ignoring grazing effects, let:

$$\hat{P}^1_{max} = \hat{P}\,e^{-bn}$$

$$c\frac{dB}{dt} = B\hat{P}^1_{max}\Delta\left\{e^{-bn}\frac{L(\phi,\eta)}{\eta} - \frac{24r}{\Delta}\right\} \qquad \text{mg } a \text{ m}^{-2}\,\text{d}^{-1}$$

By choice for suitable constants, this expression may be iteratively solved for B_{max} for various combinations of Δ/r and extinction depth. The results for $\Delta/r = 240$ and $\varepsilon_q = 1\cdot0$ are shown in Figure 5.10. The continuous line in the biomass estimates represents the simple derivation used previously. As $\varepsilon_q = 1\cdot0$ was chosen, the extinction depth becomes equivalent to the actual mixed

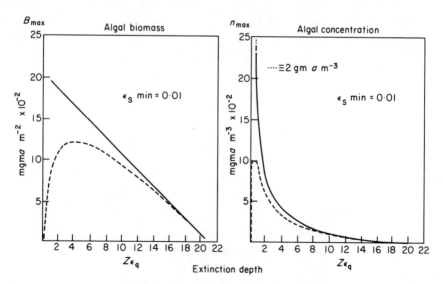

Figure 5.10 Maximum chlorophyll a concentrations and biomasses as functions of extinction depth, using either a simple photosynthetic behaviour (———) or a population density dependent response (– – –)

depth, and it may be seen that the modified equation produces a depth dependent biomass response, shown dotted. With the extinction depth being equal to the depth through which the algae are mixed, it is possible to convert the biomasses into the appropriate concentrations and these are illustrated as well. Although in very shallow water the quantitative results are more suspect, it would seem that the type of response indicated by the results from the modified net photosynthesis equation is more reasonable than an unrestrained response, to potential nutrient limited concentrations.

By similar manipulations curves of equal algal concentration may be constructed for various values of the extinction depth. The results of such calculation are shown in Figure 5.11 with the extinction depth having been split into its two components. It may be seen that the potential for the production of large algal concentrations is much greater in shallow water. This potential may be off-set to some extent by an increase in the non-algal extinction coefficient. In deep, clear water considerable further increase in depth is required before any very great reduction in concentration occurs, but slight increase in the non-algal extinction coefficient can produce a dramatic decrease.

Super-imposed on the figure are three hypothetical sets of conditions. Assume a reservoir of about 17 m deep is fed from a surface water source such that the extinction depth is approximately 13–14, immediately after winter (Point 1). The Spring increase in radiation intensity may reduce the dissolved colour material sufficiently to affect the extinction depth somewhat (Point 2). With decrease in the surface wind-work, the incident radiation may cause a

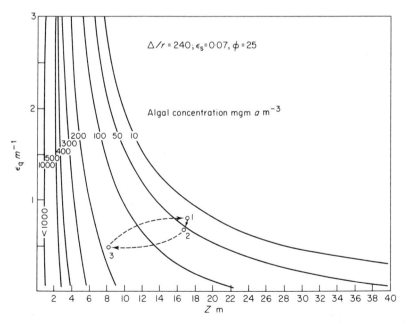

Figure 5.11 Maximum chlorophyll a concentrations as functions of basal optical property and mixed depth, in conditions favourable to growth

thermal stratification. This may produce a shallow mixed depth, over-lying a stagnant hypolimnion (Point 3). If some mechanism exists whereby the reservoir may be made sufficiently turbulent so as to destroy the stratification and to distribute the algae through the full original depth, then the results for potential algal production are clear and are implied by the arrow returning to '1'. The increase in non-algal extinctions in these circumstances will be due to non-algal particulate material being retained in suspension by the increased turbulence.

In water masses only 10–12 m in depth, the severe limitation full depth circulation imposes is not available without a concomitant increase in the non-algal particulate concentration. It is convenient if, due to the optical properties of such particles and the form of the extinction depth and algal concentration relation, this limitation is caused by very much less material than the algal population being limited.

The three points could also be taken as some indication of potential concentrations in a mixed reservoir whose mean depth was significantly reduced throughout the growing season. Full-depth turbulent redistribution would then not decrease the algal concentrations unless also causing, as above, a very marked increase in the non-algal extinction coefficient. With the depths illustrated the required turbidities to reduce the potential maximum concentrations to less than 50 mg a m^{-3} would probably be difficult to accept even if they could be produced.

5.8 CONCLUSION

It seems probable that those concerned in the management of water will need to become much more quantitative in the consideration of algal and related problems. I have attempted to show how even simple mathematical models may be used to gain quite considerable insights into the production dynamics of reservoir eco-systems. In those situations requiring educated guesswork, these insights may be extremely valuable. It is to be hoped that present development will markedly reduce the frequency with which such guesswork has to be resorted to.

I have deliberately kept the model I have used for illustration as simple as possible whilst retaining some degree of reality, in order to attempt to clearly illustrate the manipulations which have produced the relationships suggested. It will be quite obvious that even this simple model would benefit from the capacity to time-vary many of the parameters, and to be able to take into account the more complex biological, chemical and physical interactions. Even such straightforward development would confer quite a sophisticated level of biological judgement on situations or circumstances.

Acknowledgements

This contribution is by permission of Dr. E. Windle Taylor, C.B.E., Director of Water Examination. Such permission, however, in no way implies that the views expressed necessarily coincide with those of the Metropolitan Water Board (now the Thames Water Authority).

5.9 REFERENCES

Bindloss, M. E., Holden, A. V., Bailey-Watts, A. E., and Smith, I. R. (1970). Phytoplankton production, chemical and physical conditions in Loch Leven. *UNESCO-IBP Symposium on productivity problems of freshwaters.* Kazimierz-Dolny, Poland.
Ganf, G. G. (1970). 'The regulation of net production in Lake George, Uganda, E. Africa', *UNESCO-IBP Symposium on productivity problems of freshwaters.* Kazimierz-Dolny, Poland.
Levins, R. (1966). 'The strategy of model building in population biology'. *Amer. Scient.,* **54**, 421–431.
Ridley, J. E. A., Cooley, P., and Steel, J. A. (1966). 'Control of thermal stratification in Thames valley reservoirs', *Proc. Soc. Wat. Treat. Exam.,* **15**, 225–244.
Riley, G. A., Stommel, H., and Bumpus, D. F. (1949). 'Quantiatie ecology of the plankton of the western North Atlantic', *Bull. Bingham. Oceanogr. Coll.,* **12**, 1–169.
Steel, J. A. (1972a). 'The application of fundamental limnological research in water supply system design and management', *Sym. zool. Soc. Lond.,* **29**, 41–67.
Steel, J. A. (1972b). 'Factors affecting algal blooms', *Symp. Soc. Appl. Bact.* **1**: *Microbial Aspects of Pollution.*
Sverdrup, H. U., Johnson, M. W., and Fleming, R. H. (1942). *The Oceans, their physics, chemistry and general biology.* Prentice Hall, New York.
Sverdrup, H. U. (1953). 'On conditions for the vernal blooming of phytoplankton', *J. Cons. int. Explor. Mer.,* **18**, 287–295.
Symons, J. M., Irwin, W. H., Robinson, E. L. and Robeck, G. G. (1967). 'Impoundment destratification for raw-water quality control using either mechanical or diffused-air pumping', *J. Am. Wat. Wks. Ass.,* **59**, 1268–1291.
Talling, J. F. (1957a). 'The phytoplankton population as a compound photosynthetic system', *New Phytol.,* **56**, 133–149.
Talling, J. F. (1957b). 'Photosynthetic characteristics of some freshwater plankton diatoms in relation to underwater radiation, *New Phytol.,* **56**, 29–50.
Talling, J. F. (1960). 'Self-shading effects in natural populations of a planktonic diatom', *Wett. Leben.,* **12**, 235–242.
Talling, J. F. (1965). 'The photosynthetic activity of phytoplankton in East African lakes', *Int. Rev. Hydrobiol.,* **50**, 1–32.
Talling, J. F., Wood, R. B., Proseer, M. V., and Baxter, R. M. (1973). 'The upper limit of photosynthetic productivity by phytoplankton: evidence from Ethiopian soda lakes', *Freshwat. Biol.,* **3**, 53–76.
Vollenweider, R. A. (1965). 'Calculation model of photosynthesis-depth curves, and some implications regarding day rate estimates in primary production', *Mem. Ist. Ital. Idrobiol.,* 18 Suppl., 425–457.

6

Modelling of Dissolved Oxygen in a Non-Tidal Stream

M. B. Beck

6.1 INTRODUCTION

Since the classical study of Streeter and Phelps (1925) the modelling and analysis of dissolved oxygen (DO) levels in a river has occupied a major portion of the literature on water quality models. This is appropriate for two reasons: on the one hand DO is a good index of the general healthy state, or otherwise, of an aquatic environment and on the other hand, because many variables of pollution and quality interact with the DO, its variations in time and space are particularly difficult to characterize. One of the more well known measures of waste pollution, the biochemical oxygen demand (BOD), is primarily responsible for the consumption of DO in a river and, hence, for the degradation of water quality. Owing to the significance of the interaction between the DO and the BOD it has become common parlance to talk about DO–BOD models.

The emphasis in this paper lies upon the use of control and systems analysis techniques for the modelling of DO–BOD interaction in a non-tidal reach of river. For this case, then, the 'system' (see Andrews, Chapter 1) is defined as the body of water flowing through a given stretch of river, which can usually be suitably specified with upstream and downstream boundaries marked by the location of weirs, effluent discharges and confluences between rivers, and so forth. Defining the system in this manner affords a convenient picture of the river with visually, and conceptually, well-defined inputs and outputs.

The control aspects of the study are largely concerned, at this stage, with the derivation of *dynamic* models for the process of DO–BOD interaction. For, although it may be desirable to operate a process at a chosen *steady-state*, or reference, value, it is by no means a reasonable assumption that the inputs to the system are time-invariant. Therefore, some form of dynamic control is required to maintain predetermined satisfactory levels of DO, for instance, in a reach of river. Alternatively, it may be desirable to utilize the time-varying nature of water quality in order to meet, and balance, the competing demands made on a river's self-purification capacity by the assimilation of waste effluent loads and by the abstraction of potable water; this is especially so if the river is viewed as a network for the transfer of water between supply and demand centres. Thus, a reach of river, while it may be considered as a separate entity for certain studies, is really an integral part of a much larger water quality system which embraces other mutually interacting systems, viz. potable water treatment and supply, a sewer network, and wastewater treatment plant (Beck, 1976a). Moreover, the successful operation and design of a water quality system is heavily dependent, in certain critical areas, upon an intimate under-standing of the dynamics of water quality in the river.

The contents of the paper are organized as follows. Since a degree of 'jargon' has already been imprinted on the discussion, e.g. dynamic, steady-state, time-invariant, Section 6.2(1) attempts to clarify the ideas behind the terminology of mathematical modelling and Section 6.2(2) gives an introduction to the various assumptions which are made in the formulation of different mathematical representations of a systems behaviour. Section 6.3 reviews briefly the historical development of mathematical models for DO–BOD interaction. Starting with the contribution of Streeter and Phelps, it is noticeable that the past five years have seen a sudden development in relatively complex models; indeed, there are models which may lead us to question the hitherto (assumed) over-riding importance of BOD as the principal variable interacting with the DO.

Sections 6.2 and 6.3 are deliberately general in nature such that the latter part of the paper, which has a clear control and systems bias, can be studied in the broader context of a much wider variety of scientific disciplines. Section 6.4 outlines the use of system identification and parameter estimation in *dynamic* mathematical modelling and Section 6.5 describes the results of a case study of the River Cam in Eastern England. It is our intention that this latter section should both illustrate the concepts discussed in Section 6.2 and make some additional contribution to the previous investigations reviewed in Section 6.3. In the final short section, Section 6.5(4), we remark upon the application of the models in simulation and control system synthesis studies.

6.2 MATHEMATICAL MODELS FOR DO–BOD INTERACTION

Many of the general concepts of mathematical modelling are discussed elsewhere in this book, e.g. (Andrews, Chapter 1; Petrie, Chapter 2), and it is not the pur-pose of an 'applications' paper to reiterate what has been said before.

(1) *Some definitions of terminology*

We may start by defining the component features of the system in a manner similar to that of Andrews (Chapter 1), where, from Figure 6.1., we have

$\mathbf{x}$ = a vector of state (of the system) variables,
$\mathbf{u}$ = a vector of deterministic input (or forcing) variables
$\mathbf{y}$ = a vector of output variables,
α = a vector of parameters for the system characterization,
ξ = a vector of stochastic disturbances of the system, i.e. system 'noise',
η = a vector of stochastic measurement errors associated with the outputs $\mathbf{y}$, i.e. measurement 'noise'.

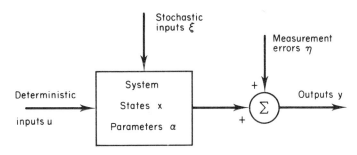

Figure 6.1 Schematic representation of the system variables

For the most general case it may be assumed that all these quantities display variations both in time, t, and space, z. (Here z is assumed to be one-dimensional with respect to the particular application of river water quality modelling.) In other words, the system's behaviour is described in mathematical terms, i.e. the model, by a set of dependent variables, x, u, y, ξ, and η and parameters α, which are functions of the independent variables, t and z.

Now let us make a clear distinction between a *dynamic* and a *steady-state* model: thus,

(a) a *dynamic* model relates time-varying inputs $u(t,z), \xi(t,z)$ to a time-varying state $x(t,z)$ of the system; furthermore, the noisy observations of the system $y\,(t,z)$ are also time-varying;

(b) a *steady-state* model relates time-invariant inputs $u(z)$, $\xi(z)$ to a time-invariant state $x(z)$, which is observed through time-invariant measurements $y(z)$.

In this sense, then, it is unfortunate that the classical DO–BOD study of Streeter and Phelps (1925) refers to an independent variable called the 'time of flow'. In most practical applications of their work it is assumed that the velocity of river flow is time-invariant and, hence, the independent variable 'time of flow' is tantamount to the independent variable of distance along the stream, i.e. z. Certainly, this nomenclature can lead to the confusing appearance of their model as a dynamic model when, in reality, it is merely a steady-state model. The implications of proposing a steady-state model for the description of DO–BOD interaction are mainly the assumptions that concentrations of DO and BOD, i.e. x, and the flow-rate of the river are time-invariant at all points z along the river.

Secondly, the distinction between *stochastic* and *deterministic* models is made by noticing that Figure 6.1 represents a stochastic system whereas, if it is assumed that

$$\xi(t,z) = 0 \qquad \text{(for all } t, z)$$
$$\eta(t,z) = 0 \qquad \text{(for all } t, z)$$

the system representation reduces to the class of deterministic models. Thus, for example, if a deterministic characterization of DO–BOD interaction is made, it implies that experimental observations of DO and BOD contain no errors of measurement and that there are no chance disturbances of the system such as those which may result from the meteorological conditions in the system's environment.

Finally, the choice of a mathematical model structure can take one of two polarized forms. A *black box* model is considered here to be one in which the system description merely relates the inputs **u** to the outputs **y** in the manner of 'causes' and 'effects', respectively. For this type of model no information is assumed for the physicochemical and biological mechanisms which govern the internal description, i.e. **x**, **α**, of the system; in this case the black box is literally a fair reflection of our knowledge of the system block in Figure 6.1. On the other hand, an *internally descriptive* model structure exploits much more, if not all, of the available a priori information on the phenomena determining the system's behaviour. To illustrate these points, consider the situation shown in Figure 6.2(a). A black box model assumes that, from empirical observation, the output DO, BOD concentrations at the downstream boundary of the reach are related by a cause–effect mechanism to the DO and BOD concentrations at the upstream end of the reach. However, the black box model does not assume any explicit knowledge of the phenomena of reaeration and BOD

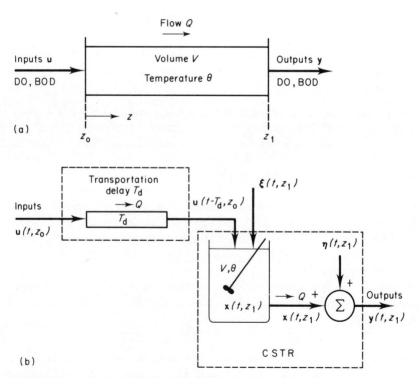

Figure 6.2 A simple model for DO–BOD interaction in a reach of river: (a) a single reach of river; (b) CSTR representation

decay, for example, although these factors would be included in an internally descriptive model.

Notice now that this definition of a black box model is at some variance with that of Petrie (Chapter 2). What Petrie has termed a black box model, the control engineer would prefer to call an internally descriptive model. Of course, we should always remember that no model of a real system is perfect: if it were we might call it a 'white box' model. It seems, therefore, that a better view of mathematical models would be one in which there is a whole spectrum of models ranging from the black box to the 'white box' and where most internally descriptive models fall within a 'grey box' band! The terminology is, nevertheless, convenient since it mirrors the level of the investigator's insight into the physical mechanisms of the system's behaviour.

(2) Dynamic mathematical model forms

This presentation of dynamic mathematical model forms is a brief survey of the aforementioned spectrum of model structures. It is arranged in an ascending order of complexity: that is to say, the order of presentation is a logical sequence which could be followed in the derivation of a final model structure with sufficient complexity and accuracy to meet the needs of a specific application.

(a) Black box models

The simplest type of model is a black box model, mainly because it assumes no a priori knowledge of the system's characteristics. Let us consider the problem where the time variation of a water quality index, e.g. DO, BOD, chloride concentration, temperature, etc. is to be modelled at a *fixed* point location, z_p say, on the river. A general functional relationship for the black box model is given by,

$$y(k, z_p) = f_1 \{ y(k - r, z_p), U(k - r, z), N(k - r) \} \qquad (6.1)$$

where

$$\left. \begin{array}{l} Y(k - r, z_p) = [y(k - 1, z_p), y(k - 2, z_p), ..., y(k - r, z_p)] \\ U(k - r, z) = [u(k, z), u(k - 1, z), ..., u(k - r, z)] \\ N(k - r) = [n(k - 1), n(k - 2), ..., n(k - r)] \end{array} \right\} \qquad (6.2)$$

$\mathbf{u}(k, z)$ and $\mathbf{y}(k, z_p)$ are respectively the observations of the multiple system inputs and the system output at the kth sampling instant of time; $n(k)$ is a sequence of independent, normal $(0, \lambda^2)$ random variables, and f_1 is a scalar function in the variables, as shown.

The black box model, or input/output model, is a time-series model in which the prediction of the output $\mathbf{y}(k, z_p)$ is a function of previous observations of that output, i.e. $Y(k - r), z_p)$, and the inputs, i.e. $U(k - r, z)$, together with past realizations of the 'lumped' stochastic noise sequence $N(k - r)$. Inspection of equation (6.1) and Figure 6.1 shows that the assumed knowledge of the system behaviour rests solely upon the empirical observations of the input and output variables; further it is assumed that all the stochastic effects of the quantities ξ and η of Figure 6.1 are lumped together in a single noise process N in equation (6.1).

The model can be used in two ways. Firstly, it may be desirable to model DO time-variations at a point z_p on the basis of measurements of DO, temperature, BOD, and so on, all at the *same* point z_p on the river; thus for equation (6.1),

$$U(k - r, z) = U(k - r, z_p) \qquad (6.3)$$

Alternatively, and this is probably a more useful application (see Section 6.5) the output DO, BOD variations at the downstream boundary of a reach of river, $z_p = z_1$ in Figure 6.2a, are predicted from data on the input DO, BOD disturbances at the upstream boundary of the reach, i.e. for equation (6.1),

$$\begin{aligned} \mathbf{y}(k, z_p) &= \mathbf{y}(k, z_1); \ Y(k - r, z_p) = Y(k - r, z_1); \\ U(k - r, z) &= U(k - r, z_0) \end{aligned} \qquad (6.4)$$

(b) *Discrete-time internally descriptive models.*

Because of the problems in finding unique representations of the model structure, see e.g. Åström and Eykhoff (1971), the black box model of equation (6.1) is often limited in practice to multiple input/single output systems. (Note, however, that under certain assumptions an array of black box models can be used to model a multivariable system, see section 5.) A more complex model for multivariable (multiple input/multiple output) systems is given by a discrete-time, state-space model,

$$\mathbf{x}(k, z_p) = \mathbf{f}_2' \{ \mathbf{x}(k - 1, z_p), \mathbf{x}(k - 1, z) \}$$

with noisy observations,

$$\mathbf{y}(k, z_p) = \mathbf{g} \{ \mathbf{x}(k, z_p), \mathbf{n}(k) \}$$

$$\left.\begin{aligned} & \\ & \\ & \\ & \end{aligned}\right\} \qquad (6.5)$$

Equation (6.5) is a particular form of the model in which ξ and η of Figure 6.1 are lumped together into a single noise vector $\mathbf{n}(k)$; this form of the model is useful for comparison with the later continuous-time internally descriptive models. Alternatively, substituting for $\mathbf{x}$ from the observations process, we obtain the form

$$\mathbf{y}(k, z_p) = \mathbf{f}_2 \{ \mathbf{y}(k - 1, z_p), \mathbf{u}(k - 1, z), \mathbf{n}(k) \} \qquad (6.6)$$

which permits comparison with the black box model above. $\mathbf{f}_2, \mathbf{f}_2', \mathbf{g}$ are all vector functions in the variables as shown.

Like the black box model of equation (6.1), the discrete-time, internally descriptive model of equation (6.6) can be employed under conditions similar to those of equations (6.3) and (6.4). Note that this model, while it retains some of the simplicity of the black box model, since it does not require the solution of differential equations, is conceptually more complex than equation (6.1). It not only deals with a multivariable system, but it also requires a priori knowledge of the *internal* causality of the system by virtue of the explicit inclusion of $\mathbf{x}$ in equation (6.5); such knowledge of $\mathbf{x}$ is included implicitly in the model form of equation (6.6).

(c) *Continuous-time, internally descriptive models*

The class of continuous-time, internally descriptive models are all derived from the application of mass balance principles to a reach of river. The analysis of these models is well-suited to a background of chemical process theory (see, e.g. Himmelblau and Bischoff, 1968), since, after all, the behaviour of a reach of river can be viewed along lines similar to that of a chemical reactor.

Theoretically, the quantity of fluid mixing provides a convenient criterion by which to distinguish between these models as follows:

(i) *Intimate mixing*; a first-order ordinary differential equation with associated transportation delay T_d to allow for transportation effects between spatial points z_0 and z_1 (see Figure 6.2b).

$$\left. \begin{aligned} \frac{d\mathbf{x}}{dt}(t,z_1) &= \mathbf{f}_3\{\mathbf{x}(t,z_1), \mathbf{u}(t-T_d, z_0), \boldsymbol{\xi}(t,z_1)\} \\ \text{with noisy observations } \mathbf{y}(t,z_1) &= \mathbf{g}_1\{\mathbf{x}(t,z_1), \boldsymbol{\eta}(t,z_1)\} \end{aligned} \right\}$$

In chemical engineering terminology the model of equation (6.7) corresponds to a continuously stirred tank reactor (CSTR) (Himmelblau and Bischoff, 1968) plus transportation delay idealization of a reach of river.

(ii) *No mixing*; a first-order partial differential equation model represented by,

$$\left. \begin{aligned} \frac{\partial \mathbf{x}}{\partial t}(t,z) &= \mathbf{f}_4\left\{\frac{\partial \mathbf{x}}{\partial z}(t,z), \mathbf{x}(t,z), \mathbf{u}(t,z_0), \boldsymbol{\xi}(t,z)\right\} \\ \text{with noisy observations } \quad \mathbf{y}(t,z) &= \mathbf{g}_2\{\mathbf{x}(t,z), \boldsymbol{\eta}(t,z)\} \end{aligned} \right\} \qquad (6.8)$$

Equation (6.8) is the well-known description of a plug flow reactor (PFR) model in which $\partial \mathbf{x}(t,z)/\partial z$ is the bulk advective flow component of material flux.

(iii) *Partial mixing*; a second-order partial differential equation model represented by,

$$\left. \begin{aligned} \frac{\partial \mathbf{x}}{\partial t}(t,z) &= \mathbf{f}_5\left\{\frac{\partial^2 \mathbf{x}}{\partial z^2}(t,z), \frac{\partial \mathbf{x}}{\partial z}(t,z), \mathbf{x}(t,z), \mathbf{u}(t,z_0), \boldsymbol{\xi}(t,z)\right\} \\ \text{with noisy observations } \quad \mathbf{y}(t,z) &= \mathbf{g}_2\{\mathbf{x}(t,z), \boldsymbol{\eta}(t,z)\} \end{aligned} \right\} \qquad (6.9)$$

As a more complex extension of equation (6.8), the model of equation (6.9) describes the behaviour of a PFR with the additional effects of partial mixing represented by a term for longitudinal dispersion, i.e. $\partial^2 \mathbf{x}(t,z)/\partial z^2$.

The three types of internally descriptive models are arranged in ascending order of complexity in the sense that ordinary differential equations are more easily solved than partial differential equations; likewise, the higher the order of the differential equations the more difficult is their solution. From the following section it is apparent that most investigations of DO–BOD interaction have used models of the continuous-time, internally descriptive type. Indeed, previous studies have been almost exclusively involved with various forms of the models given by equations (6.8) and (6.9).

This is perhaps understandable in view of the conceptual ease of considering a river as a continuous-time process with properties which vary both in time and space. While being much simpler, equation (6.7) may appear to be a strange idealization of the river system, whereas, with equation (6.6), it is not natural to describe DO–BOD interaction as a discrete-time, or digital, process. Notice, however, that if equation (6.7) is integrated between the instants $k - 1 \leqslant t \leqslant k$, an equation of the form equation (6.6) results.

6.3 THE DEVELOPMENT OF INTERNALLY DESCRIPTIVE MODELS—A REVIEW

For the internally descriptive models much work has been done on the postulation of theories which describe the 'internal' phenomena governing the behaviour of DO–BOD interaction in a reach of river. In this section we review the development of mathematical expressions for the physico-chemical, biological production/consumption processes which affect the levels of DO and BOD in the system. That is to say, once an expression is obtained for these *source* terms, $S\{\mathbf{x}(t,z)\}$ say, where

$$S\{\mathbf{x}(t,z)\} = \begin{bmatrix} S_1\{\mathbf{x}(t,z)\} \\ S_2\{\mathbf{x}(t,z)\} \end{bmatrix}$$

and S_1 is a set of production/consumption terms for DO, S_2 is a set of production/consumption terms for BOD, it can then be inserted into the desired model structure, i.e. equation (6.7), (6.8), or (6.9).

The interaction between DO and BOD is indeed a complex process which depends upon a multitude of factors, some or all of which may be important for any given river system. Figure 6.3 provides a reasonably concise and complete picture of those processes which are believed to govern DO–BOD interaction. (For the reader previously unacquainted with the physical nature of the system, a good introduction to the subject is given by Downing, 1967.) It is impossible to review all of the models which have been proposed for DO–BOD interaction, since there is truly a plethora of published papers on the subject. However, it is possible to state clearly that the classical study of Streeter and Phelps, 1925 has been a dominant feature in the evolution of DO–BOD models. They assumed that the balance between DO and BOD concentrations is the result of two processes only, the reaeration of the stream and the consumption of DO in the oxidation (or decay) of BOD, i.e.

DO: $\qquad S_1 = K_1(C_s - x_1) - K_2 x_2 \qquad$ (a)

BOD: $\qquad S_2 = -K_2 x_2 \qquad\qquad\qquad$ (b) $\qquad$ (6.10)

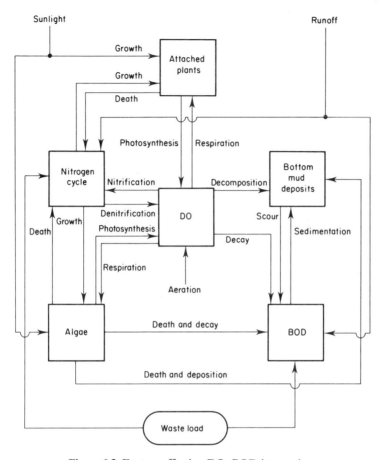

Figure 6.3 Factors affecting DO–BOD interaction

Here, x_1 = concentration of DO (mg/l)
x_2 = concentration of BOD (mg/l)
C_s = saturation concentration of DO (mg/l)
K_1 = reaeration rate constant for DO (day^{-1})
K_2 = BOD decay rate constant (day^{-1}),

and the arguments, t, z, are omitted for notational convenience.

Streeter and Phelps employed their equations in the form of a steady-state solution to the first-order partial differential equation model, equation (6.8). No significant improvement in the detail of the Streeter–Phelps equations was made until the mid 1960's and, in practice, equations (6.10) may still afford a useful description of DO–BOD interaction for certain applications, see, for example, Tarrasov and colleagues (1969), Grantham and colleagues (1971). But, judging by Figure 6.3, it is evident that the Street–Phelps equations are a considerable simplification of the true nature of DO–BOD interaction. A more realistic representation of the system is proposed by Dobbins (1964) who gives the following expressions for the source terms S,

DO: $\qquad S_1 = K_1(C_s - x_1) - K_2 x_2 + D_B \qquad$ (a)

BOD: $\qquad S_2 = - K_2 x_2 - K_3 x_2 L_A \qquad$ (b)

$$(6.11)$$

in which the additional parameters are defined as

K_3 = rate constant for BOD removal (from the flowing system) by sedimentation and/or adsorption (day^{-1}),

D_B = net rate of addition of DO to the reach of river by the combined effects of the decomposition of bottom mud deposits, and the photosynthesis/respiration of plants [(mg/l)/day],

L_A = net rate of addition of BOD to the reach of river, e.g. by local surface runoff [(mg/l)/day].

Incorporating and rationalizing some of the contributions from the intervening years since 1925 (e.g. Thomas, 1948; O'Connor, 1961), Dobbins presents his model in the form of second-order partial differential equations, i.e. equations (6.11) substituted in equations (6.9). However, in this earlier paper (Dobbins, 1964) he chooses merely to analyse the simpler steady-state solution of the model and the concomitant problems of theoretical and empirical methods for assessing the magnitudes of the various model parameters. Dobbins also investigates some fundamental aspects of the effects of longitudinal dispersion on the steady-state DO and BOD profiles and he postulates a theoretical mechanism for the process of atmospheric reaeration of the stream. A similar application of the steady-state version of the Dobbins model has been used by Fan and colleagues (1971) for the study of longitudinal dispersion effects in a large-scale multi-reach river system.

In a subsequent paper Dobbins, with his coworker Dresnack (Dresnack and Dobbins (1968)), treats the problem of numerically solving the second-order partial differential equations with finite difference schemes for the more general case of dynamic conditions in the river. The term D_B is also modified such that we have

DO: $\qquad S_1 = K_1(C_s - x_1) - K_2 x_2 - D'_B - R + P \qquad$ (a)

BOD: $\qquad S_2 = K_2 x_2 - K_3 x_2 + L_A \qquad$ (b)

$$(6.12)$$

where now

D'_B = rate of removal of DO by decomposition of bottom mud deposits [(mg/l)/day],

R = rate of removal of DO by plant respiration [(m/l)/day],

P = rate of addition of DO by plant photosynthesis [(mg/l)/day].

The 'Achilles heel' of the Dresnack–Dobbins model is the assumption of a steady, i.e. time-invariant, flow-rate for the river which can prove to be quite restrictive for their solution of the dynamic model (see, e.g. Beck, 1973). [Modelling the time-variations (dynamics) of river flow is of considerable importance for any water quality model since it determines both the retention time and dispersive properties of the reach of river; several authors have tackled this problem with respect to DO–BOD quality models, e.g. Whitehead and Young, 1975a; Grantham and colleagues, 1971.]

Equations (6.11) or equations (6.12), are, nevertheless, a good initial approximation to the true nature of DO–BOD interaction, although like the Streeter–Phelps equations, they can be shown to be inadequate in specific circumstances (see Section 6.5). The statement that they are a good initial approximation can be argued thus: the equations are sufficiently general to be applied across a broad spectrum of varying types of river, yet they are sufficiently detailed in their (dynamic) structure to illuminate the more specific modes of DO–BOD interaction which may attach to a specific reach of river.

Working along lines similar to those of Dobbins, and at almost the same time, Camp (1965) presents an alternative set of proposals for rectifying the inadequacies of the Streeter–Phelps model. Although Camp only discusses the existence of steady-state conditions in a river, his equations may be expressed as,

DO: $\quad S_1 = K_1(C_s - x_1) - K_2 x_2 + P'$ (a)

BOD: $\quad S_2 = -K_2 x_2 - K_3 x_2 + p$ (b)

$$(6.13)$$

with

P' = rate of addition of DO by algae through photosynthesis $[(mg/l)/day]$,

p = rate of addition of BOD (to the flowing system) by resuspension of bottom mud deposits $[(mg/l)/day]$.

The significance of the Dobbins/Camp models for DO–BOD interaction is evidenced both by the large number of applications they have found in later studies and by the fundamental role they have played in nearly all the subsequent contributions to this field of research. A particularly good discussion of their models is given by Hansen and Frankel (1965); they suggest a further important modification whereby DO–BOD interaction is described in a form equivalent to the following,

DO: $\quad S_1 = K_1(C_s - x_1) - (K_{21} + K_{22})x_2 - D'_B + P_m \cos(\omega t - \phi)$ (a)

BOD: $\quad S_2 = -(K_{21} + K_{22})x_2 - K_3 x_2 + p$ (b)

$$(6.14)$$

When comparing equations (6.14) with equations (6.12), (6.13), it is observed that the diurnal variations in DO are now introduced by the term $P_m \cos(\omega t - \phi)$, where

P_m = amplitude of diurnal oscillations in the rate of addition of DO by the photosynthetic/respiratory activity of the stream biota $[(mg/l)/day]$, (cf. the definition of P for equation (6.12): clearly, the definitions of photosynthetic/respiratory effects on the DO are generally identical, although each individual author may choose to suggest particular biological sources for the effects, e.g. plants, algae.)

ω = period of the oscillations $(2\pi \text{rad}/\text{day})$,

ϕ = a phase lag (rad)

In addition, the previous definition of the rate constant K_2 is disputed in the sense that Hansen and Frankel believe there should be an explicit distinction between the 'gross' rate of BOD oxidation observed 'in situ' in the river, i.e. $K_2 (= K_{21} + K_{22})$, and that rate which is measured by a laboratory analysis of a sample of river water, i.e. K_{21}. K_{22} in this case, then, describes the effects of attached aquatic growths and slimes in converting oxygen-demanding organic matter into stable end-products. This kind of modification is more interesting than it seems at first sight, for it raises the question of how to assign values for the parameters of the model; we shall consider this point again in Section 6.4(3).

Hansen and Frankel's model is also significant for another rather different reason. As expressed in equation (6.14), the symbol p is an inadequate representation of their proposals: p would be more correctly represented by $p\{\mathbf{x}(t,z)\}$, thus indicating that the effects of bottom deposits are a function of other state variables $(x_3, x_4, \ldots)$ in the system description, i.e. the concentration of BOD in the mud deposits. Now, if we recall the pictorial description of the river water quality system in Figure 6.3, we can observe that, hitherto, our definition of the state $\mathbf{x}$ covers only the concentrations of DO, BOD as dissolved and suspended material in the *flowing* river. Hansen and Frankel's model is, as it were, the first in a rapidly developing line of 'second-generation' DO–BOD models where the order of the state $\mathbf{x}$, and hence the definition of the system, is extended to embrace a number of additional variables which are found to be relevant for the description of DO–BOD interaction. For instance, in addition to the chemical composition of the mud deposits, we might include the concentrations of live and dead algae, or the concentration of nitrogen-bearing compounds, as other state variables (see below and section 5).

By the late 1960's the idea of using these second-generation DO–BOD models, together with more complex descriptions for the diurnal DO variations, was developed and expanded, albeit perhaps unintentionally, by O'Connor and his colleagues (O'Connor, 1967; O'Connor and Di Toro, 1970). The O'Connor model of DO–BOD interaction takes the following form,

$$\text{DO:} \quad S_1 = K_1(C_s - x_1) - K_{2c}x_{2c} - K_{2n}x_{2n} - D'_B - R + P \quad \text{(a)}$$
$$\text{BOD:} \quad S_2 = -K_{2c}x_{2c} - K_{2n}x_{2n} \quad \text{(b)}$$
$$\left. \right\} \quad (6.15)$$

where the state variable x_2, representing BOD concentration, has been differentiated into two types,

x_{2c} = concentration of carbonaceous BOD (mg/l),
x_{2n} = concentration of nitrogenous BOD (mg/l),

with the additional parameters defined as

K_{2c} = carbonaceous BOD decay rate constant (day^{-1}),
K_{2n} = nitrogenous BOD decay rate constant (day^{-1}).

In contrast to the Dobbins model of equation (6.12), the terms R and P of equation (6.15) refer to the effects of respiration and photosynthesis, respectively, of algae. And, as Hansen and Frankel also indicate, O'Connor (O'Connor and

Di Toro, 1970) treats the term P in a more general sinusoidal form using a Fourier series expansion.

With the presentation of the work of O'Connor the review of the simpler, low-order, state-space models of DO–BOD interaction is complete. Looking back over the contributions that have been made since the original study of Streeter and Phelps, we can see that the various models are, by and large, quite similar. From an applications point of view these later models have been employed in a wide variety of investigations. Broadly speaking, those who have followed the approach of O'Connor have tended to substitute equations (6.15) into a first-order partial differential equation model structure, equation (6.8), as does O'Connor himself. Providing the assumption of no fluid mixing is legitimate, this model structure can be a useful approximation to the true nature of the system. One particular advantage of the dynamic version of the model is that it can be reduced to an ordinary differential equation form, whereby it is easier to solve, by the method of characteristics (Di Toro, 1969), and it does not require the assumption of a time-invariant flow-rate for the river. Possibly for these reasons, the approach of O'Connor has been followed for studies in the formulation of state and parameter estimation problems in the DO–BOD system, Koivo and Phillips (1971), (1972); Koivo and Koivo (1973). On the other hand, the Dobbins/Camp approach has been used by Goodman and Tucker (1971), as a repeated steady-state solution for the simulation of a dynamic model, by Lee and Hwang (1971) for steady-state parameter estimation, and, in a reduced form, by Kendrick and colleagues (1970) for the synthesis and evaluation of an optimal control scheme for in-stream DO.

Over the past five years the second-generation DO–BOD model has become well established. And with this development there appears to be a trend towards the introduction of the nonlinear kinetic model of Monod (1942) into DO–BOD interaction. In one form the Monod function has been postulated for the oxidation of BOD (Shastry and colleagues, 1973), and in many other forms it is suggested as a description for the growth of algae (Chen, 1970; Rich and colleagues, 1972; Thomann and colleagues, 1974) and for the growth of bacteria and protozoa in river sediments (Rutherford and O'Sullivan 1974). This last example of a higher-order state-space formulation may be considered as an extension of the earlier work of Hansen and Frankel on the activities of decomposing mud deposits.

The most comprehensive of these contemporary models for the pictorial system in Figure 6.4, although we should note that strictly speaking, it refers to estuarine reaches of a river, is that proposed by Thomann and colleagues, (1974).

It is, in fact, appropriate that the work of Thomann should be mentioned here, since he has made an invaluable contribution in the field of modelling and control of water quality. In particular, he has done much to introduce the systems analysis approach (Thomann, 1972) and is one of the first authors to propose the use of time-series analysis for the forecasting of water quality data, e.g. DO and temperature (Thomann, 1967). Notice that the use of these types of model, a black box model, represents quite a departure from the more 'traditional' method of using internally descriptive models for predicting DO variations, as witnessed by this review. [For further applications of time-series analysis for the stochastic modelling and prediction of water quality data we may refer to the review of Thornes and Clark, 1975.] Clearly, the black box approach is a more empirical approach, although it does not contain the same kind of empiricism which is apparent in the studies of, for example, Owens and colleagues (1969). More recently, some of the sophisticated algorithms for black box model identification and parameter estimation have been proposed by Huck and Farquhar (1974) for DO and chloride prediction. Both Huck and Farquhar and Thomann (Thomann, 1967) address the problem of time-series models for water quality variations at a fixed spatial location, i.e. equation (6.1) with the conditions of equation (6.3); we shall discuss in section 4 some results for the other type of black box model given by the substitution of equation (6.4) into equation (6.1).

6.4 MODEL OBJECTIVES, IDENTIFICATION, PARAMETER ESTIMATION, AND VALIDATION

(1) Objectives, instrumentation

Bearing in mind that, as control and systems analysts, we are often concerned with time as the *single* independent variable, we now address the problem of identifying and verifying specifically *dynamic* models of DO–BOD interaction from a given set of experimental data. For, although the subject of modelling DO–BOD interaction has received much attention in the literature, few investigators have used their models in a truly dynamic context or verified them against field data. This is understandable since the intensity and duration of sampling and monitoring required is often prohibitive, especially in the case of 5-day BOD tests.

However, for many forecasting, parameter estimation, or control applications two factors are of considerable importance. Firstly, a river system is unlikely to be in a steady state, since this implies that the DO and BOD inputs and outputs, as well as the volumetric flow-rate, are time-invariant; therefore, a dynamic model should be assumed.

Of course, that is not to say that a steady-state model is inappropriate for all studies. For instance, a steady-state model is more useful than a dynamic model in quantifying the effects on river water quality of possible strategies for meeting expected demands for water and for disposing of expected domestic and industrial effluent in future years. A good example of this kind of *long term* planning and management (steady-state) model is given by Fawcett (1975). Alternatively, a steady-state model of DO–BOD interaction may be coupled with an economic analysis in order to formulate a capital investment policy for wastewater treatment facilities (see e.g. Dysart, 1969). In contrast, for the *short term* day-to-day operation of an on-line, automatic control scheme for water quality, or as a more accurate reflection of the true nature of the system, a dynamic model is more suitable. Indeed, we might remark that almost any system in the real world is naturally both dynamic and stochastic.

Secondly, it is not constructive to use a model unless it has been shown to be a reasonable description of the observed behaviour of the system. Normally, in the case of water quality modelling field data are at a premium. And, if the available manpower is limited, as is generally the case, then only small portions of a river system can be investigated in detail. Yet despite the current focus of attention on environmental technology, which is broadening the scope of commercial probes and instruments required for on-line measurement and control, there continue to be difficulties in developing the techniques for direct access to, or correlative measures of, some of the most important biochemical and biological variables, e.g. BOD. These problems are compounded by the hostile environment in which sensors are fouled rapidly and require frequent calibration, cleaning and maintenance in order that they retain the sensitivity needed to monitor minute changes in the concentrations of dissolved and suspended substances. Suffice it to say, that, in the majority of cases, the observation of water quality variables is very much a labour-intensive affair of manual or automatic sampling followed by laboratory analysis.

(2) Experimental design

Because the premium on water quality data is so high, it is desirable that experimental work should be well planned in advance so that the amount of information derived can be maximized. Any a priori knowledge of the system's dynamic behaviour is an advantage, since this knowledge can be used to establish, for example, the relevant state variables to be measured, the sampling rate at which the measurements should be made, the duration of the experiment, and the desired nature of the input forcing signals.

With respect to system identification and parameter estimation, these conditions have been discussed fully elsewhere, and, for the sake of brevity, we merely refer to the work of Gustavsson (1975). Notice, however, that the forcing disturbances **u**, if they can be artificially manipulated, e.g. step, pulse, pseudo-random binary sequence (PRBS) testing, must be constrained in order to retain process stability at all times. Clearly, for the DO–BOD interaction system it is only possible to observe the system as it behaves naturally and artificial disturbances are precluded. This means that the maximum benefit may not be obtained from an experiment as, indeed, is the case where subsequent modelling shows that the experiment should have included measurements of additional state variables. One such experiment for the identification of DO–BOD interaction in the River Cam is described in Beck (1973): its success can be judged by the results of Section 6.5 and by the fact that a more comprehensive experiment has subsequently been carried out on the same river.

(3) Identification, parameter estimation and model validation

It follows that, with the retrieval of experimental data, a model must be chosen to represent the information in those data. As we have mentioned in Section 6.2(1), there is essentially a choice between two model types for the description of the system's dynamics:

 (i) a black box (or input/output) model;
 (ii) an internally descriptive (or mechanistic) model.

Briefly, once the type of model is chosen to meet the specific conditions of any given problem, the construction and validation of the model is achieved in two stages:

 (iii) identification
 (iv) parameter estimation

At the identification stage the causal relationships between the inputs and outputs, together with the order and structure, of the model are defined. The separation of the identification and estimation stages is essentially one of concept. Identification is often carried out with a parameter estimation scheme in which the analyst is more concerned with assessing the adequacy of the model structure than with estimating the parameters accurately. It is at this stage where the analyst's own understanding of the physical system is fully exploited in the interpretation of the results. For, despite the attempts to put the identification stage on a firm mathematical basis, the choice of the final model structure remains very much an art.

Thus, having identified the model structure, the parameters of the model are estimated according to any one of a number of procedures which fall into the two categories of,

 (v) off-line methods
 (vi) recursive (or on-line) methods.

By off-line we mean the estimation of parameters by repeated updating of the estimates after each iteration through a given block of N data samples, whereas the recursive methods adapt the estimates at each sampling instant t_k (where $k = 1, 2, \ldots, N$) *within* that set of N sampled observations.

Clearly it is beyond the scope of this paper to review the whole subject of identification and estimation; it is sufficient to refer to the work of Åström and Eykhoff (1971) and Eykhoff (1974), who give a more complete discussion of the underlying theoretical aspects of this fast-developing field. Alternatively, a concise and clear exposition of recursive approaches to time-series analysis is given by Young (1974).

When the final model is obtained, its validity should be assessed by carrying out various statistical tests on the errors between the observed system and model output responses. Essentially, the question must be answered: does the model simulate the system adequately in accordance with the proposed objectives and application of the model? We should prefer a positive reply, but then no model is perfect and the modelling exercise itself is a continuing process of evolution which depends upon the interplay between experiment, analysis, and synthesis. We might also ask the question: are the estimated parameter values physically meaningful? Yet in resolving this question, we must bear in mind the danger in comparing estimates of the parameters α obtained by applying parameter estimation techniques to the set of 'in situ' field observations y, as here, with those obtained from empirical relationships (e.g. for K_1, the reaeration rate constant, see Owens and colleagues, 1964) or from laboratory analyses of bottled river water samples (e.g. for K_2, the BOD decay rate constant, see the discussion of Hansen and Frankel, 1965). There is no entirely objective means of stating that one method is better than another. Finally, let us also remark that, if the same vector of parameters α in the source terms S is estimated from the same set of data, but using different internally descriptive model structures, i.e. equations (6.7), (6.8), (6.9), then the estimates of α will differ according to the assumptions made about the mixing properties of the reach.

6.5 A CASE STUDY: THE RIVER CAM

The case study described in this section forms a part of a systems analysis project aimed at the application of system identification, parameter estimation, and control methods to the solution of certain problems of river water quality management (Beck, 1976a). Several dynamic models have been identified and verified against field data collected once daily over an 80-day period, 6 June 1972 to 25 August 1972, from a 4·7 km stretch of the River Cam immediately downstream of the Cambridge Sewage Works effluent outfall (Figure 6.4). The main purpose here is to summarize the modelling results in a manner which illustrates the points discussed in the preceding sections. The presentation is according to the ordering of the model types listed in Section 6.2(2) and is not necessarily the order in which these results have been obtained. Each aspect of the identification of the models is reported more fully elsewhere.

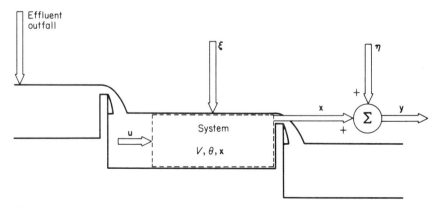

Figure 6.4 The experimental reach of river: a part of the River Cam outside Cambridge

154

(1) Black box models

Referring to Figures 6.2a and 6.4, and equation (6.1) with conditions given by equation (6.4), explicit forms of black box models can be obtained for the input/output relationships between the upstream and downstream DO, BOD using an *off-line* method of *maximum likelihood* identification (Åström and Bohlin, 1965). The two multiple input/single output models are given by (Beck, 1974).

$$
\begin{aligned}
&\text{DO:} \quad y_1(k) = 0\cdot715y_1(k-1) + 0\cdot174u_1(k-1) + 0\cdot057u_3(k) \\
&\qquad\qquad + 0\cdot044u_3(k-1) + 0\cdot554n_1(k) \qquad\qquad\text{(a)}
\end{aligned}
$$

$$
\begin{aligned}
&\text{BOD:} \quad y_2(k) = 0\cdot751y_2(k-1) + 0\cdot102u_2(k-1) + 0\cdot048u_3(k-2) \\
&\qquad\qquad + 0\cdot060u_3(k-4) + 0\cdot618n_2(k) - 0\cdot313n_2(k-1) \quad\text{(b)}
\end{aligned}
$$

(I)

where y_1, y_2 = downstream observations of DO, BOD respectively, i.e. $y_1(k,z_1), y_2(k,z_1)$
u_1, u_2 = upstream observation of DO, BOD respectively, i.e. $u_1(k,z_0), u_2(k,z_0)$
n_1, n_2 = stochastic noise sequences
$u_3(k)$ = an observation of the sunlight incident on the system during the kth day (hr/day)

and the sampling interval, $t_k - t_{k-1}$ equals one day. The results of models (I) are shown in Figure 6.5 for the purely deterministic case where $n_1(k) = n_2(k) = 0$ for all k.

Notice here that the dynamics of the DO and BOD are assumed to be independent, i.e. there is no interaction between DO and BOD; on the basis of these results we might even tentatively ask how significant, in dynamic terms, is the interaction between the DO and the BOD? The sunlight input u_3 is included to account for the significant effects of floating algal populations on the DO and BOD (Beck, 1975). This is an important feature of the Cam study where algae are observed to produce pronounced peaking effects in both the DO and the BOD between days 36–48 and from day 60 onwards. It is further evident from model I that this algal effect expresses asynchronous time-dependency relationships with the DO and BOD: equation I (a) implies that the photosynthetic production of DO in the reach at time k is dependent upon the sunlight during that same day, as might be expected, while equation I(b) implies that the peak BOD response to the algal effect follows some one or two days behind the peak DO response. Finally, recalling that BOD is a notoriously difficult variable to implement in any practical on-line control scheme, since it takes 5 days to measure, the DO model of equation I(a) has a good potential for control applications: in the first instance, it is a very simple model, and in the second instance it does not require any information on BOD conditions in the reach. (Therefore, such a model can not be used to evaluate control schemes which manipulate effluent BOD levels for the maintenance of in-stream DO; it can, of course, be used to investigate the operation of artificial aeration systems.)

155

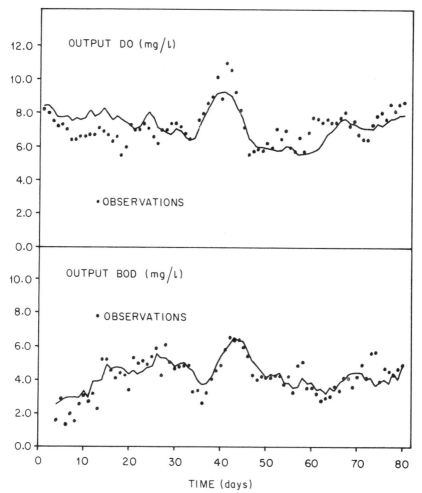

Figure 6.5 Deterministic output (downstream) DO, BOD responses given by Model I

(2) A discrete-time, internally descriptive model

Using a multivariable extension of the *recursive instrumental variable-approximate maximum likelihood* (Ivaml) technique for system identification (Young, 1974), Young and Whitehead (1975) have obtained a discrete-time, internally descriptive model from the Cam data,

$$\begin{bmatrix} x_1(k) \\ x_2(k) \end{bmatrix} = \begin{bmatrix} \alpha_{11}\,V/Q(k-1) & \alpha_{12} \\ 0 & \alpha_{22}\,V/Q(k-1) \end{bmatrix} \begin{bmatrix} x_1(k-1) \\ x_2(k-1) \end{bmatrix} +$$

$$+ \begin{bmatrix} \beta_{11}\,0\ \ \beta_{13}\,\beta_{14} \\ 0\ \ \beta_{22}\beta_{23}\,0 \end{bmatrix} \begin{bmatrix} u_1(k-1) \\ u_2(k-1) \\ I(k-1) \\ C_s(k-1) \end{bmatrix} \quad (\text{II})$$

with the observations process

$$\begin{bmatrix} y_1(k) \\ y_2(k) \end{bmatrix} = \begin{bmatrix} x_1(k) \\ x_2(k) \end{bmatrix} + \begin{bmatrix} n_1(k) \\ n_2(k) \end{bmatrix} \quad (6.16)$$

(The particular nature of the statistics of n_1 and n_2 in this model are, in view of the multivariable characterization, considerably more complex than those given for model I, see Young and Whitehead, 1975.)

Here x_1, x_2 = concentrations of DO, BOD respectively at the downstream boundary of the reach of river (mg/l)

Q = volumetric flow-rate in the reach of river (m³/sec)

V = constant volumetric hold-up in the reach ($= 1\cdot51\ (10)^5\ \mathrm{m}^3$)

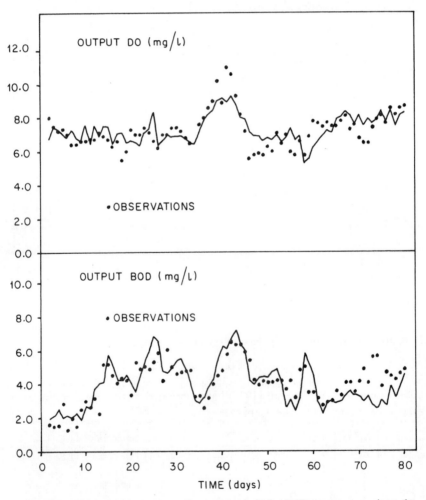

Figure 6.6 Deterministic output (downstream) DO, BOD responses given by Model II

$I(k-1)$ is a function of the input sunlight u_3 and we have called this term a 'sustained sunlight effect' (Beck and Young, 1975), for reasons which will become clear in section 6·5(3); it accounts for the effects of algae on the DO and BOD, as observed with model I.

Figure 6.6 shows the deterministic response of model II with the following estimated parameter values,

α_{11}	$-0·26$	β_{11}	$0·33$	β_{22}	$0·27$
α_{12}	$-0·25$	β_{13}	$0·84$	β_{23}	$0·43$
α_{22}	$0·36$	β_{14}	$0·1$		

In general, the results of this model are similar to those of model I, except for the elimination of the initial error in the predicted DO output, compare Figures 6.5(a) and 6.6(a). This is probably a result of the manner in which flow-rate variations are included in model II, although the physical mechanisms underlying the mathematical explanation of the observed variations are not altogether clear. Recent work, in which a black box model similar to model I is reappraised using a different identification technique (recursive IVAML), also indicates that the inclusion of flow-variations can improve the DO prediction over the initial period of the experiment (Beck, 1976b).

Model II is a good example of how the discrete-time, internally descriptive model bridges a 'conceptual gap' between the black box and the continuous-time, internally descriptive models. It includes more a priori knowledge of the physical system, e.g. the coupling between the DO and BOD dynamics, the inclusion of the saturation concentration of DO and the sustained sunlight effect as inputs to the system, yet it retains much of the simplicity of the discrete-time, difference equation approach of the black box model. There are, however, difficulties in interpreting the parameters and their estimates within the physical framework of the system.

(3) Continuous-time, internally descriptive models

The argument supporting the use of models based on the CSTR idealization of a reach of river (Figure 6.2b) has been discussed elsewhere (Beck, 1973; Young and Beck, 1974; Beck and Young 1975); we shall not, therefore, enlarge upon it here. It suffices merely to point out that it combines simplicity, i.e. *one* independent variable, time t, with a capability for accurate simulation of the observed longitudinal dispersion effects, i.e. the mixing properties, of a reach of river, see Whitehead and Young (1975a). In the Cam, where we are concerned with a short reach of river, it is unnecessary to include any explicit representation of the transportation delay, T_d, which is indicated in equation (6.7). Thus, a first version of the continuous-time, internally descriptive model for DO–BOD interaction is given by,

DO:
$$\dot{x}_1(t) = -[k_1 + Q(t)/V)x_1(t)] - K_2 x_2(t) + [Q(t)/V)]u_1(t) \\ + K_4[I(t_k) - \bar{I}] + K_1 C_s(t) - D'_B(t) + \xi_1(t)$$
(a)

BOD:
$$\dot{x}_2(t) = -[K_2 + K_3 + Q(t)/V]x_2(t) + [Q(t)/V]u_2(t) \\ + K_5[I(t_k) - I] + L_A(t) + \xi_2(t)$$
(b)

where
$$I(t_k) = (t_{k-1}) + \frac{1}{\tau}\left[u_3(t_k)\left\{\frac{\theta(t_k) - \bar{\theta}}{\theta}\right\}I(t_{k-1})\right]$$
(c)   III

and
$$[I(t_k) - I] = 0 \text{ for } I(t_k) < I$$
(d)

where the dot notation refers to differentiation with respect to time t and the argument t_k replaces the argument k for notational consistency. The equations of modal III are based upon the proposals of Dobbins, i.e. equations (6.11) or (6.12) substituted into the first-order ordinary differential equation structure of equation (6.7). The major modification is the inclusion of a low-pass filter mechanism, equation III (c), for the sustained sunlight effect $I(t_k)$ in which

τ = time constant of the low-pass filter (days),
θ = temperature of the river water (°C),
$\bar{I}$ = a threshold level of the sustained sunlight effect,
$\bar{\theta}$ = an arbitrary mean river water temperature (°C),
K_4, K_5 = proportionality constants for the sunlight effect in the DO, BOD
   equations respectively. No specific units are assigned to $I, \bar{I}, K_4, K_5$,
   in view of the dimensional anomaly of equation III (c).

In other words, $\bar{I}$ indicates the requirement for a certain minimum level of $I(t_k)$, i.e. equation III (d), before significantly-sized algal populations can become established, while τ expresses the dependence of this effect on *prolonged* periods of warm, sunny weather, which stimulate the growth of algae.

The derivation of model III is reported in an earlier paper (Beck and Young, 1975). Given that we have noisy, sampled observations of the output (downstream) DO and BOD,

$$\left.\begin{array}{l} y_1(t_k) = x_1(t_k) + \eta_1(t_k) \\ y_2(t_k) = x_2(t_k) + \eta_2(t_k) \end{array}\right\}$$
(6.17)

the structure of model III can be identified, with simultaneous estimation of the parameters, using a *recursive* technique known as the *extended Kalman filter* (EKF) (see, e.g. Jazwinski, 1970) (We might alternatively consider the EKF as a means of statistically validating the dynamic model structure.) The results of this identification procedure applied to the Cam data are discussed in Beck and Young (1976).

The deterministic output DO and BOD responses predicted by model III, with $\xi_1 = \xi_2 = 0$, are plotted in Figure 6.7 with the following estimates of the parameters,

K_1	0·17	K_5	0·32	$L_A(t)$	0·0 for all (t)
K_2	0·32	$\bar{I}$	6·0		
K_3	0·0	$\bar{\theta}$	8·0	$D'_B(t)$	$\begin{cases} 2{\cdot}7 \text{ for } 0 \leqslant t \leqslant t_{19} \\ 0{\cdot}4 \text{ for } t > t_{19} \end{cases}$
K_4	0·31	τ	4		

The dashed lines in Figure 6.7 show the responses that are obtained if the Dobbins equations are used for model III, i.e. $K_4 = K_5 = 0$. It is now possible to see the importance of including explicitly the effects of algae in the model, although the pseudo-empirical expression, I, still fails to explain completely the observed variations in the BOD over the latter period of the experiment. The estimates obtained for D'_B are a reflection of the initially marked depression of DO levels which have been noted previously with respect to models I and II.

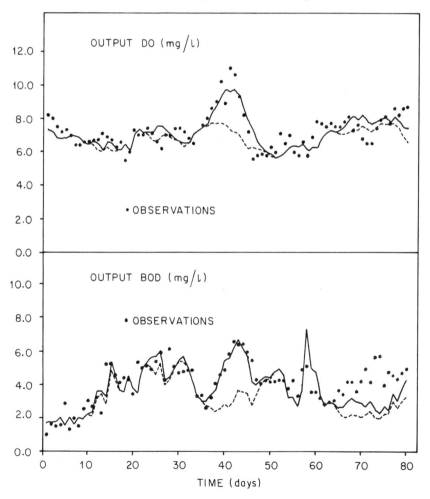

Figure 6.7 Deterministic output (downstream) DO, BOD responses given by Model III (dashed lines denote responses with $K_4 = K_5 = 0$ in Model III)

While model III is, generally speaking, an adequate description of DO–BOD interaction, it does not characterize completely some of the observed variations in the BOD data. And, since the effects of algae are clearly important disturbances of the DO and BOD dynamics, it is interesting to hypothesize a second-generation DO–BOD model which we call a DO–BOD-algae model.

Referring once again to Figure 6.3 for the representation of the internal mechanism of the system, a DO–BOD-algae model for the Cam might take the following form (Beck, 1975),

$$
\text{DO: } \dot{x}_1(t) = -[K_1 + Q(t)/V]x_1(t) - K_2 x_2(t) + K_6 x_3(t)[u_3(t_k)]^s
$$
$$
\qquad\qquad - K_7 x_3(t) + [Q(t)/V]u_1(t) + K_1 C_s(t) - D'_B(t) \qquad (a)
$$

$$
\text{BOD: } \dot{x}_2(t) = -[K_2 + K_3 + Q(t)/V]x_2(t) + K_8 x_4(t)
$$
$$
\qquad\qquad + [Q(t)/V]u_2(t) + L_A(t) \qquad (b)
$$

$$
\text{Live algae: } \dot{x}_3(t) = -[K_9 + Q(t)/V]x_3(t) + \left\{ \frac{\hat{\mu}u_3(t_k - T_d)}{K_{10} + u_3(t_k - T_d)} \right\} \qquad (c)
$$

$$
\text{Dead algae: } \dot{x}_4(t) = -[K_8 + Q(t)/V]x_4(t) + K_9 x_3(t) - R_s \qquad (d)
$$

(IV)

in which K_6 = proportionality constant for the production of DO by the photosynthetic activity of live algae,

$\quad\;\; K_7$ = proportionality constant for the consumption of DO by the respiration of live algae,

$\quad\;\; K_8$ = rate constant for the production of BOD by the redissolution of dead algal material,

$\quad\;\; K_9$ = specific death/decay rate constant of algae,

$\quad\;\; K_{10}$ = saturation coefficient for the rate-limiting nutrient of algal growth,

$\quad\;\; s$ = exponential power for the dependence of algal photosynthetic production of DO on sunlight conditions,

$\quad\;\; \hat{\mu}$ = maximum specific growth rate of algae,

$\quad\;\; R_s$ = rate of sedimentation of undissolved dead algal material.

The synthesis of model IV from model III is quite a complex process (Beck, 1975) involving the use of off-line, maximum likelihood procedures for identification (Beck, 1974).

The experimental basis for the model rests largely upon the evidence that the time-dependency of the DO and BOD dynamics on the algal population dynamics is *asynchronous*, see model I. Thus, by proposing two additional state variables, the concentrations of live and dead algae, x_3, x_4, respectively, it is possible to resolve some of the inadequacies of model III, which expresses *synchronous* relationships for the algal effects with the DO and BOD through the term $I(t_k)$ in equation III(c). This omission in the pseudo-empirical term $I(t_k)$ is perhaps best demonstrated by model II, where, for the period $t_{36} \rightarrow t_{48}$, the predicted DO model response *lags* behind the observed peak, while the model BOD output *leads* the observed BOD peak.

Most of the additional parameter definitions are self-explanatory in terms of the additional physical and biochemical mechanisms described by the model. Equation IV(c) includes a Monod (1942) function in which the sunlight incident on the system is assumed to be the rate-limiting 'nutrient' for the growth of algae. The pure time delay, T_D (= 1 sampling interval, i.e. one day), affords a better prediction by the model and this may, in fact, be due to a third form of algae which can be termed an 'inert' or stored-mass, phase in their life cycle. It is assumed that no significant quantities of either live or dead algae enter the system at the upstream boundary of the reach of river.

Because of the absence of any measurements of the state variables x_3 and x_4, which represent the concentrations of the algal population, this last model for DO–BOD-algae interaction cannot be identified in any really precise statistical

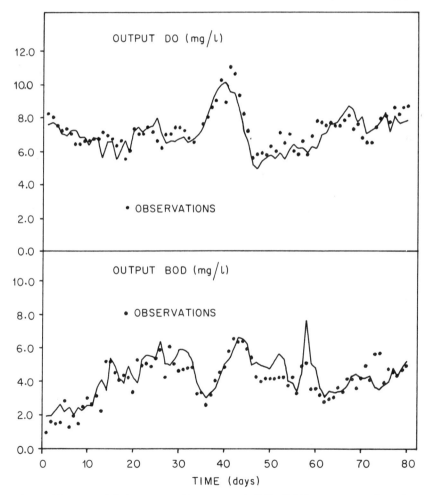

Figure 6.8 Deterministic output (downstream) DO, BOD responses given by Model IV

fashion. For this reason model IV is presented in a purely deterministic form, and, given the relevant parameter estimates of model III, a suitable prediction of the output DO and BOD concentrations is shown in Figure 6.8. (See Beck, 1975 for approximate values of the parameters.) The most important features of the model are that it attempts to eliminate some of the pseudo-empiricism of model III, namely $I(t_k)$ and that it is capable of describing the observed BOD variations for the final period of the experiment, compare Figures 6.7b and 6.8b. Of course, it should be emphasized that the model is a qualitative, rather than quantitative, description of DO–BOD-algae interaction and, in view of the data limitations, it remains a somewhat hypothetical model of the real system. The physicochemical, biological processes of the death, decay and production of a BOD by algae are, in particular, not well understood.

However, we may surmise that these effects are sensitive to flow-rate, Q, conditions in the river, although the reasons for this require further clarification. More recent studies of the Cam data do not reveal the same kind of dependence on Q for BOD-algae interaction, yet they do show a significant dependence on Q for DO-algae interaction (Beck and Whitehead, 1976b). This latter evidence suggests that the production of DO by algae decreases with increasing flow-rate Q, which would be plausible if it can be assumed, for example, that turbidity, and hence the reduction of light penetration into the river, increases with an increase in Q.

Thus, although it can be concluded that the four models presented here are capable of describing the major observed time-variations of DO–BOD interaction, and very similar models have been likewise identified elsewhere (Whitehead and Young, 1975a), much work remains to be done. Where discrepancies are already known to exist, experimental studies are currently in progress in the hope that additional information can be obtained for a better model of the system, e.g. a second-generation DO–BOD–algae/nitrogen cycle model.

(4) Applications of the models

The fact that this paper deals only with the development, identification, and validation of mathematical models for DO–BOD interaction, and in particular with the dynamics of the system, is indicative of the major effort which has been required to obtain a model that can be usefully, and confidently, employed in an applications context. Apart from the derivation of a better understanding of this complex system, which would be important for any class of river, the application of dynamic models of DO–BOD interaction is perhaps best suited for the analysis of stretches of river where critically poor quality conditions exist. For it is here that the dynamic model would seem to be of greatest benefit in the synthesis and evaluation of control schemes with a view to implementing some form of automation for the day-to-day maintenance of a satisfactory water quality. Alternatively, it may be useful to employ the dynamic model in a simulation context in order to assess the effects of severe *transient* disturbances of the system, i.e. thunderstorms, de-oxygenation by sudden death and decay of algae, discharges of storm-water overflows, high nitrate loads in surface water runoff.

Unfortunately, there is no space to pursue any of these matters in detail here. All manner of interesting studies could be proposed for applying the model in a purely simulations capacity. However, for control applications some very pertinent questions must be answered before it is likely that we shall see a proper on-line automation innovated in the control of river pollution.

And not the least of these questions will be the question of economics: do the benefits which might accrue from the automatic control of water quality outweigh the costs of implementing such a technology? It is probable that this problem, which also implies the resolution and specification of desirable stream quality standards, will be much more difficult to solve than the technical problem of designing a control system, for which some preliminary proposals for the automatic control of DO levels in a reach of river have already been made. Using the Cam models *simple* feedback controllers for the downstream DO by regulation of an upstream effluent discharge are evaluated in Young and Beck, 1974. This investigation pays particular attention to the problem of avoiding the need for BOD measurements in the control scheme. An alternative approach to effluent regulation is the use of artificial aeration, as discussed by Whitehead and Young (1975b), (see also Tarrasov and colleagues 1969), although this latter method is essentially an interim, but cost-effective, expedient. Nevertheless, let us hope that the long and winding path from the 1972 Cam experiment, via the identification of the mathematical models and the synthesis of control laws, to the point where controlled mechanical aerators are currently being assessed in the same stretch of river, will be a fruitful exercise in the much-needed application of control and systems analysis techniques to water quality control.

6.6 CONCLUSIONS

A brief review of mathematical models for DO–BOD interaction in a non-tidal reach of river has been presented. We have attempted to combine a control and systems analysis approach with an understanding of the physical, chemical and biological phenomena that govern the nature of the system. In particular, we have been at pains to point out the various types of model which might be used in a given situation, while we have treated the historical development of DO models in a manner which facilitates comparisons between each author's contribution irrespective of his objectives for, or his application of, the model. It is evident firstly that, with only a few notable exceptions, most previous models have been of the internally descriptive type, and secondly that there is a current trend towards the construction of a second-generation of DO–BOD models which take a much broader view of the characterization of water quality.

In the latter half of the paper there is a more obvious focus of attention on the dynamic (and stochastic) aspects of DO–BOD models. Some concepts of system identification and parameter estimation are introduced as background material for the results of the case study of the River Cam.

These results are used primarily to illustrate the point that several types of mathematical model can be obtained for the prediction of DO–BOD variations in a river, but they also give a more detailed picture of what is meant by a higher-order, second generation model. The case study is, if nothing else, one of the few (reported) attempts at validating *dynamic* DO–BOD models against field data; this is a feature of DO–BOD modelling research which has hitherto been sadly lacking in the literature.

Acknowledgements

This paper summarizes some work which has been continuing over the past four or five years. During that time the author has benefited considerably from the guidance and comments of Professor P. C. Young and to Dr. P. G. Whitehead of the Centre for Resource and Environmental Studies, National University, Canberra.

6.7 REFERENCES

Åström K. J., and Bohlin, T. (1965). 'Numerical identification of linear dynamic systems from normal operating records', *Proc. IFAC symposium on the Theory of Self-adaptive Control Systems*, Teddington, England, (also in 'Theory of Self-adaptive Control Systems', P. H. Hammond (Ed.), Plenum Press, New York, 1966).

Åström, K. J., and Eykhoff, P. (1971). 'System identification—a survey', *Automatica*, **7**, No. 2, 123–162.

Beck, M. B. (1973). *Ph.D. thesis*, University Engineering Department, Cambridge.

Beck, M. B. (1974). 'Maximum likelihood identification applied to DO-BOD-algae models for a freshwater stream', *Report 7431(C)*, Lund Institute of Technology, Division of Automatic Control, Sweden.

Beck, M. B. (1975a). 'The identification of algal population dynamics in a freshwater stream', in *Computer Simulation of Water Resources Systems*, G. C. Vansteenkiste (Ed.), North-Holland Publishing Co., Amsterdam, pp. 483–494.

Beck, M. B. (1976a). 'Dynamic modelling and control applications in water quality maintenance', *Water Research*, **10**, 575–595.

Beck, M. B. (1976b). 'Identification and parameter estimation of biological process models', in *System Simulation in Water Resources* (ed. G. C. Vansteen Kiste), North-Holland Publishing Co., Amsterdam.

Beck, M. B., and Young, P. C. (1975). 'A dynamic model for DO-BOD relationships in a non-tidal stream', *Water Research*, **9**, 769–776.

Beck, M. B., and Young, P. C. (1976). 'Systematic identification of DO-BOD model structure', *Proc. A. S. C. E., J. Env. Eng. Div.*, **102**, EE5, 909–927.

Camp, T. R. (1965). 'Field estimates of oxygen balance parameters', *Proc. A. S. C. E., J. Sanit. Eng. Div.*, **91** (SA5), 1–16.

Chen, C. W. (1970). 'Concepts and utilities of ecologic model', *Proc. A. S. C. E., J. Sanit. Eng. Div.*, **96** (SA5), 1085–1097.

Di Toro, D. M. (1969). 'Stream equations and Method of characteristics', *Proc. A. S. C. E., J. Sanit. Eng. Div.*, **95** (SA4), 699–703.

Dobbins, W. E. (1964). 'Bod and oxygen relationships in streams', *Proc. A. S. C. E., J. Sanit. Eng. Div.*, **90** (SA3), 53–78.

Downing, A. L. (1967). 'The effects of pollution on the oxygen balance of rivers', Paper presented at a meeting of the Royal Society of Health, London.

Dresnack, R., and Dobbins, W. E. (1968). 'Numerical analysis of BOD and DO profiles', *Proc. A. S. C. E., J. Sanit. Eng. Div.*, **94** (SA5), 789–807.

Dysart, D. C. (1969). 'The use of dynamic programming in regional water quality planning', Paper presented at the A. A. P. S. E. Systems Analysis Conference.

Eykhoff, P. (1974). 'System identification—Parameter and state estimation', John Wiley and Sons, London.

Fan, L. T., Nadkarni, R. S., and Erickson, L. E. (1971). 'Dispersion model for stream with several waste inputs and water intakes', *Water Resources Bulletin*, **7**, No. 6.

Fawcett, A. (1975). 'A Management model for river water quality', Paper presented at a symposium on Water Quality Modelling of the Bedford Ouse, Anglian Water Authority, Churchill College, Cambridge.

Goodman, A. S., and Tucker, R. J. (1971). 'Time-varying mathematical model for water quality', *Water Research*, **5**, 227–241, 1971.

Grantham, G. R., Schaake, J. C., and Pyatt, E. E. (1971). 'Water quality simulation model', *Proc. A. S. C. E., J. Sanit. Eng. Div.*, **97** (SA5).

Gustavsson, I. (1975). 'Survey of applications of identification in chemical and physical processes', *Automatica*, **10**, No. 1, 3–24.

Hansen, W. D., and Frankel, R. J. (1965). 'Economic evaluation of water quality—a mathematical model of dissolved-oxygen concentration in freshwater streams', *SERL Report No. 65–11*, Sanitary Engineering Research Laboratory, University of California, Berkeley.

Himmelblau, D. M., and Bischoff, K. B. (1968). *Process analysis and simulation*, John Wiley, New York, 1968.

Huck, P. M., and Farquhar, G. J. (1974). 'Water quality models using the Box-Jenkins method', *Proc. A. S. C. E., J. Env. Eng. Div.*, **100** (EE3), 733–752.

Jazwinski, A. H. (1970). *Stochastic processes and filtering theory*, Academic Press, New York.

Kendrick, D. A., Rao, H. S., and Wells, C. H. (1970). 'Optimal operation of a system of waste water treatment facilities', *I. E. E. E. Symposium on Adaptive Processes, Decision, and Control*, Austin, Texas, 1970.

Koivo, A. J., and Phillips, G. R. (1971). 'Identification of mathematical models for DO and BOD concentrations in polluted streams from noise currupted measurements', *Water Resources Research*, **7** (4), 853–862.

Koivo, A. J., and Phillips, G. R. (1972). 'On determination of BOD and parameters in polluted stream models from DO measurements only', *Water Resources Research*, **8** (2), 478–486.

Koivo, H. N., and Koivo, A. J. (1973). 'Optimal estimation of polluted stream variables', *Proc. I. F. A. C. Conference on Identification and System Parameter Estimation, paper PE-2*, The Hague/Delft, the Netherlands.

Lee, E. S., and Hwang, I. (1971). 'Stream quality modeling by quasilinearization', *Journal W. P. C. F.*, **43** (2), 306–317.

Monod, J. (1942). *Recherches sur la croissance des cultures bacteriennes*, Hermann et Cie, Paris.

O'Connor, D. J. (1961). 'Oxygen balance of an estuary', *Trans. A. S. C. E.*, **126** (III), 556.

O'Connor, D. J. (1967). 'The temporal and spatial distribution of dissolved oxygen in streams', *Water Resources Research*, **3** (1), 65–79.

O'Connor, D. J., and Di Toro, D. M. (1970). 'Photosynthesis and oxygen balance in streams', *Proc. A. S. C. E., J. Sanit. Eng. Div.*, **96** (SA2), 547–571.

Osens, M., Edwards, R. W., and Gibbs, J. W. (1964). 'Some reaeration studies in streams', *International Journal of Air and Water Pollution*, **8**, 469–486.

Owens, M., Knowles, G., and Clark, A. (1969). 'The prediction of the distribution of dissolved oxygen in rivers', in *Advances in Water Pollution Research*, S. H. Jenkins (Ed.), Pergamon Press, Oxford, pp. 125–137.

Rich, L. G., Andrews, J. F., and Keinath, T. M. (1972). 'Diurnal pH patterns as predictors of carbon limitation in algal growth', *Water and Sewage Works*, **119**, 126–130.

Rutherford, J. C., and O'Sullivan, M. J. (1974). 'Simulation of water quality in Tarawera River', *Proc. A. S. C. E., J. Env. Eng. Div.*, **100** (EE2), 369–390.

Shastry, J. S., Fan, L. T., and Erickson, L. E. (1973). 'Nonlinear parameter estimation in water quality modelling', *Proc. A. S. C. E., J. Env. Eng. Div.*, **99** (EE3), 315–331.

Streeter, H. W., and Phelps, E. B. (1925). 'A study of the pollution and natural purification of the Ohio River', *Bulletin No. 146*. U. S. Public Health Service.

Tarrasov, V. J., Perlis, H. J. and Davidson, B. (1969). 'Optimization of a class of river aeration problems by the use of multivariable distributed parameter control theory', *Water Resources Research*, **5** (3), 563–573.

Thomann, R. V. (1967). 'Time series analysis of water quality data', *Proc. A. S. C. E., J. Sanit. Eng. Div.*, **96** (SA1), 1–23.

Thomann, R. V. (1972). *Systems analysis and water quality management*, Environmental Research and Applications Inc., New York.

Thomann, R. V., Di Toro, D. M., and O'Connor, D. J. (1974). 'Preliminary model of Potomac estuary phytoplankton', *Proc. A. S. C. E., J. Env. Eng. Div.*, **100** (EE3).

Thomas, H. A. (1948). 'Pollution load capacity of streams', *Water and Sewage Works*, **95**, 409.

Thornes, J., and Clark, M. W. (1975). 'Non-sequential water quality project—Paper 2: A review of some applications of stochastic empirical modelling in water quality investigations', *Technical Report*, Department of Geography, London School of Economics and Political Science.

Whitehead, P. G., and Young, P. C. (1975a). 'A dynamic stochastic model for water quality in part of the Bedford-Ouse River system', in G. C. Vansteenkiste (Ed.), *Computer Simulation of Water Resources Systems*, North-Holland Publishing Co., Amsterdam, pp. 417–438.

Whitehead, P. G. and Young, P. C. (1975b). 'The Bedford-Ouse study dynamic model— fourth report to the Steering Group of the Great Ouse Associated Committee', *Technical note CN/75/1*, Control Division, University Engineering Department, Cambridge.

Young, P. C. (1974). 'A recursive approach to time-series analysis', *Bulletin of the Institute of Mathematics and its Applications (IMA)*, **10** (5/6), 209–224.

Young, P. C., and Beck, M. B. (1974). 'The modelling and control of water quality in a river system', *Automatica*, **10** (5), 455–468.

Young, P. C., and Whitehead, P. G. (1975). 'A recursive approach to time-series analysis for multivariable systems', in G. C. Vansteenkiste (Ed.), *Computer Simulation of Water Resources Systems*, North-Holland Publishing Co., Amsterdam, pp. 39–52.

7

Nitrification in the River Trent

J. H. N. GARLAND

7.1 INTRODUCTION

During the course of an investigation aimed at finding quantitative relations between the flow of pollutants and variations in water quality at any point in the Trent system (Garland and Hart, 1972), it became apparent that the total quantity of ammoniacal nitrogen admitted to the River Trent by the tributaries draining the catchment area above Nottingham consistently exceeded the quantity observed flowing in the river at Nottingham. From calculations made for mass balances of inorganic nitrogen it was observed that the deficit in the mass flow of ammoniacal nitrogen was virtually accounted for by an increase in the mass flow of oxidized nitrogen, thus indicating that nitrification was the most likely cause of the loss of ammonia from the river water.

Because nitrification is a contributory factor to the demand exerted on the oxygen resources of a stream undergoing self-purification, detailed consideration is given in this chapter to the results of analyses of Trent water (made for the purposes of the Trent Research Programme) and the results of a series of monthly intensive surveys of the mid-Trent subsequently undertaken to study the dynamic features of the nitrification process.

7.2 FEATURES OF THE TRENT BASIN

The Trent is an industrial river. It rises above and drains the City of Stoke and is soon joined by the grossly polluted River Tame which drains the heavily industrialized urban area of the West Midlands. Below its confluence with the Tame the Trent carries the treated domestic and industrial effluent from a population of about 3 million persons, having received about 70% of the total polluting load discharged in the basin. The flow of water however is about 30% of the eventual fresh-water flow of the Trent and as a consequence the river is in its poorest condition at this point.

Before reaching Nottingham the river is joined by the Rivers Dove and Derwent, both of which are major tributaries used for public water supply purposes. The volume of water brought in by these rivers, which rise in the uplands of Derbyshire, substantially improves the quality of the River Trent by way of dilution. Additional pollution above Nottingham occurs in the smaller Rivers Soar and Erewash, the former conveying the waste water from Leicester and the latter draining the industrial Erewash valley which is located midway between Derby and Nottingham. However the dilutions afforded by the Trent at its junctions with the Soar (about 8:1) and Erewash (50:1) are such as to result in only very small increases in concentrations of pollutants in the main river. Below Nottingham the Trent is a good coarse fishery and remains of good quality until it receives a number of polluting industrial discharges to its estuarial reach before joining the Yorkshire Ouse to form the Humber. The history, uses, quality of the waters, and problems of management in the Trent basin have been described by Lester (1971).

7.3 THE OXIDATION OF AMMONIACAL NITROGEN AND ITS EFFECT ON THE MASS FLOW OF OXYGEN IN THE TRENT

The behaviour of inorganic nitrogen in the River Trent system has been deduced by constructing annual mean flow corrected mass balances using river water quality data made available by the Trent River Authority (now part of the Severn–Trent Regional Water Authority).

When the annual mean daily mass flows of ammoniacal-N from the tributaries draining to the river above Nottingham are summed and compared with the mass flows surviving in the river at Nottingham, as in Figure 7.1, it is evident that ammonia is being lost from the system and that both mass flows have been declining over the period. Lester (1971) has attributed the decline in the mass flow of ammonia to the building of new and the extension of existing sewage-treatment plants, the diversion of trade effluents to sewerage systems, and the decline in the use of the carbonization of coal as a source of town gas. While it is difficult to quantify these events individually it was deduced that the mass flow of ammoniacal-N to the sewage works from coal carbonization effluents declined from 8·7 t/d in the period 1962–4 to 3·4 t/d in the period 1968–70; a reduction of well over 50%. However abatement in the mass flow in tributaries over the period 1964–70 has been about 10 t/d, suggesting that in addition to the reduc-

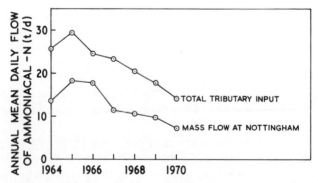

Figure 7.1 Trends in the mass flow of ammoniacal–N in the Trent basin to Nottingham, 1964–1970

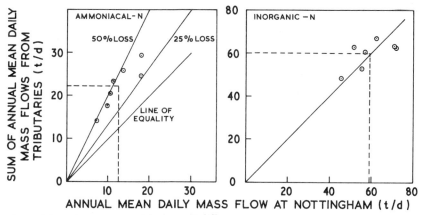

Figure 7.2 Relationships in the Trent basin between sum of annual mean daily flows of ammoniacal-N and inorganic-N from the tributaries above Nottingham and the corresponding flows at Nottingham, 1964–1970
(average mass flows shown by dashed line)

tion in the quantity of ammonia in crude sewage an increasingly effective nitrifying regime has been established at the sewage-treatment works. Speculatively this might be attributed to the concomitant lessening in the flow of cyanide and sulphur compounds thereby facilitating the growth of nitrifying bacteria at these works.

Figure 7.2 affords a comparison between the sum of mass flows of ammoniacal and inorganic N input to the Trent by its tributaries and the mass flows observed at Nottingham. The dashed lines represent the average mass flows over the period and show that consistent losses of ammonia occur. On average the flow of ammonia was reduced from 22·3 t/d to 12·7 t/d in the Trent, a loss of 43%. In contrast Figure 7.2 also shows that the mass flow of inorganic nitrogen compounds at Nottingham is over-estimated about as frequently as it is underestimated by summation of the tributary inputs, the average flow from the tributaries (59·8 t/d) being very close to that at Nottingham (59·0 t/d), indicating that inorganic nitrogen is largely conserved. The conservation of inorganic nitrogen, embracing a loss of ammoniacal-N points to nitrification as the phenomenon of importance to the nitrogen budget of the Trent.

Field observations on the behaviour of ammonia in bodies of water labelled with radioactive tracers suggest that a model based on first-order kinetics would suffice for the purpose of accounting for decay in the annual mean mass flows. For the purpose of constructing such a model it is assumed that an average decay constant is applicable to the river as a whole between the point at which the load is first measured in the Upper Trent and Nottingham and it is convenient to work in terms of distance travelled rather than time, and in loads rather than concentrations.

The mass balance for ammoniacal-N would be written

$$M_N = \sum_1^n M_i \exp - K_N x_i \tag{7.1}$$

when M_N is the total mass flow passing Nottingham from the tributaries each of which are x_i distant from Nottingham and impose an initial load on the river of M_i; k_N is the universally applicable decay coefficient (per unit distance travelled). Solution of equation 7.1 is easily accomplished by trial and error methods and for the 1970 data (given in Table 7.1 below) the value of k_N was found to be 0.0142 km^{-1}, which at an average river water velocity of 55 km d^{-1}

Table 7.1 Annual mean mass flows in the River Trent in 1970 of NH$_3$ and carbonaceous oxygen demand

	Nitrogenous oxygen demand		Carbonaceous oxygen demand		
	Calculated mass flow of ammoniacal nitrogen (t/d)	Observed mass flow of ammoniacal nitrogen (t/d)	Observed mass flow of carbon-aceous BOD (t/d)	Calculated mass flow of carbon-aceous BOD (t/d)	Calculated mass flow of carbon-aceous UOD (t/d)
Upper Trent	1·14	1·14	4·09		6·93
Above Tame	0·99			3·95	6·69
Tame input	8·83		20·88		35·37
	9·82	9·97	23·85	24·82	42·06
Above Dove	7·64			23·33	39·54
Dove input	0·51		3·47		5·88
Below Dove	8·15	7·35	27·52	26·80	45·42
Above Derwent	5·65	5·00	26·34	24·49	41·50
Derwent input	1·12		6·81		11·54
Below Derwent	6·77	6·76	33·34	31·30	53·04
Above Soar	6·32			30·78	52·16
Soar input	2·04		3·50		5·93
	8·36			34·28	58·09
Above Erewash	7·80			33·71	57·12
Erewash input	0·45		1·37		1·32
	8·25			35·08	58·44
Trent at Nottingham	7·19	7·14	33·92	33·92	57·4
Decay constant	0·0142 km $^{-1}$ (0·781 d^{-1})				0·0035 km $^{-1}$ (0·193 d^{-1})
Total tributary input		14·09			67·97
Load oxidized	6·90				10·51
Oxygen consumed (t/d)	29·88				10·51

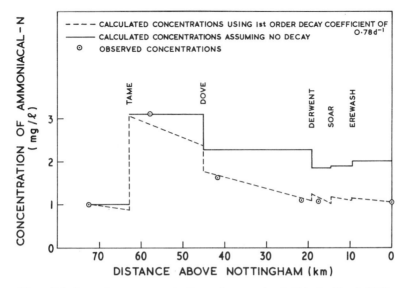

Figure 7.3 Annual mean concentrations of ammoniacal–N in the Trent, 1970

corresponds to a value of $0.78\,\mathrm{d}^{-1}$. Table 7.1 shows that observed and calculated mass flows are in reasonable agreement. Figure 7.3 shows the corresponding concentrations.

The effect of nitrification is examined by calculating a mass balance of dissolved oxygen taking into account additions of oxygen to the river by the tributaries and reaeration through the surface, and losses as a result of the oxidation of ammonia and exertion of a carbonaceous BOD. The rates of consumption of oxygen by carbonaceous matter are derived from BOD measurements made in the presence of a nitrification inhibitor (allylthiourea) using an equation of the same form as equation 7.1).

The results of the carbonaceous BOD mass-flow calculations are given in Table 7.1 and displayed in terms of concentrations in Figure 7.4. The value of the decay coefficient was found to be $0.0035\,\mathrm{km}^{-1}$ ($0.193\,\mathrm{d}^{-1}$).

The data listed in Table 7.1 show that $6.90\,\mathrm{t/d}$ of NH_3 was oxidized and $10.51\,\mathrm{t/d}$ carbonaceous BOD was satisfied, corresponding to a total oxygen consumption of $40.39\,\mathrm{t/d}$ of which $29.88\,\mathrm{t/d}$ is due to the oxidation of ammonia (at 4.33 parts O_2 per part N w/w). Thus 74% of the oxygen consumed in the self-purification process is accounted for by nitrification.

The overall differential equation for the change in oxygen mass flow in any sub-reach is given by

$$\frac{-\mathrm{d}[O_2]}{\mathrm{d}x} = 4.33\,M_0 k_N \exp(-k_N x) + k_L L_0 \exp(-k_L x) - k_A.\bar{Q}C_s + k_A.[O_2]$$

$$(7.2)$$

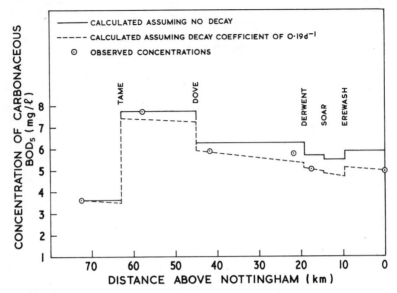

Figure 7.4 Annual mean concentrations of carbonaceous BOD in the Trent, 1970

where $\bar{Q}$ is the average daily mean flow of water, C_S the solubility of oxygen in water at the average temperature, and k_A and k_L the reaeration and carbonaceous decay coefficients (per unit distance), and M_0 and L_0 the mass flows of ammoniacal-N and carbonaceous UOD at the head of the reach. This equation is a modification of the well known Streeter–Phelps equation in which the oxygen demand term has been separated into its carbonaceous and nitrogenous components.

Upon integration by standard methods equation 7.2 becomes

$$[O_2] = [O_2]_{SAT} - \frac{4 \cdot 33 M_0 k_N \exp(-k_N x)}{k_A - k_N} - \frac{L_0 k_L \exp(-k_L x)}{k_A - k_L}$$

$$+ \left[[O_2]_0 - [O_2]_{SAT} + \frac{4 \cdot 33 M_0 k_N}{k_A - k_N} + \frac{L_0 k_L}{k_A - k_L} \right] \exp - k_A x \quad (7.3)$$

where $[O_2]$ is the mass flow of oxygen at a distance x below the head of the sub-reach, $[O_2]_{SAT}$ the mass flow under saturated conditions, and $[O_2]_0$ the mass flow entering the reach.

The value of the reaeration constant (k_A) was calculated using the method of Tsivoglou and Wallace (1972) who showed that in over 400 determinations of the reaeration coefficient in five rivers of widely differing characteristics, the stream reaeration capacity was directly proportional to the energy expended by the flowing water. The rate coefficient was shown to be directly proportional to the rate of energy expenditure by the water as it falls between two points (effectively equal to the rate of loss of potential energy) and was given by an equation which can be rewritten as

$$k_A = E.\Delta h/t \qquad (7.4)$$

where Δh is the change in surface elevation between two stations, t the time of travel of water between the stations, E a constant of proportionality called

Table 7.2 Mass-flow of dissolved oxygen (t/d)

	Nitrification only $k_L = k_A = 0$	Carbonaceous oxidation only $k_N = k_A = 0$	Nitrification + carbonaceous oxidation $k_A = 0$	Reaeration + nitrification + carbonaceous oxidation	Observed mass flow of dissolved oxygen
Upper Trent	9·65	9·65	9·65	9·65	9·65
Above Tame	9·01	9·42	8·78	9·95	
Tame input	10·69	10·69	10·69	10·69	
Below Tame	19·70	20·11	19·47	20·64	20·34
Above Dove	10·25	17·58	7·50	20·48	
Dove input	14·26	14·26	14·26	14·26	
Below Dove	24·51	31·84	21·76	34·74	38·36
Above Derwent	13·72	27·94	7·06	35·91	35·91
Derwent input	15·30	15·30	15·30	15·30	
Below Derwent	29·02	43·24	22·36	51·21	50·89
Above Soar	27·09	42·35	19·54	51·70	
Soar input	8·79	8·79	8·79	8·79	
Below Soar	35·88	51·14	28·33	60·49	
Above Erewash	33·50	50·17	24·98	60·27	
Erewash input	1·16	1·16	1·16	1·16	
Below Erewash	34·66	51·33	26·14	61·43	
Trent at Nottingham	30·06	49·35	19·56	61·44	61·42
Decay constant	0·0142 km^{-1} (0·781 d^{-1})	0·0035 km^{-1} (0·193 d^{-1})	k_L, k_N	k_L, k_N $k_A = 0·06$ km^{-1} (3·30 d^{-1})	
Total tributary input	59·85	59·85	59·85	59·85	
Oxygen consumed (total tributary input minus Trent at Nottingham)	29·79	10·50	40·29		

174

the escape coefficient, and k_A the reaeration rate constant (base e), all in consistent units.

Using an escape coefficient of $0.142 \, \text{m}^{-1}$ (at 25°C) and a difference in elevations in the average water level at the gauging stations defining the head of the Upper Trent reach and Nottingham of 38·5 m, the reaeration coefficient was calculated to be $0.0567 \, \text{km}^{-1}$ at an annual average water temperature of 13·1°C. This value agrees well with a value of $0.06 \, \text{km}^{-1}$ ($3.3 \, \text{d}^{-1}$) which was found to give the best fit to the observed mass-flow data.

The results of applying equation 7.3 to the separate processes governing the flow of oxygen are given in Table 7.2, where the final calculated mass balance of oxygen may be compared with that observed. The agreement is regarded as satisfactory and permits the total oxygen balance to be written in tonnes per day as

| Mass flow of oxygen at Nottingham | = | Mass flow of oxygen input from the tributaries | + | Mass flow of oxygen entering by reaeration | − | Mass flow of oxygen consumed by nitrification | − | Mass flow of oxygen consumed by carbonaceous oxidation |

i.e. $61.44 = 59.85 + 41.88 - 29.79 - 10.50$

Thus, of a total oxygen resource of 101·76/d, about 30% is consumed by nitrification and a further 10% by oxidation of carbonaceous matter. Concentration data corresponding to the mass flows of oxygen given above are shown in

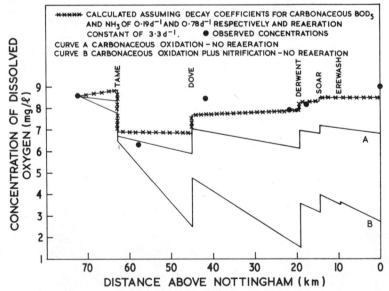

Figure 7.5 Effects of nitrification and carbonaceous oxidation on the annual mean concentration of dissolved oxygen in the Trent, 1970

Figure 7.5, where it is seen that in the absence of reaeration nitrification would cause a severe depletion of the oxygen content of the stream.

In the discussion so far, nitrification has been consider empirically, it being assumed that the process is a first-order reaction, the rate parameters of which can be deduced from annual mean mass-flow data. The next step is to consider methods of estimating concentrations of nitrifying bacteria from the results of routine water-quality surveillance programmes since these, if successful, would afford further insight into the phenomena of nitrification and allow a tentative approach to the problem of accounting for a substantial portion of the BOD exerted in samples of river water.

7.4 APPLICATION OF MICHAELIS–MENTEN AND MONOD GROWTH RELATIONSHIPS TO RESULTS OF BOD MEASUREMENTS

Mass flow of *Nitrosomonas* in the Trent

It is generally accepted that the growth of bacteria occurs exponentially according to the differential equation

$$\frac{dx}{dt} = \mu X \tag{7.5}$$

where dx/dt is the growth rate of bacteria, μ is the specific growth rate and x the instantaneous concentration of bacteria at time t. The specific growth-rate parameter μ is a function of the substrate concentration (S) and is given by the Michaelis–Menten equation

$$\mu = \mu_{max}\frac{S}{S + K_S} \tag{7.6}$$

where μ_{max} is the maximum growth-rate constant which is approached when $S \gg K_S$ and K_S is the Michaelis or saturation constant.

The integral of equation 7.5 is

$$X = X_0\exp\int_0^t \mu.dt \tag{7.7}$$

where X_0 is the initial concentration of activity growing bacteria. Eliminating X between equation 7.7 and the Monod growth relationship (equation 7.8)

$$X = X_0 + Y(S_0 - S) \tag{7.8}$$

where Y is the yield constant and S_0 the initial substrate concentration yields an equation for determining the initial concentration of bacteria, X_0, in terms of the specific growth-rate parameter and changes in the substrate concentration, i.e.

$$X_0 = \frac{Y(S_0 - S)}{\left(\exp \int_0^t \mu.dt\right) - 1} \qquad (7.9)$$

Equation 7.9 is applied to the results of BOD measurements in order to estimate initial concentrations of *Nitrosomonas* contributing to the nitrogenous oxygen demand (NOD).

The NOD is the difference between the oxygen demands exerted by samples of water in the presence and absence of a nitrification inhibitor (allylthiourea) and may be related to the quantity of substrate (i.e. ammoniacal nitrogen) oxidized under the conditions of the test. Thus knowing the initial concentration of ammoniacal nitrogen in the water as sampled, the final concentration at the end of the BOD test may be estimated from BOD measurements and using literature values of the saturation constant and yield constants, the integral in equation 7.9 may be evaluated approximately and hence X_0 estimated.

It is assumed that the NOD exerted in the BOD test over 5 days is due to the complete oxidation to nitrate of part or all of the ammoniacal nitrogen and the usually very small quantity of nitrite initially present.

Thus

$$NOD = 4.33\left[[NH_3]_0 - [NH_3]_5 \right] + 1.11[NO_2']_0 \qquad (7.10)$$

and in the presence of a nitrification inhibitor (specific to *Nitrosomonas*) the BOD exerted is given by

$$BOD_I = BOD_C + 1.11[NO_2']_0 \qquad (7.11)$$

where the subscripts I and C refer to nitrification inhibited and carbonaceous BOD, respectively. Thus the difference between the unsuppressed and inhibited BOD determinations is equivalent to the oxygen demand

$$BOD - BOD_I = 4.33\left[[NH_3]_0 - [NH_3]_5 \right] \qquad (7.12)$$

exerted by the oxidation of ammonia to nitrate, the initial and final concentrations of ammoniacal nitrogen being indicated by the subscripts 0 and 5, respectively. The values of 4.33 parts of oxygen per part of $NH_3 - N$ and 1.11 per part of $NO_2' - N$ oxidized to nitrate are the net oxygen requirements for the oxidation of these substrates by *Nitrosomonas* and *Nitrobacter* respectively, the theoretical values deduced by Montgomery and Borne (1966) having been experimentally verified by Wezernak and Gannon (1967) who also found that the yield constant, Y, was 0.09 mg cells/mg $NH_3 - N$ oxidized. When the BOD equivalent of the ammoniacal nitrogen consumed is substituted and the yield constant of 0.09 inserted in equation 7.9, the estimating equation for the initial, concentration of *Nitrosomonas* becomes

$$X_0 = \frac{0 \cdot 0207 \left[BOD - BOD_1 \right]}{\left[\exp \int_0^t \mu.dt \right] - 1}$$ (7.13)

BOD determinations usually involve the addition of a volume of the sample of polluted water to dilution water which contains $0 \cdot 4$ mg/1 $NH_3 - N$, the dilution given depending upon the experience of the operator and the source of the sample. The procedure used by the Trent River Authority (Cooke, 1973) is to dilute samples from grossly polluted sources by the addition of 4 volumes of dilution water, from less polluted sources by the addition of 1–2 volumes of water and to treat samples from unpolluted sources without further dilution. These additions are assumed to dilute the nitrifying bacteria present in the sample when collected in the same proportion as the water added, and it is further assumed that nitrifying bacteria are not inadvertently introduced at the time the test solutions are prepared.

The growth-rate term, i.e. the denominator in equation 7.13, can only be evaluated approximately, since substrate concentrations are not measured during the course of or at the end of the BOD test and an assumption has to be made about the way in which the specific growth-rate parameter (μ) changes with time. It is assumed that the integral $\int_0^t \mu.dt$ is approximately given by the product of the average of the specific growth-rate constants at the beginning and end of the test and the duration of the test, i.e.

$$\int_0^5 \mu.dt = \frac{5(\mu_0 + \mu_5)}{2}$$ (7.14)

This assumption will clearly be erroneous where prevailing bacterial numbers are large and substrate concentrations low, thereby permitting the bulk of the nitrogenous oxygen demand to be exerted very rapidly and to be substantially complete well before the test is ended. As a rough guide the method should not be used where the calculated nitrogenous oxygen demand accounts for more than 80% of the oxygen required to oxidize the initial content of $NH_3 - N$ to $NO'_3 - N$. Using values of μ_{max} and K_s given by Knowles, Downing, and Barrett (1965) ($0 \cdot 762$ and $0 \cdot 728$, respectively, at 20°C),

$$\int_0^5 u.dt = 0 \cdot 762 \times 2 \cdot 5 \left[\frac{[NH_3]_0 + 0 \cdot 4 D}{[NH_3]_0 + 0 \cdot 4 D + 0 \cdot 728 (1 + D)} + \frac{[NH_3]_5}{[NH_3] + 0 \cdot 723} \right]$$ (7.15)

where D is the dilution factor used in the test and $0 \cdot 4$ mg N/l the NH_3 content of the dilution water.

The equation used to estimate the concentration of *Nitrosomonas* in the river (X_R) when the sample was collected becomes, in terms of experimentally measured values of the BOD and deduced values for ammoniacal nitrogen concentrations,

$$X_R = (1 + D)X_0 \qquad (7.16)$$

where

$$X_0 = \cfrac{0 \cdot 0207 \cfrac{(BOD - BOD_l)}{1 + D}}{\left[\exp 1 \cdot 905\left[\cfrac{[NH_3]_0 + 0 \cdot 4D}{[NH_3]_0 + 0 \cdot 4D + 0 \cdot 728(1 + D)} + \cfrac{[NH_3]_5}{[NH_3]_5 + 0 \cdot 728}\right]\right] - 1}$$

$$(7.17)$$

and

$$[NH_3]_5 = \frac{[NH_3]_0 + 0 \cdot 4D}{1 + D} + \frac{BOD - BOD_l}{4 \cdot 33(1 + D)} \qquad (7.18)$$

It is emphasized that this method of estimating *Nitrosomonas* was devised to make use of existing water quality surveillance data. Given a free choice it would be rejected in favour of the results of experiments in which ammoniacal-nitrogen concentrations were determined during the course of the 5-day incubation period, and the concentrations observed as a function of time fitted to the appropriate fully integrated rate equation. Such experiments do not yet fall within the scope of water quality surveillance programmes, but the occasional routine experiment would be worth performing to seek information about the occurrence of *Nitrosomonas* in river waters and thereby account more fully for the measured BOD.

With this limitation in mind, it is possible to estimate the mass flow of *Nitrosomonas* in river basins from BOD measurements assuming that Michaelis–Menten kinetics apply to the process of autotrophic nitrification in the BOD bottle and that published values of the kinetic constants apply. Equation 7.17 has been applied to average concentrations of $NH_3 - N$ and BOD with and without ATU at a number of stations in the Trent basin over the period 1968–1972 and enables estimates of the production of *Nitrosomonas* to be made. Mass flows of *Nitrosomonas* have been estimated as the product of the concentrations of bacteria and annual mean daily flow of water and are thus not necessarily an accurate measure of the annual mean mass flow. Nevertheless the results display approximate additivity about confluence points, and comparison between the sum of the approximate mass flows of *Nitrosomonas* entering the Trent from its tributaries and the mass flow of these bacteria 'observed' at Nottingham indicates a consistent production within the main river (Table 7.3). Only about one-third of the mass flow at Nottingham is attributable to the input from the tributaries. This is qualitatively consistent with the premise that bacteria can only grow and multiply at the expense of an appropriate substrate. Quantitatively, however, there are large discrepancies between the bacterial masses expected to be produced by the oxidation of the quantities of ammoniacal nitrogen calculated to have been removed from the

Table 7.3 Estimates of production of *Nitrosomonas* in suspension in the River Trent and proportion of load of ammoniacal nitrogen oxidized attributable to *Nitrosomonas* in suspension, 1968–1972

Year	Mass flow of *Nitrosomonas*		Ammoniacal-N oxidized in river			
	From tributaries (kg/d)	At Nottingham (kg/d)	Produced in river (kg/d)	Total (t/d)	Attributable to *Nitrosomonas* in suspension* (t/d)	%
1968	51	150	99	9·9	1·1	11·1
1969	50	139	89	7·8	1·0	12·8
1970	61	147	86	7·0	1·0	14·3
1971	35	92	57	9·9	0·6	6·1
1972	72	108	36	5·7	0·4	7·0

*At a yield of 0·99 units of bacterial mass per unit mass of NH_3-N oxidized.

river and the 'observed' increases in bacterial masses which are only about 10% of those expected. Consistent discrepancies of this magnitude indicate that nitrification in the Trent is largely dependent on nitrifying bacteria attached to the bed of the river.

The above supposition is supported by results obtained independently by Curtis, Durrant and Harman, who conclude from numbers of bacteria measured in the system that the average quantities of ammoniacal nitrogen removed in the Trent between February and December 1972 attributable to bacteria in suspension was 7%, a result which is identical to that deduced for 1972 from the approximate mass-balance approach presented above (Table 7.3).

Clearly the interpretation of the extent to which benthic processes are involved depends upon the value of the yield constant used to estimate the quantity of bacterial mass produced by the quantity of substrate oxidized. Estimates of the value of the yield constant in the literature vary but a range of 0·05–0·10 appears acceptable to the majority of workers. Thus Knowles, Downing, and Barrett (1965) concluded that their choice of a value of 0·05 might have been slightly low. Wezernak and Gannon (1967) observed a yield constant of 0·097 in their study of the oxygen requirements of nitrifying bacteria. On the other hand, Stratton and McCarty (1967) and Lawrence and McCarty (1970) quote a value of 0·29 which was deduced by McCarty (1965) on theoretical grounds when considering the thermodynamics of bacterial growth and is regarded by these authors as the 'true' constant applicable to the production of living cells. Considerable difficulty might attend the determination of the proportion of total bacterial mass which is 'alive' at any time after the start of a culture, and such a distinction is probably unnecessary in the case of nitrifying bacteria since the 'mortality' constant used by Stratton and McCarty ($0.05 \, d^{-1}$) implies a half-life of survival of *Nitrosomonas* of nearly 14 days. This observation when applied to the results of Wezernak and Gannon would increase the value of the yield constant from 0·097 as reported to 0·108. In the context in which yield constants are used in this present study, such a difference is of negligible impact and the value of the yield constant of 0·09 selected earlier is regarded as appropriate to the Trent situation.

7.5 AN INTUITIVE MODEL OF THE NITRIFICATION PROCESS IN A RIVER

Since nitrification in the Trent probably takes place largely on the bed of the river, any kinetic model will be strongly influenced by hydraulic considerations. The model that is proposed is shown diagrammatically in Figure 7.6. In any reach the lower layer of water, that fraction of the total flow which impinges on the bed, is represented as a volume (V_L) swept out by a fraction α of the flow Q while water is detained in the reach for a time Δt.

$$V_L = \alpha Q . \Delta t \qquad (7.19)$$

The remainder of the flow is then confined to an upper layer of volume

$$V_U = (1 - \alpha) Q . \Delta t \qquad (7.20)$$

The river is considered to be divided into consecutive reaches of length such that the residence time (Δt) within each is short, 2 h being convenient for computational purposes. Nitrification is assumed to proceed independently in each layer in accordance with Monod kinetics at rates governed by the appropriate values of the kinetic parameters and effective concentration of bacteria.

Thus in the upper layer, when the concentration of ammoniacal nitrogen

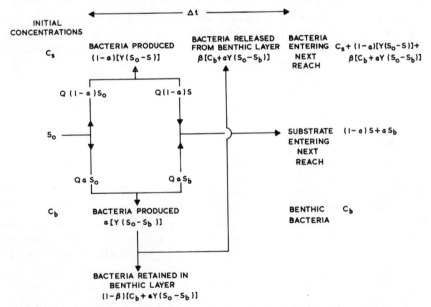

Figure 7.6 Model for simultaneous nitrification in the benthos and overlying water

entering the reach is S_0, the concentration S at the end of the reach is obtained by solving the integrated Monod equation

$$\mu_{max}t = \frac{YK_S}{X_S + YS_0}\ln\frac{S_0}{S} + \left(1 + \frac{YK_S}{X_S + YS_0}\right)\ln\left(1 + \frac{Y(S_0 - S)}{X_S}\right) \quad (7.21)$$

for S, where X_S is the initial concentration of bacteria freely suspended in the upper layer of water in the reach and other symbols are as previously defined. As a result of substrate utilization, the concentration of bacteria (X) at the end of the reach will have increased in proportion to the amount of ammoniacal nitrogen oxidized, i.e.

$$X = X_S + Y(1 - \alpha)(S_0 - S) \quad (7.22)$$

so that the mass flow of bacteria leaving the reach resulting from the mass flow entering the reach and growth in the upper layer within the reach will be

$$QX_S + QY(1 - \alpha)(S_0 - S)$$

and the mass flow of substrate in the upper layer at the end of the reach will be

$$Q(1 - \alpha)S$$

In the benthic layer it is assumed that bacteria are attached to the material comprising the bed of the stream at an effective concentration X_b. This may be regarded as the concentration of bacteria which would be produced if the nitrifying sediment were homogeneously suspended throughout the lower layer while a particular body of water was in the reach and was instantaneously settled and retained in the reach as the body of water moved out.

The concentration of substrate (S_b) in the lower layer at the end of the reach is calculated as before using equation 7.21 with the value of X_b inserted as the initial bacterial concentration, and at the end of the reach instantaneous mixing of the two layers of water occurs to produce a uniform concentration of substrate in the water entering the next reach which is instantaneously divided as before.

Thus the mass flow of substrate (S_1) entering the second reach is given by

$$Q(1 - \alpha)S + \alpha S_b$$

and its concentration is thus

$$(1 - \alpha)S + \alpha S_b$$

constant-flow conditions being assumed.

The bacterial concentration (X') in the benthic layer will be increased in proportion to the amount of substrate removed in that layer

$$X' = X_b + \alpha Y(S_0 - S_b) \quad (7.23)$$

and it is assumed that a small fraction β of the total nitrifying bacteria in the sediment is released at the end of the reach to the upper layer, thus admitting the possibility that at certain times of the year (for example winter flood) β could have a value such that bacteria were washed out of the reach at rates exceeding their rate of growth in the benthos, thus avoiding the difficulty of having to concede that the concentration of bacteria on the bed of the river would increase perpetually as long as substrate was present in solution in the river water.

Thus the concentration of bacteria in suspension entering the second reach will be the sums of the concentration of bacteria entering the first reach, of bacteria formed in suspension, and of bacteria released from the bed of the river, i.e.

$$X_S + (1 - \alpha) Y (S_0 - S) + \beta [X_b + \alpha Y (S_0 - S_b)].$$

The substrate and bacterial concentrations entering any reach are thus calculated from the extent of nitrification in the preceding reach and the procedure may thus be repeated for any number of consecutive reaches.

In applying the model to conditions thought to be typical of the mid-Trent in summer, three cases have been examined in which the effective concentrations of bacteria in the hypothetical benthic layer were 400, 1000, and 2000 $\mu g/l$ respectively. These values are arbitrary but were chosen to reflect the observation (Curtis, Durrant, and Harman) that the concentration of bacteria in the sediment greatly exceeds on a volumetric basis the concentration of suspended bacteria. Initial concentrations of substrate and suspended bacteria were chosen to be 2·0 mg/l and 20 $\mu g/l$, respectively. Values of the kinetic parameters (μ_{max}, K_S) were those appropriate to a temperature of 20°C. Six consecutive reaches of retention time each of 2 h were considered, which in terms of the velocities of flow of water in the mid-Trent would be equivalent to a length of about 4 km.

The choice of values for the parameters α (the proportion of water impinging on the bed, i.e. comprising the lower layer) and β (the proportion of benthic nitrifying bacteria released to the upper layer from the lower layer) is arbitrary. The value for the parameter α was taken to be 0·4 on the basis that not all the ammonia entering the reach is oxidized.

In the case of the parameter β it was considered that in order to prevent washout of the bacteria attached to the bed and to allow their concentration to grow to a peak under the summer conditions of low flow (when by observation nitrification occurs most rapidly in the Trent) its value had to be small compared with unity. A value satisfying these criteria under the conditions in which the model was operated was 0·005.

In principle both parameters could be measured by tracer techniques if the exacting requirements of a suitable tracer material could be met. Such a water-soluble tracer would need to be selectively, instantaneously, and irreversibly adsorbed by biological material on the bed of the stream and at the same time be released after attachment along with particles of the biomass abraded by turbulent action. Under these circumstances the proportion of tracer surviving in true solution after traversing n reaches would be p^n where $p = 1 - \alpha$ and the proportion of the tracer adsorbed by bed material and subsequently released by abrasive action (determined by assay of suspended matter) would be

$$\beta(1 - p^n)$$

It is unlikely that anything other than a radioactive tracer would suffice and a possible choice is phosphorus-32. The estimates of the parameters made in this way would thus be average values since it would be assumed that reaches were uniform in character.

The onerous task of fitting the model to observations made in the mid-Trent has not been undertaken, but the results of three simulations using benthic concentrations of bacteria of 2000, 1000, and 400 μg/l presented in Figure 7.7 are encouraging; the effective rate equations are found to be statistically indistinguishable from first-order equations, the fitted equations being

$$2 \cdot 1591 \exp - 4 \cdot 3511 \, t (r = 0 \cdot 9978),$$
$$2 \cdot 0621 \exp - 1 \cdot 6794 \, t (r = 0 \cdot 9969),$$
and
$$2 \cdot 0106 \exp - 0 \cdot 6174 \, t (r = 0 \cdot 9992)$$

for the three benthic concentrations respectively, where t is the time of travel in days. As will be shown, this is qualitatively in accord with observations and values of the rate-constants are comparable to those found to apply in the Trent (see Table 7.4 later).

In the simulation most of the nitrification is attributable to the benthic process. Thus by putting $\alpha = 0$ with an initial concentration of bacteria in

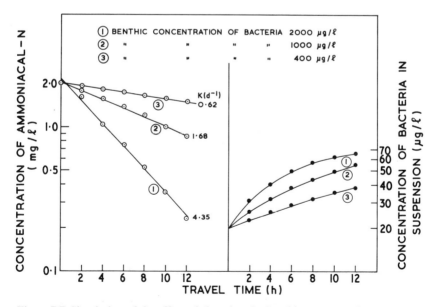

Figure 7.7 Simulation of the effect of changing the benthic concentration of *Nitrosomonas* on the extent to which ammoniacal–N is removed from a river and the extent to which overlying water is enriched with *Nitrosomonas* as a result of benthic release and growth

suspension of 20 μg/l the quantity of ammoniacal nitrogen removed was only 0·07 mg/l after 12 h travel, whereas when the benthic layer comprised 40% of the flow (i.e. $\alpha = 0·4$) and benthic concentrations of bacteria were 400, 1000, and 2000 μg/l, respectively, the quantities of ammoniacal nitrogen removed in 12 h were 0·52, 1·13, and 1·77 mg N/l, respectively. The quantities of ammoniacal nitrogen removed by the growth of bacteria in suspension when the benthic concentrations of bacteria were 400, 1000, and 2000, were 0·0497, 0·0591, and 0·0615 mg/l, representing about 9·3, 5·2, and 3·5% of the total ammonia removed, respectively. Release of bacteria from the benthos appears to be a factor of importance in explaining increases in bacterial concentration over the mid-Trent reach (Figure 7.7). In the simulations, increases in bacterial concentration due to growth in suspension were found to be almost negligible in comparison with the increases resulting from benthic release which are of the order required to be in qualitative agreement with the results obtained by a consideration of the concentrations of *Nitrosomonas* deduced from BOD measurements.

7.6 RESULTS OF INTENSIVE NITRIFICATION SURVEYS CONDUCTED ON A 23-KM LONG REACH IN THE MID-TRENT IN SOUTH DERBYSHIRE BETWEEN WILLINGTON AND SHARDLOW

A series of monthly surveys was carried out with few interruptions between June 1971 and April 1973. Samples of water were collected at 7 stations in the reach, sampling being initiated when the time of arrival of the peak concentration of the radiotracer (potassium bromide-82, half-life 35·4 h), released about 6 km upstream of the first station, could be estimated by consulting a record of radioactive counts with time displayed as a continuous trace on a portable potentiometric recorder. Samples were collected at frequent intervals for about 10 min either side of the arrival of the peak, the object being to discern changes taking place in the chemical composition of a labelled body of water as it traversed the reach. After collection, samples were stored in ice-boxes and analysed within two days for $NH_3 - N$, $NO_2' - N$, and oxidized $- N$ by standard auto-analytical techniques. As an illustration, results of two surveys are shown in Figure 7.8.

During the July 1971 survey at an average water temperature of 19·9°C and velocity of flow of 0·594 ms^{-1}, 52·6% of the ammonia entering the reach was removed, whereas during the December 1972 survey at an average temperature of 9·4°C and velocity of 1·116 ms^{-1}, only 9% was removed. The velocity of flow was so high in December that Stations 3 and 4 could not be sampled in time. Despite a clearly observable loss of 0·7 mg/l ammoniacal nitrogen during July and an imputed loss of 0·1 mg ammoniacal nitrogen during December, there was no marked change in the concentration of nitrite-nitrogen, presumably because it is removed almost as rapidly as it is formed. Data obtained on other surveys are summarized in Table 7.4 together with the first-order rate constants and correlations coefficient obtained for the decay of ammoniacal

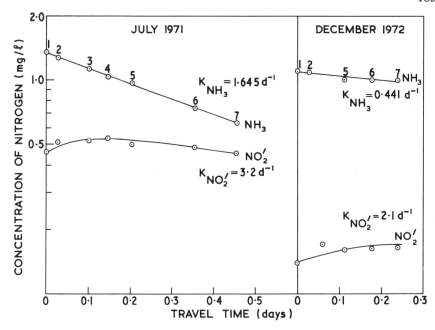

Figure 7.8 Concentrations of ammoniacal and nitrite nitrogen observed in the mid–Trent at Stations 1–7 on two intensive surveys

nitrogen and the rate constants for the decay of nitrite-nitrogen assuming first-order consecutive reaction kinetics. These data show the range in concentrations of ammoniacal nitrogen and the variability in river conditions (temperature and velocity) encountered in the period of observations, and the high correlation between the observed changes in the concentrations of ammoniacal nitrogen and those predicted by first-order kinetics. Concentrations of *Nitrosomonas* given in Table 7.4 are those reported by Curtis, Durrant, and Harman.

The reaction sequence

$$NH_3 \xrightarrow{K_N} NO_2' \xrightarrow{K_Y} NO_3'$$

may be considered as classical first-order consecutive reactions in which the respective reaction rate constants K_N and K_Y depend upon the hydraulic and other conditions of the stream.

The differential equations for the rate processes are

$$-\frac{d[NH_3]}{dt} = K_N[NH_3] \tag{7.24}$$

and

$$\frac{d[NO_2']}{dt} = -\frac{d[NH_3]}{dt} - K_Y[NO_2'] \tag{7.25}$$

Table 7.4. Results of intensive surveys of the River Trent

Year/month	Concentration of NO₂'-N at Entry (mg/l)	Exit (mg/l)	Concentration of NH₃-N at Entry (mg/l)	Exit (mg/l)	Average water temperature (°C)	Average water velocity (cm/s)	Rate constants for decay of NH₃ (K_N) (d^{-1})	NO₂' (K_Y) (d^{-1})	Concentration of Nitrosomonas in suspension at entry to reach* (µg/l)	Correlation coefficient for first-order decay of NH₃-N
1971										
June	0.38	0.37	1.05	0.64	16.9	79	1.48	3.5	—	0.97
July	0.46	0.45	1.33	0.64	19.9	59	1.65	3.2	—	0.99
Aug	0.56	0.49	1.26	0.62	19.7	60	1.50	2.0	—	0.93
Sept	—	—	—	—	—	—	—	—	—	—
Oct	0.69	0.58	2.03	0.45	17.5	44	2.42	3.4	—	0.97
Nov	0.37	0.44†	1.44	1.02†	10.1	65	1.05	2.8	—	0.84
Dec	0.35	0.46	1.93	1.22	11.2	71	1.05	3.3	—	0.94
1972										
Jan	0.46	0.46	1.23	0.85	9.2	95	1.39	3.1	—	0.99
Feb	0.11	0.13	1.04	0.90	6.8	93	0.50	3.8	0.3	0.99
Mar	0.13	0.13	1.12	0.94	6.1	120	0.71	5.4	0.4	0.98
Apr	0.22	0.26	0.94	0.70	10.2	89	0.97	3.2	0.2	0.96
May	0.51‡	0.40	1.31‡	0.35	19.2	56	2.95	4.6	16.0	0.99
June	0.51	0.46	1.10	0.48	15.5	54	2.28	3.6	4.6	0.99
July	0.71	0.28	0.83	0.13	22.9	44	2.99	3.8	6.0	0.94
Aug	0.40	0.24	1.07	0.26	20.2	53	2.71	5.2	6.2	0.99
Sept	0.62	0.68	1.70	0.48	16.8	44	2.06	3.3	2.7	0.99
Oct	0.79	0.80	1.99	0.55	15.9	44	2.06	2.6	—	0.98
Nov	0.23	0.23	1.14	0.75	9.5	72	1.09	4.6	2.0	0.98
Dec	0.14	0.17	1.10	1.00	9.4	112	0.44	2.1	1.2	0.91
1973										
Jan	—	—	—	—	—	—	—	—	—	—
Feb	0.20‡	0.32	1.8†	1.13	10.8	74	1.34	6.0	—	0.99
Mar	—	—	—	—	—	—	—	—	—	—
Apr	0.24	0.45	2.42	1.18	10.4	60	1.64	7.1	—	0.98

* Assuming a cell mass of 10⁻¹² g/cell   † At Station 6   ‡ At Station 2

where the rate of change of concentration of nitrite-nitrogen is given by the difference between its rate of formation from ammoniacal nitrogen and its rate of removal by oxidation to nitrate-nitrogen.

Integration of equation 7.24 and substitution into equation 7.25 yields

$$\frac{d[NO_2']}{dt} = [NH_3]_0 K_N e^{-K_N t} - K_Y [NO_2'] \qquad (7.26)$$

$$[NO_2'] = [NO_2']_0 e^{-K_Y t} + [NH_3]_0 \frac{K_N}{K_Y - K_N}\left(e^{-K_N t} - e^{-K_Y t}\right) \qquad (7.27)$$

where the concentrations of the nitrogen species are in square brackets, the subscript 0 refers to the concentrations of reactants entering the reach and t is the time of travel of water in the reach under steady-flow conditions.

By differentiation of equation 7.27 and equating the differential to zero, the time taken for the concentration of nitrite-nitrogen to reach a maximum is given by

$$t_{max} = \frac{1}{K_Y - K_N} \ln \frac{K_Y}{K_N}\left\{ 1 - \frac{[NO_2]_0}{[NH_3]_0}\left(\frac{K_Y}{K_N} + 1\right)\right\} \qquad (7.28)$$

and this equation may be used to estimate a starting value of K_Y for iterative fitting to equation 7.27, since K_N is known from a least-squares fit to the regression of ammoniacal-nitrogen concentrations against time and t_{max} can be roughly estimated (see for example Figure 7.8) from a plot of nitrite-nitrogen concentrations against time. The upper limit to the value of K_Y is given by the relationship

$$K_Y < K_N\left[\frac{[NH_3]_0}{[NO_2]_0} - 1\right]$$

this condition being necessary to prevent the occurrence of a negative term in the logarithmic part of equation 7.28.

If t_{max} is to be a positive quantity, i.e. if the peak concentration of nitrite-nitrogen is to occur in or below the reach considered, then

$$K_Y > K_N$$

In all cases it was found that the value of the rate constant for the removal of nitrite-nitrogen (K_Y) exceeded the value of the rate constant for its formation from ammoniacal-nitrogen (K_N), the minimum, average, and maximum values of the ratios of the two values being 1·26, 3·05, and 7·60, respectively. High values of this ratio tend to occur in the winter and low values in the summer and early autumn, this effect being largely due to a pronounced seasonality in the values of K_N. The average values of the two rate-constants K_N and K_Y during the 1972 surveys were 1·68 and 3·78 d^{-1}, respectively.

Examination of the values of the decay constant for the removal of ammoniacal-nitrogen shows that the seasonal effect manifests itself as a dependence on both the velocity and temperature of the river water. A significant reduction in the variance of estimates of K_N was found by postulating the multiple regression model

$$K_N = A + BT + \frac{C}{V} \qquad (7.29)$$

where A, B, and C are constants, T is the temperature in °C, V the velocity of flow in km/h and K_N the decay constant (d^{-1}) in preference to either of the simple regression models,

$$K_N = A + BT, \text{ or } K_N = A' + \frac{C}{V}.$$

Thus the equation

$$K_N = -0.6272 + 0.0789\,T + \frac{2.5773}{V} \qquad (7.30)$$

explained 79·6% of the variation in K_N, while the simple models each explained about 70%. The calculated residuals about equation 7.30 were not random but were positively correlated with bacterial numbers in suspension found by Curtis, Durrant, and Harman at the head of the reach. They were however independent of the ammoniacal-nitrogen concentrations. Multiple regression of the rate-constant K_N on temperature $(T, °C)$, velocity (V, kmh^{-1}), and bacterial number $(N, thousands/ml)$ yielded the equation

$$K_N = 0.0833\,T + \frac{2.1234}{V} + 0.0646N \qquad (7.31)$$

which explains 97·2% of the variation in K_N, the standard error of estimate being 0.211 d^{-1}. The value of the rate-constant given by equation 7.31 is dominated by the temperature and velocity of the flow of water, a low numerical but statistically significant dependence on bacterial number being consistent with a benthic process which is governed more by the physical state of the water than by the concentrations of bacteria in suspension.

In contrast, an examination of the value of the decay-constant (K_Y) for the removal of nitrite yielded results which were statistically not significant. There was a slight indication of a negative correlation with temperature $(r = -0.073)$ (K_Y tending to be higher in value during the winter months) and a slight positive dependence on reciprocal velocity $(r = 0.103)$. No correlation could be established between values of K_Y and numbers of *Nitrobacter* either in suspension or in the sediment, initial concentrations of either ammoniacal or nitrite–N, and values of the decay-constant K_N. No marked seasonal variation in the values of K_Y was apparent. It should however be remembered that estimates of values of this decay constant are subject to much greater errors than those found for the decay of ammoniacal nitrogen. A characteristic feature of nitrite-nitrogen concentrations in the reach is an absence of a marked change with time which in many instances is difficult to distinguish within the scatter of the analytical results, particularly when low concentrations are encountered. (For example see the results of the December 1972 survey presented in Figure 7.8.)

It is probable therefore that the values found for this constant are dominated by error terms, a conclusion supported by the observation that the values appear to be derived from a log-normal distribution having a geometric mean of $3.7\ d^{-1}$ and a standard deviation $\log_{10}\ 1.378$.

In general terms it is clear that an adequate if somewhat empirical representation of the nitrification process is available for the Trent without recourse to kinetic models of complexity greater than first order. Concentrations of ammoniacal nitrogen may be predicted from a knowledge of temperature and velocity of flow and bacterial numbers if known as well. Concentrations of nitrite do not change markedly.

7.7 SUMMARY AND CONCLUSIONS

1. An annual mean mass balance of the oxygen flow in the Trent system between Yoxall and Nottingham showed that nitrification exerts the major demand on the oxygen resources of the Trent. In averaged round figures, in 1970 of the total resource of 100 t/d, about 60 t/d enter in solution in tributary waters, 40 t/d enter through the 'surface' of the main river, while 30 t/d and 10 t/d are consumed by the oxidation of ammoniacal nitrogen and the exertion of the carbonaceous BOD, respectively. The decay of both ammoniacal nitrogen and ultimate carbonaceous BOD between intermediate points was well represented by the classical first-order processes, and the independently calculated value of the reaeration constant based on the average fall in the elevation of water surfaces between Yoxall and Nottingham (38·5 m) and an average velocity of 55 km/d was found to be in good agreement with that necessary to account for the mass flow of oxygen at Nottingham.

2. A method was devised whereby the results of routine BOD measurements yielded estimates of the concentrations of *Nitrosomonas* bacteria in the waters of the Trent system. The concentrations calculated were in reasonable agreement with those found using the most probable number method of evaluating the results of incubating diluted samples of river water, but are insufficient to account for more than about 10% of the observed losses of ammoniacal nitrogen. The discrepancy indicates that nitrification in the Trent is predominantly a benthic process.

3. An intuitive model employing Michaelis–Menten kinetics to represent processes occurring on the bed and in the overlying water of the river yielded results which were qualitatively in agreement with observations. The model predicted that ammoniacal-nitrogen concentrations would be logarithmically dependent upon time of travel and that increases in bacterial concentration would be observed not so much as a result of growth in suspension but more as a result of benthic release. The model was extended to simulate seasonal changes in the nitrification process, but it is clear that formidable problems attend the fitting of observed data to a model even as simple as the one proposed.

Further insight into the actual hydraulic behaviour of the system is needed. This might be obtained from tracer and other simulative studies.

4. The results of 20 monthly surveys show that ammoniacal-nitrogen removal from the river is a pseudo-first-order process. Values of the correlation coefficients for the first-order fitted equations were (with one exception) uniformly high, exceeding 0·9, the majority being greater than 0·97. Values of the rate constant for the removal of ammoniacal nitrogen displayed some seasonal dependence, being high in the summer and low in the winter. This effect was manifest as a direct dependence on the temperature and an inverse dependence on velocity. Statistically significant but numerically slight dependence of the value of the rate constant on bacterial number in suspension was also demonstrated, 97·2% of the variation in the values of rate constant during 1972 being accounted for.

5. By analogy with the observed behaviour of ammoniacal nitrogen, it was assumed that the hydraulic processes leading to a pseudo-first-order reaction in the case of ammoniacal nitrogen applied with equal effect to the oxidation of nitrite by *Nitrobacter*. Values of the decay constants for the removal of nitrite-nitrogen were calculated using classical first-order consecutive reaction kinetics as a model for the process. These were found to be uniformly greater in value than the values of the first-order decay constants for the removal of ammoniacal nitrogen, indicating that the second stage in the consecutive reaction

$$NH_3 \longrightarrow NO_2' \longrightarrow NO_3'$$

proceeds rapidly under all conditions so far encountered. This conclusion is supported by the observation that while substantial quantities of ammoniacal nitrogen are removed from the river, increases in nitrite-nitrogen concentrations are slight. Difficulties in explaining the variation in the values of the decay constant for nitrite-nitrogen removal can in part be attributed to difficulties in discerning accurately small changes in the low concentrations of nitrite in the reach, with the result that considerable error attends the values of the constant found.

Alternative models for the second-stage reaction have not been explored. These will require much more computational effort since the equations involving processes other than first order will have to be solved numerically. It is also clear that very great care and attention to detail in sampling and analysing for nitrite is required if any significance is to be attached to changes in concentrations, since these are likely to be very small. It is quite possible that routine analytical methods, although sensitive, are not capable of giving the precision required to detect subtle variations in nitrite concentrations which would be implied by more complex models of the oxidation process.

Acknowledgements

Mr I. C. Hart was responsible for much of the data processing and field work during intensive surveys, and was assisted by Mrs S. Williams. Mr W. F. Lester,

Director Scientific Services, Severn–Trent Water Authority is warmly thanked for the supply of data on water quality.

This paper is published by permission of the Director of the Water Research Centre.

7.8 REFERENCES

Cooke, R. (1973). (Trent River Authority) Private communication.

Courts, E. J. C., Durrant, K., and Harman, M. M. I. 'Nitrification in rivers in the Trent Basin', *Water Research* (in press).

Garland, J. H. N., and Hart, I. C. (1972). 'Effects of pollution on quality of water in the river system'. In *Report of Trent Steering Committee*, **4**. Water Resources Board, Reading.

Knowles, G., Downing, A. L., and Barrett, K. J. (1965). 'Determination of kinetic constants for nitrifying bacteria in mixed culture, with the aid of an electronic computer', *J. gen. Microbiol.*, **38**, 263–278.

Lawrence, A. W., and McCarty, P. L. (1970). 'Unified basis for biological treatment design and operation', *J. sanit. Engng. Div., Am. Soc. civ. Engrs*, **96** (SA3), 757–778.

Lester, W. F. (1971). 'The River Trent and the economic model research programme'. In *Proceedings of Symposium. The Trent Research Programme*. Inst. Wat. Pollut. Control, Maidstone.

McCarty, P. L. (1965). 'Thermodynamics of biological synthesis and growth', *Proc. 2nd int. Conf. Wat. Pollut. Res., Tokyo*, 1964, **2**, 169–199.

Montgomery, H. A. C., and Borne, B. J. (1966). 'The inhibition of nitrification in the BOD test', *J. Proc. Inst. Sew. Purif.*, **4**, 357–368.

Stratton, F. E., and McCarty, P. L. (1967). 'Prediction of nitrification effects on the dissolved-oxygen balance of streams', *Envir. Sci. Technol.*, **1**, 405–410.

Tsivoglou, E. C., and Wallace, J. R. (1972), 'Characterization of stream re-aeration capacity', *U. S. envir. Protect. Ag., Ecol. Res. Ser. Rep. EPA-R3-72-012*, U. S. Govt Printing Office, Washington, D. C.

Wezernak, C. T., and Gannon, J. J. (1967). 'Oxygen-nitrogen relationships in autotrophic nitrification', *Appl. Microbiol.*, **15**, 1211–1215.

8

Estuarine Dispersion

L. GALLAGHER and G. D. HOBBS

8.1 INTRODUCTION

In recent years it has become increasingly necessary for engineers and scientists to be able to predict the degree to which pollutants will be dispersed when discharged to an estuarine environment. Only when the degree of dispersal is known can reliable estimates be made of their likely ecological impact. Many polluting substances dissolve in or are miscible with water and in consequence are transported from place to place by the natural motion of the water within the estuary. The dispersal of such substances is the subject of this chapter.

The chapter is not an exhaustive review. Its objective is to introduce the reader, who is assumed to have had no previous encounter with the subject, to the concept of dispersion and to provide him with sufficient background knowledge to enable him to read and understand the literature cited. In consequence much detail has been omitted and many of the arguments have been greatly simplified.

Section 8.2 is devoted to a description in physical terms of the dispersion process. Section 8.3 discusses the determination of turbulent diffusion coefficients and is followed in Section 8.4 with a review of the methods currently available for calculating effective longitudinal dispersion coefficients. The penultimate section reviews the experimental evidence available and the agreement or disagreement existing between theory and observation. The last section provides a brief summary of the 'state of the art'.

8.2 THE DISPERSION PROCESS

The movement of water in even the simplest estuaries is complex, see for example Ippen (1966). It can however be divided into two major categories—bulk motion (e.g. tidal flow, freshwater flow) and turbulent motion (e.g. eddies created by 'roughness' or irregularity in the channel bed and banks).

If a soluble material is discharged to an estuary it will be carried away from the point of discharge by the bulk water motion and will be spread out, both along and perpendicular to the direction of flow, by the diffusive effects of the turbulent motion. As the material diffuses across the flow it will become spread into regions possessing different bulk velocities. For example, if it is discharged from the channel edge, where friction maintains low velocities, it will eventually spread out towards the centre of the channel where velocities are generally much higher. The spreading into streams moving at different velocities accelerates, at least initially, the dispersal of the materials in the direction of flow. This combined effect of flow, or more specifically, spatial gradients in the flow (shear) and turbulent diffusion is termed 'dispersion'.

In order to make a quantitative estimate of the dispersion suffered by a material discharged to an estuary it is therefore necessary to have some measure of the strength of the turbulent diffusion process and a quantitative knowledge of the bulk water velocity. The latter is amenable to direct measurement and, at least in principle, poses no great problems, but turbulent transport is a much more difficult process to quantify.

By analogy with molecular diffusion (a process which is very much slower at transporting material than turbulent diffusion), it is assumed that the mass flux (mass/unit area/unit time) in a given direction is proportional to the concentration gradient in that direction, where the coefficient of proportionality E is called the turbulent or eddy diffusion coefficient. Thus if the coordinate x is directed along the estuary, the longitudinal mass flux is

$$- E_x \rho \frac{\partial c}{\partial x},$$

where $c(x)$ is the concentration, ρ is the density of the water and the sign has been chosen to yield a positive mass flux down a concentration gradient. There will be corresponding mass fluxes in the vertical (y) and transverse (z) directions, with turbulent diffusion coefficients E_y and E_z which will not necessarily be equal either to each other or to E_x. The determination of the turbulent diffusion coefficients is discussed in Section 8.3.

In order to obtain the total mass flux it is necessary to add to the turbulent contribution the flux of material transported by the bulk motion. Thus:

longitudinal mass flux $= (uc - E_x \partial c/\partial x)$
vertical mass flux $= (vc - E_y \partial c/\partial y)$
transverse mass flux $= (wc - E_z \partial c/\partial z)$

where u, v, and w are the longitudinal, vertical, and transverse bulk water velocities.

Summing these fluxes a mass balance equation can now be derived (see for example p. 567 of Ippen, 1966) of the form

$$\frac{\partial c}{\partial t} = - \frac{\partial}{\partial x}\left(uc - E_x \frac{\partial c}{\partial x} \right) - \frac{\partial}{\partial y}\left(vc - E_y \frac{\partial c}{\partial y} \right) - \frac{\partial}{\partial z}\left(wc - E_z \frac{\partial c}{\partial z} \right) + S \quad (8.1)$$

where the term $S(x, y, z, t)$ has been added to account for external sources of material. Subject to appropriate boundary conditions and a knowledge of u, v, w, E_x, E_y, and E_z, this equation can, in principle, be solved to determine the spatial and temporal distribution of the concentration, c.

Only in very exceptional circumstances would such a procedure be adopted. Firstly, a knowledge of the concentration distribution in such fine detail would be quite unnecessary for engineering applications. Secondly, the basic velocity and turbulent diffusion coefficient data would not be known with sufficient accuracy. Thirdly, field data could never be sufficiently comprehensive or detailed to verify more than just the 'macroscopic' features of the predictions. Finally the computational effort required to solve the equation for realistic estuarine conditions would be enormous and costly.

The procedure conventionally adopted to bring the problem down to an acceptable size is to average equation (8.1) over one or more of the space dimensions. Averaging vertically, i.e. with respect to y, yields a 2-dimensional 'plan' model suitable for the investigation of wide vertically mixed estuaries and bays (see for example Leendertse, 1970; and Orlob and colleagues, 1969). Averaging transversely, i.e. with respect to z, yields a 2-dimensional 'elevation' model suitable for the study of narrow stratified estuaries (e.g. Hobbs, 1970; Hobbs and Fawcett, 1973).

Averaging both vertically and transversely yields a 1-dimensional model suitable for estuaries which are well mixed both vertically and transversely (e.g. Okubo, 1964). It is this last form which is most used, perhaps more for its comparative computational simplicity, than for its precise applicability.

In its 1-dimensional average form, equation (8.1) reduces, after some re-grouping of the advective and diffusive terms, to

$$\frac{\partial}{\partial t}(A\bar{c}) = -\frac{\partial}{\partial x}(A\bar{u}\bar{c}) + \frac{\partial}{\partial x}\left(AK_x\frac{\partial \bar{c}}{\partial x}\right) + S \qquad (8.2)$$

where $\bar{c}(x,t)$ and $\bar{u}(x,t)$ are the concentration and bulk water velocity averaged over the cross sectional area $A(x,t)$. In this equation all reference to the velocity shear, i.e. the deviation of u from $\bar{u}$, has disappeared from the advective term and must reappear in the dispersive term.

The effective longitudinal dispersion coefficient K_x replaces the turbulent diffusion coefficient E of equation (8.1) and represents the combined effects of turbulent diffusion and velocity shear. Thus in general

$$K_x = K_x(u - \bar{u}, E_x, E_y, E_z)$$

It is convenient to split K_x into components $\bar{E}_x$, K_{xy} and K_{xz} where $\bar{E}_x$ is the cross-sectional mean longitudinal turbulent coefficient and K_{xy} and K_{xz} are the longitudinal dispersion coefficient due to vertical and transverse shear, respectively. Thus

$$K_x = \bar{E}_x + K_{xy} + K_{xz}$$

The calculation of K_{xy} and K_{xz} is discussed in Section 8.4.

8.3 TURBULENT DIFFUSION

In turbulent flow, measured variables, e.g. velocity, dye concentrations, and so forth exhibit random fluctuations. It is the correlations between fluctuations that give rise to the turbulent transport of mass, momentum, and heat. Assuming that flux is proportional to the appropriate gradient then in the y-direction for example

$$\text{vertical mass flux} \qquad = E_y \rho \frac{\partial c}{\partial y}$$

$$\text{vertical momentum flux} \qquad = N_y \rho \frac{\partial u}{\partial y}$$

where N_y is the vertical turbulent (eddy viscosity) coefficient.

In order to determine E_y use may be made of Reynolds' Analogy which equates E_y and N_y, i.e.

$$E_y = N_y = -\frac{\tau_y}{\rho \partial u / \partial y} \qquad (8.3)$$

where τ_y is the vertical momentum flux or shear stress in the fluid. A knowledge of the distribution of velocity and shear stress, itself a function of the flow distribution, is therefore sufficient to determine E_y.

For many purposes it is sufficient to assume that τ_y is either spatially uniform or falls linearly with distance from a rigid bounding surface (wall, river bed, etc.) to zero at a free surface or an axis of symmetry, e.g. a pipe axis. The value at the wall τ_0 can be calculated from the friction velocity $u.$:

$$\tau_0 = \rho u^2.$$

where $u.$ is related to the friction characteristics of the wall and can be calculated from

$$u. = 3 \cdot 9 (M / R^{1/6}) u$$

M is the Manning roughness, R the hydraulic radius for the channel, and $\bar{u}$ the cross-sectional average velocity.

In his extension of Taylor's (1954) classical analysis of dispersion in a cylindrical pipe, Elder (1959) calculated E_y for an infinitely wide open channel by assuming

$$\tau = \tau_0 (1 - y/h) \qquad\qquad (y = 0 \text{ at channel bed})$$

and a logarithmic velocity profile

$$u = u. k_0^{-1} \ln (y/y_0)$$

where k_0 is Von Karman's constant (0·42), y_0 is a characteristic roughness length for the bed, and h is the channel depth. This yielded

$$E_y = k_0 y/h (1 - y/h) h u.$$

Averaged over the depth of the channel

$$\bar{E}_y = 0 \cdot 07 \, h u. \qquad (8.4)$$

An alternative approach to the determination of $\bar{E}_y$ is firstly to recall its definition in terms of averaged fluctuations

$$y = -\rho <u^1 v^1>$$

where u^1 and v^1 are the fluctuations about the mean of the longitudinal and vertical velocities, $<>$ signifies an average over time, and then to utilize Prandtl's (1952) concept of the 'mixing length'.

Prandtl argued that the magnitude of the fluctuations u^1 and v^1 could be related to the vertical velocity gradient by

$$|u^1| = |v^1| = ly\frac{\partial u}{\partial y}$$

where l_y is the 'mixing length' and can be crudely identified with the size of the eddies primarily responsible for the transport processes.

Phase arguments then lead to

$$\tau_y = -l_y\frac{2\partial u}{\partial y}\left|\frac{\partial u}{\partial y}\right|$$

and

$$E_y = l_y^2\left|\frac{\partial u}{\partial y}\right| \tag{8.5}$$

Similarly

$$E_z = l_z^2\frac{\partial u}{\partial z}$$

Only if the turbulence is isotropic will l_y and l_z be equal.

The mixing length approach was adopted by Kent and Pritchard (1959) who took a result of Montgomery (1943) that for a neutrally stable flow in an open channel

$$l_y = 0\cdot4\frac{y}{h}\left(1-\frac{y}{h}\right)h$$

They then assumed

$$E_y = n^2 l_y^2\frac{\partial u}{\partial y}$$

where the constant n was to be determined by measurement, and taking $\partial u/\partial y \simeq \bar{u}/h$ obtained

$$E_y = 0\cdot16n^2\overline{uh}\left(\frac{y}{h}\right)^2\left(1-\frac{y}{h}\right)^2$$

Averaged over depth this gives

$$\bar{E}_y = 0\cdot005n^2 h\bar{u}$$

In estuaries $u_*/\bar{u}$ is typically 0·07. Hence $\bar{E}_y \simeq 0·07 n^2 hu_*$, a result comparable with equation (8.4) if $n \simeq 1$.

Kent and Pritchard also considered the situation where the flow was stably stratified, as it is in many estuaries. Under these conditions the Reynolds Analogy is no longer valid. They concluded that l_y would be reduced by a factor $(1 + \beta Ri)^{-1}$ where Ri is the Richardson Number

$$Ri = \frac{g}{\rho} \frac{\partial \rho}{\partial y} \bigg/ \left(\frac{\partial u}{\partial y}\right)^{2\cdot\cdot}$$

g is the acceleration due to gravity. By comparison with field data for the James River they estimated values of n, β both of the order of 1/4.

They also concluded that wave action induced by surface winds would increase the magnitude of l_y near the water surface. Thus stratification would be expected to reduce E_y, while surface winds would be expected to increase it.

In this section only turbulent diffusion perpendicular to the main channel flow has been discussed. In the next section it will be shown that these are the components that are of importance even for dispersion in a direction parallel to the flow and that direct turbulent diffusion in that direction, characterized by E_x, is generally negligible. Its order of magnitude, on the assumption that turbulence is approximately isotropic, is that of $\bar{E}_y$ and $\bar{E}_z$, i.e.

$$\bar{E}_x \simeq 0·1\, hu_*.$$

8.4 LONGITUDINAL DISPERSION

(1) Dispersion in steady flow

The first calculation of the longitudinal dispersion coefficient K_x in turbulent flow was that of Taylor (1954) for flow in a cylindrical pipe. He was concerned with the dispersion of a salt plug as it moved downstream from its point of injection. He concluded that

$$K_x = 10·06\, au_*. \tag{8.6}$$

where a is the pipe radius. The dispersion results from the combined effort of radial shear in the velocity pulling the plug out lengthwise (flow at walls less than average, flow in centre greater than average), and radial turbulent diffusion which transports material from radii of extreme velocity to the radius of average velocity. The most important qualitative result of this analysis is that K_x is *inversely* proportional to the radial turbulent diffusion coefficient. The greater the transverse diffusion the smaller the longitudinal dispersion.

Elder (1959) extended this result to a wide shallow channel and concluded that vertical shear and diffusion resulted in a longitudinal dispersion given by

$$K_{xy} = 5·9\, hu_*. \tag{8.7}$$

Bowden (1965) further generalized this result for irregular channels and arbitrary velocity profiles in the vertical direction. Applications of this generalization to a number of specific profiles of interest to oceanographers gave

$$K_{xy} = \alpha hu \qquad (8.8)$$

where α lay in the range 6 to 200, the lower value corresponding to Elder's logarithmic profile and the higher to a profile characteristic of the densimetric circulation in the saline reaches of an estuary. The value of α is clearly highly sensitive to the detailed form of the velocity profile.

Fischer (1967b) investigated the dispersion arising from transverse shear and concluded that in natural streams and estuaries it would be the dominant process and that in many cases dispersion due to vertical shear could be neglected. He deduced, using arguments paralleling those of Taylor and Elder, that

$$K_{xz} = -\frac{1}{A} \int_0^b q'(z') \int_0^{z'} \frac{1}{E_z d(z'')} \int_0^{z''} q'(z''') dz''' dz'' dz' \qquad (8.9)$$

where

$$q'(z) = \int_0^{d(z)} (u(y,z) - u) dy,$$

A is the cross-sectional area, b the channel width at the water surface, $d(z)$ the depth at distance z across the channel; $\bar{u}$ is the cross-sectional mean velocity and $u(y,z)$ is the actual velocity at a point in the cross-section.

The evaluation of equation (8.9) demands detailed hydrographic data which is not available in many circumstances. Fischer derived an approximation to equation (8.9) which expresses K_{xz} in terms of the bulk parameters of the estuary:

$$K_{xz} = 0.30 \frac{\overline{(u')^2}}{u.^2} \left(\frac{L}{h} \right)^2 hu. \qquad (8.10)$$

where L is the distance of the centre of flow to the nearest bank, i.e. approximately $b/2$ and $\overline{(u')^2}$ is the transverse average to the square of the deviation of the longitudinal velocity from its transverse mean.

If typical values are inserted into equations (8.8) and (8.10) it is readily seen that in general K_{xz} is very much greater than K_{xy}.

In the derivation of equations (8.6) to (8.10), two fundamental assumptions have been made:

(i) the dispersing effect of the shear perpendicular to the flow and the counterbalancing turbulent diffusion are in equilibrium,

(ii) the equilibrium concentration distribution thereby established perpendicular to the flow is such that deviations from the cross-sectional mean are small compared to the mean.

The first assumption will be invalid in circumstances where insufficient time has elapsed for the equilibrium to be established. This will be true during the period immediately following an abrupt change in conditions, e.g. in the period immediately following the start of the injection of a dye tracer. Fischer (1967b) has estimated the length of this period to be given by

$$t = 1\cdot8\frac{L^2}{hu.}$$

The second assumption will be invalid where special circumstances lead to unusually high concentration gradients, e.g. in the discharge of a buoyant effluent or discharge into a strongly stratified estuary.

(2) Dispersion in non-steady flow

The derivation of these results also assumes that the flow is steady and always in the same direction, i.e. in general 'seawards'. This will not be the case in an estuary where the ebb and flow of the tides dominate.

Holley and colleagues (1970) have analysed this situation and distinguish between two different regimes: one in which the time taken for a diffusive equilibrium to be established is small compared to the tidal period, the other where the reverse is true. In the first case no modification to the steady flow theory is required as at each instant of time the shear/diffusion balance perpendicular to the flow has time to adjust to the new values of the velocities. In the second case diffusion perpendicular to the sheared flow will have insufficient time to destroy the distortion of the concentration profile before the distortion is removed by reversal of the flow itself. Thus in contrast to the steady flow case where the absence of turbulent diffusion perpendicular to the flow permits continuous longitudinal dispersal by the shear, in oscillatory flow the absence of diffusion results in an undistorted concentration distribution, the flow restoring it to the same condition at the end of each cycle. In this regime therefore increased transverse diffusion results in increased longitudinal dispersion.

Holley and colleagues showed that if T_d is the relevant time for diffusion (i.e. L^2/E_z or h^2/E_y) and T is the tidal period then

$$K_{xy}(\text{tidal}) = K_{xy}(\text{steady flow})$$

for $T_d/T > 0\cdot1$

$$K_{xy}(\text{tidal}) = 10\left(\frac{T_d}{T}\right)^2 K_{xy}(\text{steady flow}) \tag{8.11}$$

for $T_d/T < 0\cdot1$

On the basis of this theory, in many estuaries, K_{xy} would be unaffected by the tidal motion, but K_{xz} could be significantly reduced, particularly near the mouth.

8.5 DISPERSION IN PRACTICE

The preceding sections have concentrated on the theoretical methods available for the calculation of turbulent diffusion coefficients and the resulting longitudinal dispersion. How do these predictions fare when compared with observations?

(1) Steady flow

Both Taylor (1954) and Elder (1959) verified their results under laboratory conditions and found good agreement. Taylor measured values of K_x/au. in the range 10·0 to 12·8 for straight pipes, but reported values of double this magnitude in curved pipes. Elder's measurements in a 35 cm wide water flume with flow depths of 1 to 1·5 cms yielded a value of K_x/hu. of 6·3. These results provided indirect confirmation that $\bar{E}_y/hu$. must be of the order of 0·07. Elder also measured the transverse spread of the dye tracer and obtained E_z/hu. $= 0·23$. Very similar values (0·17–0·24) have been obtained by other workers (Orlob, 1959; Sayre and Chamberlain, 1964; Fischer, 1967a) both in laboratory flumes and in small natural channels. Larger values have been obtained in big rivers. Yotsukura and colleagues (1970) measured 0·6 in the Missouri, and Glover (1964) a value of 0·72 in the Columbia River. The enhancement of transverse mixing by the presence of bends has been confirmed by Fischer (1969) who found effective values of E_z/hu. as high as 2·4 in curved laboratory flumes.

Fischer (1967b, 1968) has deduced longitudinal dispersion coefficients from data taken in a range of environments, from laboratory flumes to major rivers. He found values for K_x/hu. mainly in the range 100–500 with an extreme value of 5600 in the Missouri. In all those cases for which he had sufficient data he found that equation (8.9), which attributes all the dispersion to transverse effects, provided predictions that were accurate to within a factor of two. He concluded that equation (8.10) was less accurate, perhaps to within a factor of four.

It would seem therefore that the existing theory is capable of predicting, with a fair degree of accuracy, longitudinal dispersion coefficients under both laboratory and natural stream conditions provided bends do not induce a significant amount of transverse mixing and provided adequate data on the cross-sectional distribution of the flow is available. If this data is not available, the dependence of K_x on the bulk hydrographic parameters of the stream can still be deduced from equation (8.10), although the determination of its absolute magnitude will require additional experimentation.

(2) Non-steady flow

The situation with respect to non-steady flow is far more confused. Observationally tidal estuaries are characterized, particularly in their wider reaches, by high values of K_x/hu., typically in the range 500–1000. Vertical variations may dominate the dispersion in one estuary, transverse in another. Which, depends on the hydrographic and hydraulic characteristics of the estuary. It may also be the case that different mechanisms dominate different reaches of the same estuary.

From measurements in the Severn, Thames, and Mersey, Bowden (1963) derived values of K_x which are one to two orders of magnitude greater than those expected from the vertical shear. However since all the measurements were taken from within the region of salinity intrusion, Bowden (1965) suggested that the discrepancy was due to the presence of vertical density gradients, the consequent vertical stability reducing E_y and hence increasing K_{xy}. A more detailed analysis of the Mersey data revealed a quantitative correlation between the K_x estimated at various depths and the local Richardson Number. Bowden concluded that at Richardson numbers in the region of 0·5 to 1·0, the theoretical values of K_x should be increased by a factor of 10 to 20. Doubt has been cast on this interpretation by Fischer (1972) who argues that at the lower values of E_y suggested by Bowden there is insufficient time during the tidal cycle for the equilibrium to be established between the vertical shear and diffusion, and that the value of K_{xy} is correspondingly reduced. The overall effect is to produce a K_{xy} which still falls short of that observed by a factor of at least 7. Fischer attributes the dispersion observed in the Mersey to a transverse shear associated with the existence of a steady horizontal circulation. The origin advanced for this circulation is the asymmetry of the channel cross-section and the tendency for the tidal flow to favour one part of the channel during the flood and another on the ebb. The argument cannot at this stage be conclusive because Fischer's estimate of the consequential longitudinal dispersion is based on a theoretical calculation of the magnitude of the circulation which has yet to be confirmed by direct observations.

Several of the earlier workers (for example, see Ippen, 1966; Okubo, 1964) adopted a purely empirical approach to the determination of K_x by assuming it to have a simple functional dependence on x, the function being characterized by one or two unknown parameters. All these functional representations had one property in common: monotonic increase towards the sea, e.g. $K_x = A \exp (bx)$ for the Delaware. The representations were substituted into equation (8.2) and the parameters adjusted until satisfactory agreement was obtained between the resulting solutions and observations of salinity intrusion or tracer experiments. The present authors using a similar approach (Hobbs, 1970) have estimated K_x for the Tees estuary from an analysis of the salinity intrusion. The result (for neap tides) is shown, labelled $K_x^{(1)}$ in Figure 8.1. Also plotted are the predicted values of K_{xy} from equation (8.7), $K_{xz}^{(s)}$ calculated from the steady flow equation (8.10) and $K_{xz}^{(T)}$ calculated from equation (8.11) for non-steady tidal flow. The ratio T/T_{dz} is also shown, where T_{dz} is the transverse diffusion time. The parameter θ characterizes the transverse velocity profile which is assumed parabolic, with $\theta = (U_{max} - U_{min})/U_{mean}$. Measurements of the flow profile at $x = 12$km yield a value of θ close to the limiting value of 1·5, i.e. $U_{min} \simeq 0$.

Several, at this stage tentative, conclusions can be drawn from the results illustrated in Figure 8.1.

(i) The 'observed' values of K_x/hu. range from 50 at the landward limit to 350 at the estuary mouth. The latter value at least is too high for vertical shear to be the dominant dispersion process. Transverse shear must therefore be playing a major role over much of the estuary length.

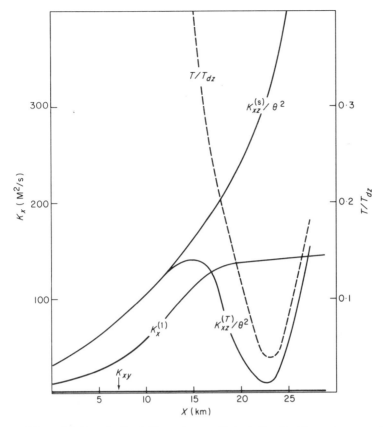

Figure 8.1 Longitudinal dispersion coefficients for the Tees estuary

(ii) In the landward reach $(0 < x < 14 \text{ kms})$ T/T_{dz} is large and transverse diffusion has time during the tidal cycle to come into equilibrium with the transverse shear. The values of K_{xz} computed from equation (8.10) exhibit the correct dependence on x but are a factor $2-6$ larger than the K_x observed.

(iii) In the seaward reach $(x > 14 \text{ kms})$ T/T_{dz} is small, the transverse processes have insufficient time to come into equilibrium during the tidal cycle, and the associated longitudinal dispersion is suppressed. The magnitude of the suppression however appears to be too large. This is perhaps not too surprising as the calculation of T_{dz} has assumed $E_z = 0.23 \, hu$. which is almost certainly an underestimate. The presence of bends, jetties, inlets, and so on will all result in a larger E_z, thereby increasing T/T_{dz} and K_{xz}.

8.6 CONCLUSION

The theory of dispersion in turbulent flow through pipes, laboratory flumes, and natural streams and channels is now well developed and can provide

204

reliable prediction of effective longitudinal dispersion coefficients. The accuracy of the prediction will depend on the amount of hydrographic and hydraulic data available.

The extension of the theory to the estuarine environment is also well advanced but, because of the intrinsically greater complexity of the water motion in estuaries in general, and because of the natural variability in characteristics between one estuary and another, the potential for deriving inaccurate estimates of K_x is still comparatively large. Nevertheless a carefully balanced combination of the qualitative conclusions of the theory and quantitative conclusions from limited field measurements can provide estimates of K_x of sufficient accuracy for the majority of engineering applications.

8.7 REFERENCES

Bowden, K. F. (1963). *Int. J. Air Water Pollution*, 7, 343–356.
Bowden, K. F. (1965). 'Horizontal mixing in the sea due to a shearing current', *J. Fluid Mech.*, 21 (2), 83–95.
Elder, J. W. (1959). 'The dispersion of marked fluid in turbulent shear flow', *J. Fluid Mech.*, 5, 544–560.
Fischer, H. B. (1967b). 'The mechanics of dispersion in Natural Streams', *J. Hydraulics Div. ASCE*, 93 (HY6), 187–216.
Fischer, H. B. (1967a). 'Transverse mixing in a sand-bed channel', *U. S. Geol. Survey Prof. Paper*, 575-D, 267–272.
Fischer, H. B. (1968). 'Dispersion predictions in natural streams', *J. Sanitary Eng. Div. ASCE*, 94 (SA5), 927–943.
Fischer, H. B. (1969). 'The effects of bands on dispersion in streams', *Water Resources Research*, 5 (2), 496–506.
Fischer, H. B. (1972). 'Mass transport mechanisms in partial stratified estuaries', *J. Fluid Mech.*, 53 (4), 671–687.
Glover, R. E. (1964). 'Dispersion of dissolved or suspended materials in flowing streams', *U. S. Geol. Survey Prof. Paper*, 433-B.
Hobbs, C. D. (1970). 'The mathematical modelling of a stratified estuary', *Proc. 5th Int. Conf. Wat. Pollut. Res. San Francisco*, 2, paper III-8.
Hobbs, G. D., and Fawcett, A. (1973). 'Two-dimensional estuarine models'. In *Water Pollution Research Technical Report No. 13*, HMSO, London, 128–138.
Holley, E. R., Harleman, D. R. F., and Fischer, H. B. (1970). 'Dispersion in homogeneous estuary flow', *J. Hydraulics Div. ASCE*, 96 (HY8), 1691–1709.
Ippen, A. T. (Ed.) (1966). *Estuary and Coastline Hydrodynamics*. MacGraw-Hill, New York.
Kent, R. E., and Pritchard, D. W. (1959). 'A test of mixing length theories in a coastal plain estuary', *J. Mar. Res.*, 18, 62.
Leendertse, J. J. (1970). 'Environmental simulation as a tool in a marine waste disposal study of Jamaica Bay', *Proc. 5th Int. Conf. Wat. Pollut. Res.*, San Francisco, 2, paper III-4.
Montgomery, R. B. (1943). 'Generalization for cylinders of Prandtl's linear assumption for mixing length', *Ann. N. Y. Acad. Sci.*, 44, 88–103.
Okubo, A. (1964). 'Equations describing the diffusion of an introduced pollutant in a one-dimensional estuary'. In K. Yoshida (Ed.), *Studies in Oceanography*, University of Washington Press, 216–226.
Orlob, G. T. (1959). 'Eddy diffusion in homogeneous turbulence', *J. Hydraulics Div. ASCE.*, 89 (HY9), 75–101.

Orlob, G. T., Selleck, R. P., Walsh, F., and Stann, E. (1969). 'Modelling of water quality behaviour in an estuarial environment', *Proc. 4th Int. Conf. Wat. Pollut. Res.*, Prague, paper III-12.

Prandtl, L. (1952). *The Essentials of Fluid Dynamics*. Blackie, London.

Sayre, W. W., and Chamberlain, A. R. (1964). 'Exploratory laboratory study of lateral turbulent diffusion at the surface of an alluvial channel', *U. S. Geol. Survey Circular 484*.

Taylor, G. I. (1954). 'Dispersion of matter in turbulent flow through a pipe', *Proc. Roy. Soc. A.*, **223**, p. 446.

Yotsukura, N., Fisher, H. B., and Sayre, W. W. (1970). 'Mixing characteristics of the Missouri River between Sioux City, Iowa and Plattsmouth, Nebraska', *U. S. Geol. Survey Water Supply Paper 1899–G*, G1-G29.

9

The Modelling of Marine Pollution

A. JAMES

9.1 INTRODUCTION

The mathematical modelling of pollution in the sea is similar in a way to the modelling procedure in fresh water. The primary models are concerned with the physical processes of dispersion of the pollutant, and the chemical or biochemical processes of removal and recycling. Secondary models are then used to interpret the biological changes which will result from the predicted changes in concentration.

There are however many differences in approach to modelling marine pollution owing to the following factors:

(a) The differences in density between seawater and the effluents. This leads to much more complex patterns of dispersion in which vertical stratification often needs to be represented.

(b) The greater range of ways in which pollutants may be added to the sea. The main possibilities are inshore on the surface (e.g. estuarine discharges), inshore on the bottom (e.g. marine pipelines), offshore on the surface (e.g. from a boat), offshore below the surface (e.g. sludge dumping).

(c) The much greater diversity of ecosystems involved. Inshore there are communities associated with the inter-tidal zone, inshore waters and inshore benthic communities. Farther out in deep water there are surface dwellers like sea birds, as well as pelagic and abyssal forms.

Work on marine pollution has tended to concentrate on the physical side particularly on dispersion from marine pipelines. Chemical and biological aspects have not been neglected but the models have considered changes in plankton and fish to the exclusion of other types of marine community. This bias is reflected in the following review which describes in the first section the major types of dispersion models and then in the second section deals in detail with some chemical and biological models.

9.2 DISPERSION MODELS

Dispersion in the sea is a complex resultant of waves, currents, density effects, and concentration gradients. It has attracted a considerable amount of attention

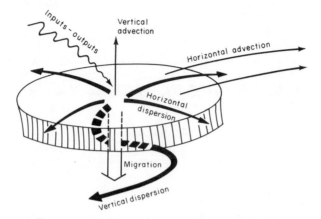

Figure 9.1 Diagrammatic representation of the relationship between different dispersive processes (after Nihoul, 1974; reproduced by permission of Elsevier Scientific Publishing Co.)

in hydraulics texts notably Ippen (1966) and the recent review by Nihoul (1974). The latter showed diagrammatically the relationships between the various processes as given in Figure 9.1. He has derived generalized equations for these relationships and showed how they may be solved for particular cases such as conservative substances.

Hydrodynamic models more directly connected with marine pollution have taken a narrower approach. They are usually specific to particular types of discharge situation and are accordingly classified in this way.

(1) Discharge from submarine pipelines

Pipelines discharging to the sea are an economically attractive method of disposing of domestic and industrial wastes for coastal communities. The

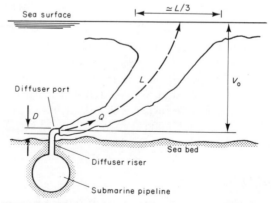

Figure 9.2 Dispersion of waste from a submarine diffuser (after Snook, 1969)

constraints placed upon their use are concerned with the degree of dispersion at or near the adjacent coastline. The dilution of a buoyant jet may be represented diagrammatically as in Figure 9.2. Various approaches have been suggested for the mathematical representation, e.g. the theoretical studies of Rawn and colleagues (1960), Abrahams (1960), Nayashi and Ito (1975).

The work by Abrahams has been the basis for most of the other studies in this field. He considered that the parameters which determine the level of dilution of waste coming from a submarine pipeline are as follows:

Rate of discharge Q
Pipe Diameter D
Jet Velocity U_j
Angle of inclination of emergent jet θ
Depth of water over pipe Y_0
Difference between jet fluid density ρ_0 and receiving water density ρ_S
Ambient water velocity U_0

Rawn, Bowerman, and Brooks (1960) expressed the dilution at the sea surface in terms of three dimensionless groups

$$\frac{Y_0}{D}, \quad \frac{V_j}{\sqrt{\dfrac{(\rho_S - \rho_0)_g D}{\rho_S}}}, \quad \frac{V_j D}{V}$$

V = Kinematic viscosity

Subsequent analysis showed that the third group (Reynolds Number) had no effect provided that the flow was turbulent ($Re > 2000$).

The above predictions assume that the jet is being discharged into a medium with zero velocity. This is unlikely since virtually all submarine pipelines are being discharged into tidal waters. The effect of the movement of ambient liquid on the dilution has been investigated by Agg and Wakeford (1972). They have developed the concept of the velocity ratio U_a/U_g as increasing the degree of mixing. Regression analysis gave a relationship of the form

$$\log\left(\frac{\text{observed dilution}}{\text{predicted dilution}}\right) = 0 \cdot 938 \log\left(\frac{V_a}{V_g}\right) + 1 \cdot 107$$

although as shown in Figure 9.3, the results are subject to a higher degree of variation, and as a result they suggested that the situation could be described by the relationship:

$$\frac{C_0}{C_m} = \phi_2\left(\frac{Y}{D}, Fr, \frac{U_a}{U_g}\right)$$

where C_0 = Observed concentration
$\quad\quad\ C_m$ = Predicted concentration
$\quad\quad\ Fr$ = Froude Number

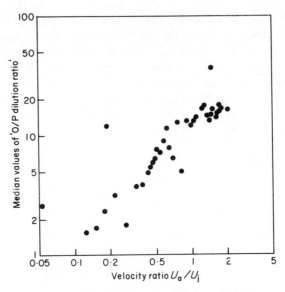

Figure 9.3 Relationship between dilution of a buoyant jet and velocity of ambient fluids (from Agg and Wakeford 1971; Reproduced by permission of The Institution of Public Health Engineers)

White and Agg (1975) later modified the above expression to a more useable form:

$$\frac{C_{\mathrm{o}}}{C_{\mathrm{m}}} = \phi_3 \left[\frac{D}{Y}, \frac{U_{\mathrm{a}}}{(g'y)^{\frac{1}{2}}}, \frac{Q_{\mathrm{S}}}{(g'y)^{\frac{1}{2}}} \right]$$

where Q = effluent discharge
Q_{S} = is a gravitational constant

(2) Dispersion of material from estuaries

Estuaries are the major pathways for waste materials to enter the sea. In many cases estuaries receive substantial discharges of domestic and industrial wastes directly, but in others the wastes and their break-down products enter via upstream discharges. Usually the discharges have been treated to minimize environmental damage, but inorganic materials like metals, oxides of nitrogen, and phosphate are not completely removed in treatment.

The materials discharged into estuaries may to some extent accumulate there depending upon bio-geochemical interactions, which may be quite complex as shown in Figure 9.4. The remainder of the material is discharged to sea in the water leaving the estuary. This may be represented as an intermittent sinusoidal pulse which builds up from Low Water to High Water and then is displaced and dispersed during the High Water to Low Water period. The

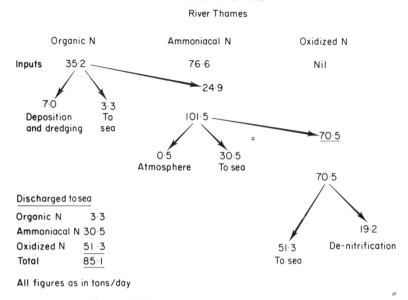

Figure 9.4 Nitrogen budget for the River Thames

shape and size of the plume and its subsequent dispersion patch can be delineated by salinity measurements as shown in Figure 9.5.

The mathematical representation of this type of marine dispersion falls into two stages:

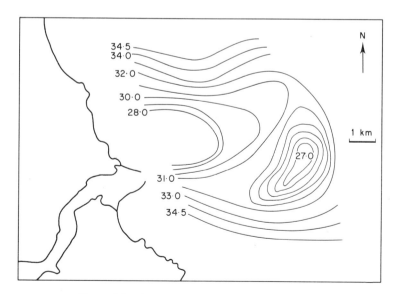

Figure 9.5 Dispersion of water leaving an estuary

(a) the representation of the pulse

(b) its subsequent dispersion.

The latter may be represented as an enlarging semi-ellipsoid as discussed in section (3). The pulse is best represented as a buoyant jet discharging into a denser liquid, with the assumption that the similarity law holds for both velocity distribution and density distribution in the diffusion region. Viscous action is presumed to have no influence on the mixing process.

The model divides up the region into four zones which may be defined as follows:

(a) Jet core. The velocity distribution and density distribution are assumed to be uniform i.e.

Velocity $= U_0$

Density $= P_0$

(b) Zone of Flow Establishment. In this zone the thickness of the jet core is assumed to be decreasing linearly

$$h = h_0 - \left(\frac{h_0}{x_0}\right)x$$

The characteristic of the mixing process of a buoyant jet is primarily volume entrainment from a lower layer.

$$\frac{dh}{dx} = E + \frac{1-x}{\alpha}\left(\frac{h_0}{h} - \frac{h_0}{x_0}\frac{X}{h}\right)E + \frac{1-\alpha}{\alpha}\frac{h_0}{X_0}$$

Where E = Entrainment constant

and $\alpha = \displaystyle\int_0^1 f(G)\,dG$

$$G = \frac{y - h_1}{h + h_1}$$

y = vertical coordinate measured from surface

$$\bar{u} = \frac{Y_0}{h}[h_1 + \alpha(h - h_1)]$$

(c) Zone of Established Flow. The average velocity in this zone due to the jet is

$$\bar{u} = \alpha U_0$$

$$\frac{d}{dx}(Uh) = EU$$

$$\frac{dP}{dx} = \frac{(P_w - P_0)E}{h}$$

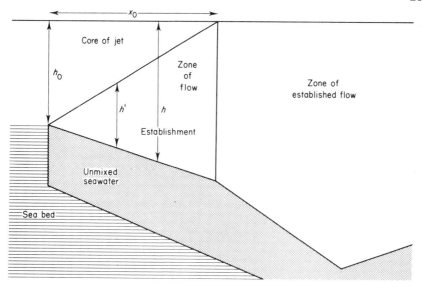

Figure 9.6 Theoretical model of dispersion from a buoyant jet

(d) Unmixed Seawater. Velocity and density are equal to the background levels. It is normally assumed that the velocity $U = 0$.

The coordinate framework and relationship of the zones are shown in Figure 9.6. Solution of the equations is by finite difference methods.

(3) Modelling of subsequent dispersion

However wastes are discharged to the marine environment, through pipelines, from estuaries, or by release from ships, the density differences usually cause the waste to form into a surface plume (or patch), from which it subsequently disperses. Because of the danger to littoral communities and the recreational interests in inshore waters, it is important to be able to simulate the dispersion of surface pollutants. As a result there have been numerous attempts at modelling this type of dispersion. This is a difficult situation to represent because of the complexity of patterns produced by combinations of vertical and horizontal mixing rates. Talbot (1975) pointed out that there were cases where any attempts to describe the diffusion process in detail is clearly hopeless. Nevertheless some measure of success has been achieved in predicting surface dispersion at sea using the formulations described below:

(a) Dispersion from a patch.
Where a waste is discharged intermittently a surface patch results. The distribution of concentrations within the patch is usually assumed to have a Gaussian form (Okubo and Pritchard, 1969).

$$\bar{C}(x,y,z,t) = \frac{M}{(2\Pi)^{3/2}\sigma_x(t)\sigma_y(t)\sigma_z(t)} \exp - \left[\frac{x^2}{\sigma_x^2(t)} + \frac{y^2}{\sigma_x^2(t)} + \frac{z^2}{\sigma_z^2(t)} \right]$$

Where $\bar{C}$ = average concentration at any point x, y, z at time t
  M = total weight of material released
  σ = standard deviation

obviously the maximum concentration $\bar{C}_{max}$ is given by the first term in the above equation, where $x = y = z = 0$

$$\bar{C}_{max} = \frac{M}{(2\Pi)^{3/2}\,\sigma_x(t)\sigma_y(t)\sigma_z(t)}$$

The divisor needs to be modified to take account of the impermeable boundary at the sea surface so that $(2\Pi)^{3/2}$ becomes $\Pi\sqrt{2\Pi}$. Field investigations have shown that patches normally elongate in the direction of travel resulting in a semi-ellipsoid. The above equation may therefore be modified to:

$$\bar{C}_{max} = \ln\frac{x^2}{\left(\dfrac{C_{max}}{C}\right)\sigma_x^2(t)} + \ln\frac{y^2}{\left(\dfrac{C_{max}}{C}\right)\sigma_y^2(t)} + \ln\frac{z^2}{\left(\dfrac{C_{max}}{C}\right)\sigma_z^2(t)}$$

(b) Dispersion from a plume.

A continuous source may be represented as the superposition of a large number of small diffusing patches each from an infinitesimally small instantaneous source and all moving with a constant velocity. Pritchard (1960) described a model for this situation based on even radial diffusion with an even layer. Whilst this model cannot account for variable trajectories as above, it does lend itself to integration and has achieved some success in prediction. For horizontal diffusion from a line source, the relationship becomes (Carter, 1975)

$$C(x,y,t) = \frac{M}{\Pi v^2 D t^2}\exp\left[\frac{-(x - Ut)^2 - y^2}{V^2 t^2}\right]$$

Where V = diffusion velocity
  U = ambient velocity
  D = thickness of surface layer

Assuming that a continuous source can be considered as N releases per unit time, then

$$C_S(x,y,t) = \int_0^t CN\,dt$$

Where C_S = concentration from a continuous release at a rate
  Q = NM, this gives for $y = 0$

$$C_S(x,0,t) = \frac{Q}{2\sqrt{\Pi VxD}}\ F(\theta)$$

where $F(\theta) \equiv \left(1 - \dfrac{2}{\sqrt{\Pi}}\right)\displaystyle\int_0^{} \exp(-\phi^2)\,d\phi = 1 - \mathrm{erf}\theta$

and
$$x \equiv Ut_1$$

$$\theta \equiv \frac{U}{V}\left(\frac{x}{x_1} - 1\right)$$

9.3 BIOLOGICAL MODELS IN MARINE POLLUTION

Developments in the simulation of biological processes and the stimulus of the International Biological Programme have generated a large number of marine models. However, most of these models have been concerned with ordinary marine processes and the following review considers only examples of models of disturbed systems.

(a) Nutrient Enrichment of the North Sea.

James and Head (1972) studied the effect of discharges on the nitrogen and phosphorus concentrations in the North Sea. They found that there was a localized fertilizing effect in waters near the mouth of an estuary. This became important during the plankton growth season because of the effect on the mass balance:

For any inshore region

$$\frac{dN}{dt} = \frac{Q}{V}N - \frac{Q}{V}N_2 - RN_2 \,(\text{Biomass})$$

where N is the concentration of the nutrient
$(Q/V)N$ is the rate of inflow
$(Q/V)N_2$ is the rate of outflow
and R is the rate of uptake per unit of biomass of the phytoplankton.

Dispersion of additional nutrients from discharges is usually very rapid due to tidal currents, and so on, so that the difference in concentration between enriched and non-enriched areas is low (see Figure 9.7). However, during the growing season the enrichment is sufficient to keep primary production at a higher level as shown in Figure 9.8.

The effect of all discharges of nutrients on the concentration in water in the North Sea was also studied by James and Head (1972). They assessed the discharges as shown in Table 9.1.

These figures were compared with the rate of entry via exchange with adjacent seas and precipitation as shown in Table 9.2.

The results of this predictive model are shown in Figure 9.9.

This type of model is useful in studying nutrient build-up but needs to be linked with biological model before any indication of change in biological productivity is obtained. An example of this type of joint chemical and biological model is the one proposed by Walsh and Dugdale (1972) to describe upwelling of nutrients on the Peruvian coast and the effect on marine communities. The physical situation is represented in the model as a two-layered series of blocks aligned in series along the axis of the plume as it moves away from the coast. Advective and non-advective transfer of material is shown in Figure 9.10

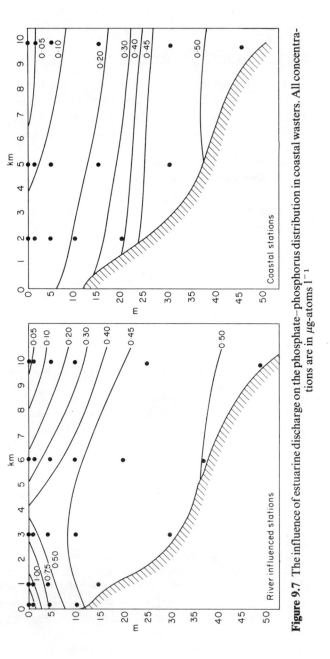

Figure 9.7 The influence of estuarine discharge on the phosphate–phosphorus distribution in coastal wasters. All concentrations are in μg-atoms l^{-1}

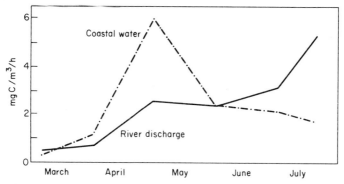

Figure 9.8 Effect of estuarine discharge of nutrients on the pattern of primary productivity

Table 9.1 Nutrient Contribution to the North Sea from Sewage and Run-off

Country	Nitrogen tons/day	Phosphorus tons/day	Source of Information
France	4	1	Volume of sewage discharged[1]
Belgium	100	24	Volume of sewage discharged[1]
Netherlands	100	24	Population[2]
Germany (via Rhine)	650	40	River concentrations
(via Ems, Weser, and Elbe)	450	100	Population[3]
Denmark	27	6	Population[2]
Norway	27	5	Population[2]
U.K. (direct)	150	35	Volume of sewage discharged[1]
(via rivers)	240	20	River concentrations
Total	1 744	255	

1. I.C.E.S. (1969).
2. 9gN and 1gP per person per day.
3. May be low as no estimate of industrial pollution included.

Table 9.2 Nutrient Inflows to North Sea from natural sources and the effect on overall concentration was modelled as if the North Sea were a stirred tank:

$$C_m = C_1 + (C_0 - C_1) e^{\frac{-ut}{v}}$$

Source	Nitrogen tons/day	Phosphorus tons/day
North Atlantic	7 700	1 230
English Channel	400	76
Baltic	180	10
Precipitation	272	27
Total	8 552	1 343

where
C_m = concentration in North Sea at time t
C_1 = Concentration of inflow
C_0 = Starting concentration in North Sea
u = Rate of inflow
v = Volume

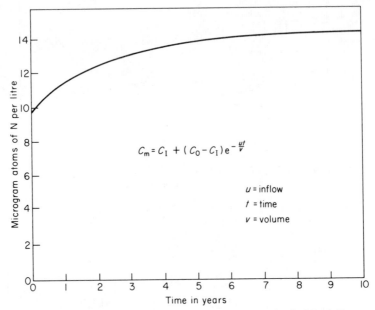

Figure 9.9 Build-up of the concentration of nitrogen in the North Sea

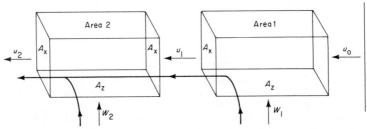

Figure 9.10 Diagrammatic representation of nutrient transfer in the Walsh and Dugdale Model (see text for explanation) (reproduced by permission of John Wiley & Sons, Inc.)

and the overall model with the biological elements is shown in Figure 9.11.

In solving such a model finite difference methods are employed. The equations are written in differential form and the rate is evaluated at each time step and then used to solve a new mass balance (for chemicals) or biomass balance (for organisms). An example of the form of the equations from Walsh and Dugdale's model is as follows:

Rate of change of phytoplankton Rate of upwelling of phytoplankton

$$\frac{dp}{dt} = \frac{(w)P_{i,j,k} - P_{i,j,k-1}}{\Delta z}$$

$$- \text{Downstream Transport}$$

$$\frac{(u)P_{i,j,k} - P_{i,j,k} - P_{i,j,k}}{2\Delta x}$$

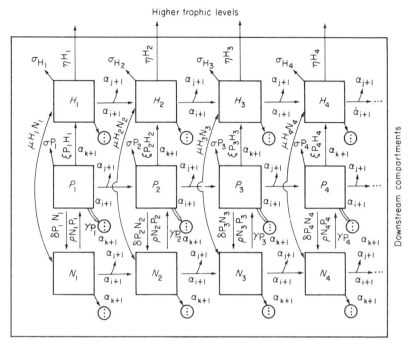

Figure 9.11 Trophic network in the Walsh and Dugdale Model (see text for explanation) (reproduced by permission of John Wiley and Sons, Inc.)

$$+ \text{ Lateral Diffusion and Mixing}$$

$$\frac{(K_y)(P_{i,j+1,k} - 2P_{i,j,k} + P_{i,j-1,k})}{(\Delta y)^2}$$

$$- \text{ Sinking} \qquad - \text{Nutrient uptake}$$

$$\frac{(W_s)P_{i,j,k}}{\Delta z} - \frac{(V_{max})(P_{i,j,k})(N_{i,j,k})}{K_t + N_{i,j,k}}$$

$$- \text{ Grazing}$$

$$\frac{(G_{max})(H_{i,j,k})(P_{i,j,k} - P^*)}{K_p + (P_{i,j,k} - P^*)}$$

where w = vertical velocity as a function of wind stress, depth, and distance offshore.

$P_{i,j,k}$ = phytoplankton concentration in surface layer with respect to X, Y and Z axis (i.e. offshore distance, lateral, and vertical distances)

u = downstream velocity as a function of W

K_y = lateral eddy coefficient as a function of i

W_s = sinking velocity of phytoplankton as a function of age

V_{max}	= maximum uptake rate of nutrients
P^*	= grazing threshold
$H_{i,j,k}$	= concentration of herbivores
K_p	= Michaelis–Menten coefficient for grazing
$N_{i,j,k}$	= concentration of nutrients
K_t	= Michaelis–Menten coefficient for uptake of nutrients
G_{max}	= Maximum rate of grazing by herbivores

A promising development in the modelling of marine pollution is in the work on biological models. These have not yet reached a stage where they can be readily applied to polluted ecosystems, but the insight they show into physical, chemical, and biological interaction is invaluable for environmental management.

An example of this type of work is the model described by Steel (1973) of a plankton ecosystem. He considered the following parameters:

(a) Vertical mixing — considers a stable thermocline situation and defines a mixing rate V as the fraction of the upper layer exchanged daily with the lower layer.

(b) Light and Temperature — both assumed to be constant and light intensity is assumed to be adequate for photosynthesis.

(c) Nutrient Uptake — considered to follow Michaelis–Menten kinetics and nitrogen is assumed to be limiting.

(d) Nutrient Concentration — it is assumed that the initial concentration is N_0 throughout. Thereafter due to stratification and uptake the concentration in the upper layer decreases to N_t and is replenished at a rate $V(N_0 - N_c)$

A fraction U of the nitrogen absorbed is considered to be returned by bacterial decay or exertion.

(e) Zooplankton Grazing Rate — considered to be density-dependent with a threshold concentration below which feeding is zero.
the rate is

$$c(p - P_1)/(D + P)$$

where P is the concentration of particulate carbon, P_1 is the threshold, and C and D are parameters defining the shift of the response curve.

The assimilation ratio is taken as a fixed value of 0·7.

(f) Zooplankton Respiration — considered as a combination of a fixed rate F and a rate E which is proportional to the food intake. The balance of gain by assimilation and loss by metabolism is then

$$(0{\cdot}7C - E)(P - P_1)/(D + P) - F$$

(g) Relation of Weight to Metabolism and Assimilation — assumed that rates of these processes vary with $W^{0{\cdot}7}$ (where W is the weight) so

$$\frac{dW}{dt} = [(0{\cdot}7C - E)(P - P_1)/(D + P) - F] \\ \times W^{0{\cdot}7}$$

and the excretion rate is

$$[E(P - P_1)/(D + P) + F]W^{0{\cdot}7}$$

(h) Zooplankton Reproduction — all zooplankton are assumed to start with a weight of $0{\cdot}2\ \mu g$ C and to grow (depending on balance between respiration and assimilation) until it reaches the adult stage. Growth then ceases and any assimilation surplus to respiration requirement is assumed to go into reproduction at a conversion efficiency of $0{\cdot}3$.

The reproduction of the female population over the life span J is therefore $0{\cdot}3\ \Sigma$ excess food. This quantity divided by $0{\cdot}2\ \mu g$ gives the number of juveniles.

(i) Herbivore mortality — the mortality pattern of zooplankton is extremely complicated. Steele thought it was due principally to predation by fish and represented it as

$$\frac{dz}{dt} = -G(Z - Z_1)(W - W_1)/(H + ZW) - \\ - GX(Z)$$

where Z is the number of copepods of individual carbon content W; G and H are parameters defining the shape of the predation curve; Z and W_1 are thresholds, and GX is coefficient of proportional predation.

(j) Initial conditions — assumed to be the beginning of a spring outburst with a plant population of P_0 and Z_0 juvenile copepods of weight W_0.

The model for one cohort then becomes

$$\frac{dR}{dt} = -ARP/(B+R) + V(R_0 - R) + UE(P - P_1)/(D + P) + FZW^{0.7}$$

Rate of change = − plant uptake + mixing + zooplankton excretion of nutrients

$$\frac{dP}{dt} = -ARP/(B+R) - VP - CZW^{0.7}(P - P_1)/(D + P)$$

Rate of plant growth = nutrient uptake − loss by mixing − zooplankton grazing

$$dW = (0.7C - E)(P - P_1)/(D + P) - FW^{0.7}$$

Rate of change of zooplankton biomass = assimilation − respiration

$$\frac{dZ}{dt} = -G(Z - Z_1)(W - W_1)/(H + ZW) - GXZ$$

Rate of change = density dependent predation − proportional predation of number

This type of model can be extremely useful in examining pollutional effects such as:

(a) Changes in nutrient level due to effluent discharges
(b) Changes in turbidity at more general effects like overexploitation or selective poisoning.

There is considerable scope for further developments in the modelling of marine pollution, particularly for extending fisheries models (e.g. Beverton, 1957) and models of primary and secondary production (e.g. Parsons and Takahashi, 1973), to take into account changes resulting from discharges of sewage and industrial wastes.

9.4 REFERENCES

Agg, A. R., and Wakeford, A. C. (1972). 'Field Studies of Jet Dilution of Sewage at Sea Outfalls', *Inst. Publ. Hlth. Engrs. J.*, **71**, 126.
Abrahams, G. (1960). 'Jet Diffusion in Liquid of Greater Density', *J. Amer. Soc. Civ. Engrs., Hyd. Div.*, **86** (HY6), 1.
Beverton, R. J. H., and Holt, S. J. (1957). 'On the Dynamics of Exploited Fish Populations', *Fishery Invest., London, Ser. II*, **19**, 1–533.
Carter, H. H. (1975). 'Prediction of Far-Field Exclusion Areas and Effects'. In A. L. H., Gameson (Ed.), *Discharge of Sewage from Sea Outfalls* Pergamon Press, p. 363.
Hayashi, T., and Ito, M. (1975). 'Initial Dilution of Effluent Discharging into Stagnant Sea Water.' In A. L. H. Gameson (Ed.), *Discharge of Sewage from Sea Outfalls* Pergamon Press, p. 253–64.
I. C. E. S. (1969). 'Pollution of the North Sea', *Report of International Council for the Exploration of the Sea*. Working Group, Series A, No. 13. Charlottenlund, Denmark.

Ippen, A. T. (1966). *Estuary and Coastal Hydrodynamics*, McGraw-Hill, New York.

James, A., and Head, P. C. (1972). 'The Discharges of Nutrients from Estuaries & Their Effect on Primary Productivity'. In M. Ruivo (Ed.), *Marine Pollution & Sea Life*, Fishing News (Books) Ltd., Surrey, England.

Nihoul, J. C. (Ed.) (1974). *Modelling of Marine Systems*. Elsevier, Amsterdam.

Okubo, A., and Pritchard, D. W. (1969). 'Summary of our Present Knowledge of the Physical Processes of Mixing in the Ocean and Coastal Waters', *U. S. Atom. Energy Commn. Rep. No. NYO-3109–40*.

Parson, N. T., and Takahashi, M. (1973). *Biological Oceanographic Processes*, Pergamon Press, Oxford.

Pritchard, D. W. (1960). 'The Application of Existing Oceanographic Knowledge to the Problem of Radioactive Waste Disposal into the Sea'. In *Disposal of Radioactive Waste*, Vol. 2. International Atomic Energy Agency, Vienna, p. 229–53.

Rawn, A. M., Bowerman, F. R., and Brookes, M. H. (1960). 'Diffusers for Disposal of Sewage into Seawater', *Proc. Amer. Soc. Civ. Engrs.*, **86** (SA2).

Steele, J. (1973). *The Structure of Marine Ecosystems*. Blackwell, Oxford.

Talbot, J. W. (1975). 'Interpretation of Diffusion Data'. In A. L. H. Gameson (Ed.), *Discharge of Sewage from Sea Outfalls* Pergamon Press, p. 321–332.

Walsh, J. J., and Dugdale, R. C. (1972). 'Nutrient Submodels & Simulation Models of Phytoplankton Production in the Sea'. In Allen, H. E., and Kramer, J. R. (Eds), *Nutrients in Natural Waters*, Wiley–Interscience, New York.

White, W. R. and Agg, A. R. (1975). 'Outlet Design'. In A. L. H. Gameson (Ed.), *Discharge of Sewage from Sea Outfalls*, Pergamon Press, p. 265–76.

III

Application to Waste Treatment

10

Modelling of Sewerage Systems

O. G. Lindholm

10.1 INTRODUCTION

In July 1971 a project for mathematical modelling of sewerage systems was started at the Norwegian Institute for Water Research (NIVA). The project is financed by the Norwegian government, and is planned to continue.

Especially, the problems of how stormwater runoff from urban areas affects the operation and efficiency of sewerage systems have been studied by the use of mathematical models describing the various technical systems. The models, which are deterministic, are programmed in FORTRAN IV and run on a UNIVAC 1108 computer (partly IBM 370/55). The FORTRAN coding and computer implementations have been worked out in cooperation with a private firm, A/S COMPUTAS.

10.2 PROGRAM CONFIGURATION

In order to analyse the sewerage system as a whole, we considered it necessary to build two models, one for the sewer network with appurtenant constructions,

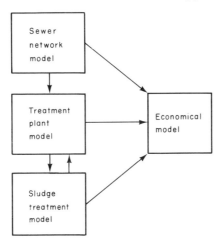

Figure 10.1 Data flow between the programs (from Deininger, 1973; reproduced by permission of Ann Arbor Science Publishers, Inc.)

and one for the sewage treatment plant. The program configuration is shown in Figure 10.1. Arrows indicate data flow between the programs. For the time being all programs are operated independently. Capital and running costs for various units are therefore computed in each individual program, in addition to the cost-analysis program.

Data from one program can be stored on file in the computer, and called from the other programs. As an example the sewer network model may compute the variables, water discharge (l/s), and BOD transportation (g/s) from an urban drainage area simulated, let's say, for one year's rain activity. These data will be called up by the sewage treatment plant model. Thus the performance of the treatment plant over the whole year can be studied when units like storm overflows, rainwater basins, and so on are varied.

10.3 PROGRAM DESCRIPTIONS

(1) Sewer network model

The main objectives of this program have been to describe the pattern of storm-water runoff in the sewer network. We wanted to know the variation in discharge at any point in the network and for each minute of a given rainfall. A hydrograph method was wanted which considered the effects of storage capacity in the sewer network and the true water velocity when the pipes were partly filled. On the other hand the method should not be too complicated, since the cost of computation should be kept on a reasonable level. The model should also be useful in practical design of sewerage systems.

Feature of network model

(a) The rainfall intensity can vary with time, i.e. for each minute of the rainfall, different intensities may be given.
(b) The runoff coefficient can vary with time, i.e. for each minute of the rainfall, different coefficients are given.
(c) Time of entry vs contributing area functions can be given. This means that different surface storage characteristics may be considered.
(d) The storage capacity of each of the sewer lines is considered.
(e) The velocity of flow is made a function of the water depth in the partially filled sewer.
(f) Storage tanks may be built in at any point of the system.
  The necessary tank volume for a given rainfall can be computed when a maximum outlet discharge from the tank is specified.
(g) Storm overflows can be considered at any point of the system. The total bypassed and diverted volumes of water and pollution are computed.
(h) The sum of industrial and domestic wastewater flow together with infiltration water is considered as a constant discharge in time, and each sewer line may have its own value for wastewater 'production'.

(i) Pumping stations and other inflow hydrographs may be given at any point of the system.

(j) Transportation of pollutants per unit time will at any specified location be computed, and as a function of time after start of rainfall. The pollutional component is given as input data in the storm water runoff, expressed as a value for each minute after start of a rainfall. Gram 'component' produced by a person per day must also be given. Total amount of the component diverted from storm overflows is computed.

(k) Besides computing water discharge (l/s) and transport of pollutants (g/s), the model can find the smallest standard pipe dimensions for the desirable sewer lines, which will avoid backwater for the particular rainfall considered.

(l) The model computes the capital costs for the total sewer network. Necessary input data are per cent of rock in each cross-sectional trench area and diameter of sewer when this is not computed by the program.

(m) When the sewer network has too small capacity, backwater may occur. The backwater level may be computed for each point of the system and is presented as a function vs time after start of a rainfall.

(n) Tunnels, canals, and pipes may be computed.

(o) The computing step between each runoff situation may be chosen.

Theory of the model

In the following a brief description is given of how the hydrographs for each pipe are computed:

(a) From the input data a rain intensity (l/s ha) and a runoff coefficient is used for each sewer line and each minute of the rainfall. The potential runoff for each sewer line is computed for every minute according to equation (10.1).

$$Q_N = \phi_N \cdot A \cdot I_N \qquad\qquad N = 1 \text{ to } M \qquad\qquad (10.1)$$

where

Q_N = Runoff (l/s) at minute N

ϕ_N = Runoff coefficient at minute N

A = Area (ha) draining to sewer line

I_N = Rainfall intensity (l/s ha) at minute N

M = Rainfall duration in minutes. (See Figure 10.2.)

(b) The delayed water discharge into the sewer inlet is computed.

$$\begin{aligned} D_1 &= Q_1 \cdot t_1 \\ D_2 &= Q_1 \cdot t_2 + Q_2 \cdot t_1 \\ D_3 &= Q_1 \cdot t_3 + Q_2 \cdot t_2 + Q_3 \cdot t_1 \qquad\qquad \text{etc.} \qquad (10.2) \end{aligned}$$

D = Discharge into sewer inlet (l/s)

t = Per cent of a sewer's potential runoff in the actual minute (See Figure 10.3.) Equations (10.1) and (10.2) lead to the inlet hydrograph shown in Figure 10.3. This hydrograph consists of superimposed 'minute-runoffs'.

230

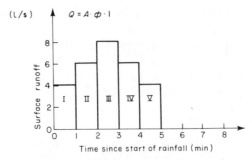

Figure 10.2 Rainwater runoff for each minute of a rainfall

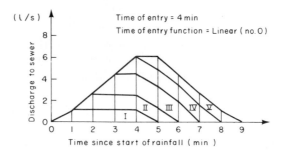

Figure 10.3 Rainwater discharge at sewer inlet (from Deininger, 1973; reproduced by permission of Ann Arbor Science Publishers, Inc.)

(c) The contributing wastewater and rainwater runoff from tributary area of the sewer are considered to adjoin at the upper end of the sewer line. The discharge from the upstream sewer which adjoins the sewer line considered at its inlet and the contribution from tributary area, is going through two additional analyses before the total outlet hydrograph of the sewer line is found.

(d) The translated hydrograph for the sewer is manipulated in a pipestorage procedure. The method is very similar to the one described by Watkins (1962) and which is developed by the British Road Research Laboratory. However, the RRL-method considers the whole network system in one operation, which may lead to inaccuracies. The method described in this paper analyses each sewer line independently. In Figure 10.4 the inflow and outflow hydrographs to a sewer line are shown. Symbols used in the following equations are defined in Figure 10.4.

$$S_2 - S_1 = [(P_2 - Q_2) + (P_1 - Q_1)]t/2$$
$$S_2 + Q_2 t/2 = (P_1 + P_2 - Q_1)t/2 + S_1$$

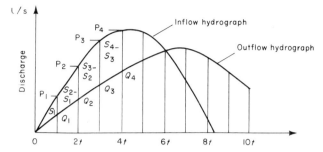

Figure 10.4 Effect of network storage volume (from Deininger, 1973; reproduced by permission of Ann Arbor Science Publishers, Inc.)

In general the equation will be:

If $Q_n < Q_{\text{capacity}}$

$$S_n + \frac{Q_n \cdot t}{2} = (P_n + P_{n-1} - Q_{n-1})\frac{t}{2} + S_{n-1} \qquad (10.3)$$

If $Q_n > Q_{\text{capacity}}$

$$Q_n = P_n$$

In equation (10.3) the right hand side is always known from the previous computation and from the inflow hydrograph. Since the relationship between the stored volume S_n at minute n, and the discharge Q_n is known from curves of partial pipe filling, both Q_n and S_n can be found. In this way the outflow hydrograph for a sewer line is found step by step.

(e) A superposing routine takes care of the branching sewers so that their hydrographs will fit correctly into the total hydrograph.

The program starts with the sewer lines in the upper end of the network and works its way down to the outlet. These data are stored on file when backwater analysis is to be done. The backwater subroutine then starts at the outlet and works its way to the upper end of the network.

(2) Sewage treatment plant model

The main objective of this model is to study the performance of a sewage treatment plant receiving both sewage and rainwater runoff over a given period or a whole year. The total program configuration, i.e. the combination of network model and treatment plant model, are to be used to analyse how a combined or a separate sewerage system will perform in a given area. On the basis of rain intensity vs duration, total precipitation duration and frequencies for a year's rain activity, some representative base rainfalls will be chosen (5–10).

These base rainfalls will be given as input to the sewer network model for a given area. The results, in terms of flow (l/s) and BOD-load (g/s), as functions of time, may further be used as input to the sewage treatment plant model.

General description of the sewage treatment plant model

Figure 10.5 shows the basis units in the model. As indicated, the flow may pass the chemical stage before the biological, and vice versa. It is also possible to let a part of the flow bypass the biological stage, flowing directly for chemical treatment. Any unit may be left out, if wanted.

Initially BOD was the only loading parameter considered, but phosphorus has now been incorporated in the model. The removal efficiency for each unit is mostly based on empirical relationships, since a more theoretical approach with extensive use of process kinetics for the present was found to be deficient and inaccurate for the prediction of removal rates for an arbitrary plant.

Capital and total annual costs are computed in the model. The capital cost functions are based on nonlinear empirical unit costs. The programs may be run from both batch and interactive time sharing terminals. However, the interactive communication between the computer and the operators is essential.

Prior to each analysis the user must choose the computing step (in minutes) between each runoff situation. By increasing the step size from 1 to 10 minutes, the computer costs will be reduced by 80–90%. For each situation we have assumed a steady state condition in the plant. The final result is obtained by superposing all the situations computed. If the main part of a yearly rainfall activity can be represented by 5 basic rainfalls, which result on an average, in a duration of say 150 minutes' runoff for each rainfall, 751 different situations have

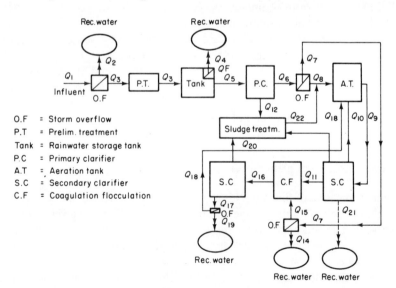

Figure 10.5 Sewage treatment plant model (from Deininger, 1973; reproduced by permission of Ann Arbor Science Publishers, Inc.)

to be analysed for a specific treatment plant. When UNIVAC 1108 needs 3/1000 sec to compute each situation, the computer will need 0·25 second to analyse the performance of a specific plant for one year, choosing 10 minutes computing steps.

Specified description of sewage treatment plant model

In the primary clarifier the removal of settleable solids is considered to follow the curve of Husmann (Munz, 1966). The BOD removal efficiency in the aeration tank is considered to follow Wuhrmann's observations (Munz, 1966). A correction for hydraulic load in the aerator is made according to Munz, (1966). A correction for temperature in the aerator is also incorporated. This correction value is taken from Eckenfelder and O'Connor (1961) and adjusted to unpublished observational data from Norway.

The amount of excess sludge from the biological stage is calculated on the basis of the sludge load factor and the temperature (Hopwood, 1965.) The performance of the secondary clarifier will vary from plant to plant. For the present we have implemented observational data (Eye, 1969), which represent a relationship between concentration of suspended solids (SS) in the supernatant and the overflow rate. The amount of BOD in the SS is a function of the sludge load factor in the aerator. This function is taken from Eckenfelder (1969). Formulation of performance for the chemical treatment stage is mostly a result of observations made in an extensive research project for physical/chemical treatment under way at our institution in Oslo.

(3) Sludge treatment plant model

The sludge treatment model is not yet sufficiently developed to be of real practical use.

10.4 OPTIMIZATION TECHNIQUE

The optimization technique used is a combination of a trial-and-error method and a gradient method. Figure 10.6 shows how the OPTI-command computes and tests the steepness of each variable's cost gradient. The variables are varied one at a time while the others are held constant.

The computer selects the variable with the least cost gradient, and increases this with the given step. Then from this new base situation the computer selects the variable again with the least cost gradient. This variable may now be different from the previous one, and it will get an increase with the given step. In this way the optimal plant is found step by step. UNIVAC 1108 used 1 sec to compute the example shown on Figure 10.6.

Assume that there are 6 unit operations with 6 possible sizes for each. An arbitrary trial-and-error method will then have to choose among 46 600 possible configurations. With this gradient technique the most probable number of

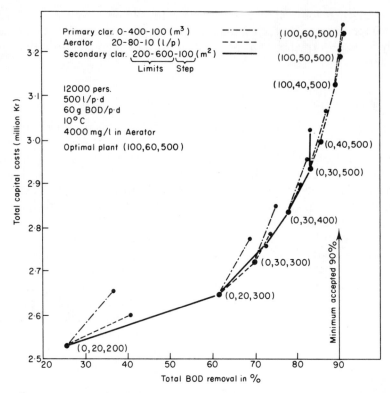

Figure 10.6 Example of an optimization (from Deininger, 1973; reproduced by permission of Ann Arbor Science Publishers, Inc.)

necessary computed configurations will be about 60 to obtain an optimal configuration.

10.5 ASSUMPTIONS

(1) Sewer network model

In a hydrodynamic aspect we assume a steady state situation at each computation. This inaccuracy is negligible when the pipe gradient is greater than 0·1% in comparison with the uncertainty of the input data. Besides, the method is tested against on-site runoff measurements and against non-stationary hydrodynamic methods with good agreement. The non-stationary hydrodynamic method will require a large number of iterations and therefore be more expensive than the method described in this paper.

Whenever there is a downstream branch in the system, we will for the time being have to split the stream at a fixed rate, or specify a maximum discharge in one sewer line and let the surplus go into the other sewer line.

(2) Sewage treatment plant model

In each computation we have assumed steady state. The time aspect is considered by superposing several situations. The removal efficiency functions are mostly based on empirical data presented in the literature.

(3) Sludge treatment model

We still know too little about the processes in each unit to follow BOD, P, N, and SS through the system with a reliable accuracy. Neither are the cost functions in the sludge model yet adequate. The project will continue and the programs will be improved as our knowledge and experience develop.

10.6 EXAMPLES AND DEMONSTRATIONS

(1) Sewer network model

To illustrate the special features of the model we have tested different flow theories on eight sewer lines linked together. We generated one inflow hydrograph only in the upper end of sewer line No. 1. This configuration enabled us to study the damping and transportation effects. Figure 10.7 shows the damping effects in the existing version of the model. The maximum discharge in sewer line No. 8 is 1/3 of maximum inflow to sewer line No. 1.

In Figure 10.8 are shown three different methods for computing storm runoff. The rational method gives a 100% higher maximum value for discharge than

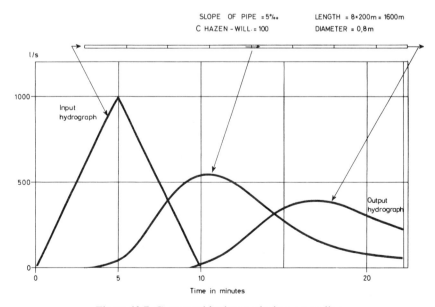

Figure 10.7 Computed hydrographs in a sewer line

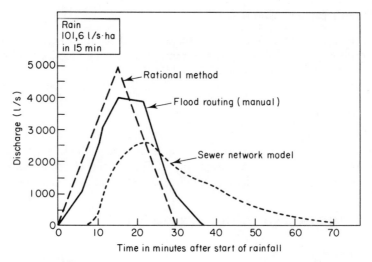

Figure 10.8 Comparison with 3 different methods (from Deininger, 1973; reproduced by permission of Ann Arbor Science Publishers, Inc.)

the method described in this chapter. The sewer network is shown in Figure 10.9.

In Figure 10.9 and 10.10 an analysis of the performance of a storm overflow is shown. Figure 10.9 shows the five outflow hydrographs from the five sewer lines. In Figure 10.10 is shown the situation if a storm overflow is located just after the joint of pipe 4 and 5. The overflow permits a discharge of 50 l/s into pipe 2, which means that maximum discharge in the overflow will be ca. 600 l/s. One would then expect that maximum outflow in pipe 1 would decrease, but this is surprisingly not the case. The reason is the required time of transportation from pipe 4 and 5 to pipe 1. This example indicates that even the hydraulic behaviour

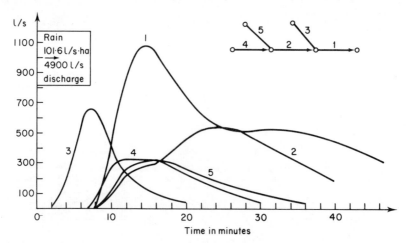

Figure 10.9 Sewer network without storm overflow (from Deininger, 1973; reproduced by permission of Ann Arbor Science Publishers, Inc.)

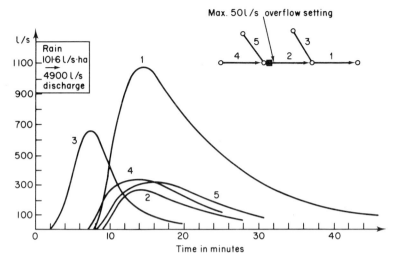

Figure 10.10 Sewer network with storm overflow

in a sewer network may be so complicated that the use of models is necessary.

An important use of mathematical models is to perform sensitivity analysis. Figure 10.11 shows the results of a sensitivity analysis of the input variables affecting the capital costs in a sewer network computation. Inaccuracy in variables (%) is shown along the horizontal axis and the corresponding decrease or increase in capital costs along the vertical axis (%). The diagram was easily obtained using the described model on a combined sewered water-shed area of area of 110 ha, 5500 persons.

The basic values of the variables which are supposed to be correct are:

Runoff coefficient, average value	=	0·4
Rain intensity, l/s ha, average value	=	89·7
Frequency of storm. Return period, years	=	2
Overland time of flow, minutes	=	10
Pipe resistance. Hazen–Williams coefficient	=	100
Hydraulic gradient, 0/00	=	4
Dry weather flow, l/pers./day	=	525
I_{ra}, index for shape of rainfall	=	1·4
I_o, indes for shape of over land storage function	=	50
I_r, index for shape of runoff coefficient function	=	1·24

Definitions
Index for shape of rainfall: I_{ra}

$$I_{ra} = \left(\frac{A}{B}\right)2$$

238

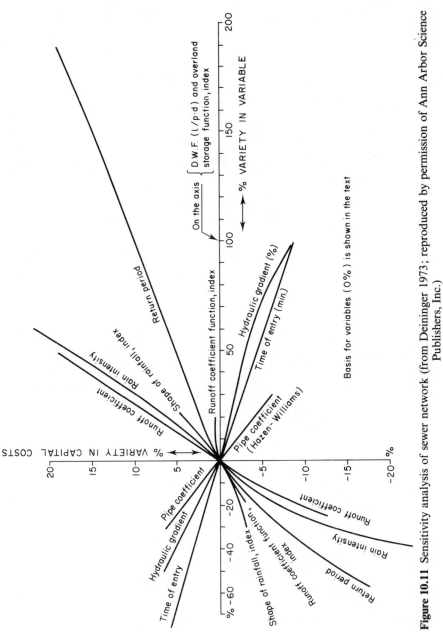

Figure 10.11 Sensitivity analysis of sewer network (from Deininger 1973; reproduced by permission of Ann Arbor Science Publishers, Inc.)

A = Volume precipitated within the most intensive 50% of rain duration
B = Total volume of rainfall

$$1 < I_{ra} < 2 \qquad \text{(usually)}$$

Index for shape of runoff coefficient function: I_r

$$I_r - \left(\frac{C}{D}\right)2$$

C = Area under the most intensive 50% of runoff coefficient curve
D = Total area under the runoff coefficient curve

$$1{\cdot}0 < I_r < 1{\cdot}5 \qquad \text{(usually)}$$

Index for shape of overland storage function: I_o
I_o = % of tributary area which contributes with water to storm inlet when 50% of time of entry has passed.

$$0 < I_o < 100.$$

(2) Sewage treatment plant model

Under Section 10.4 of this chapter an example is already shown. In addition an example showing the computation for a treatment plant in a combined sewerage area will be presented. The main data from the area are:

Number of person equivalents	= 100 000
Area	= 2000 ha
Runoff coefficient	= 0·35
Dry weather flow	= 500 1/pers/day
BOD per person equivalents	= 60 g/pers/day
Precipitation	= 0·832 m per year
Total duration of precipitation	= 792 hours per year.

From the rainfall statistics we have chosen 5 basic rainfalls to represent the yearly rainfall activity. The number of each rainfall per year is shown in Figure 10.12. The hydrographs in Figure 10.12 are the inflow hydrographs to the treatment plant caused by the 5 basic rainfalls.

Figure 10.13 shows three alternatives for BOD concentration in storm runoff. (Usually 60% from pipe deposits and 40% from surfaces.) Figure 10.14 shows the sewage treatment plant configuration. In Figure 10.15 some results are shown of the analyses for the whole year's rain activity. Curve A in Figure 10.13 is used. The influence in total costs and yearly outlet of BOD by varying the four unit operations is computed for three outgoing configurations. The conclusion is that the storage tank must be increased to obtain a substantial decrease in transport of BOD to the recipient.

The optimization routine is used to compute the results shown in Figure 10.16. Approximately 160 different plant configurations had to be computed to obtain

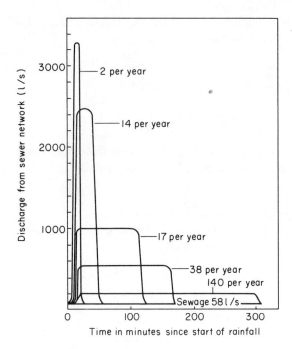

Figure 10.12 Hydrographs caused by 5 base rainfalls

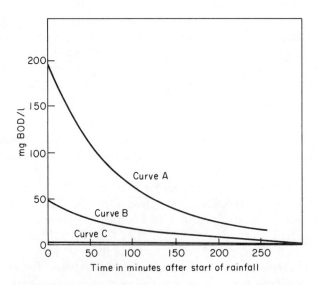

Figure 10.13 BOD concentration in urban runoff due to surface debris and deposits in combined sewer lined (from Deininger, 1973; reproduced by permission of Ann Arbor Science Publishers, Inc.)

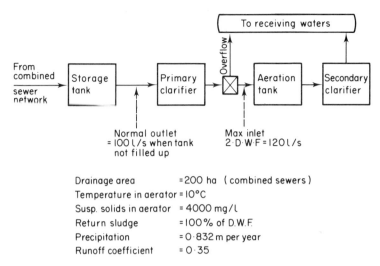

Figure 10.14 Sewage treatment plant configuration

the optimal 'path' for the three different assumptions of storm-runoff contamination. For each configuration the model had to go through 75 different flow-patterns to simulate the whole year's rain activity. It is obvious that the degree of pollution in the storm-water runoff affects seriously the cost of the optimal plant and the configurations of unit operations. The computations reveal that

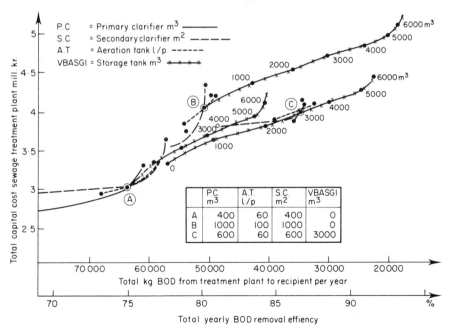

Figure 10.15 Example—treatment plant serving a combined area

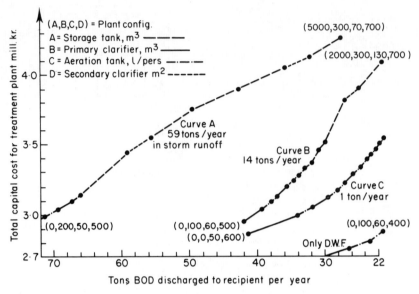

Figure 10.16 Optimal plant versus quality and quantity of influent

the money should rather be invested in storage tanks than in aerators when the storm water is heavily polluted, and vice versa when storm water is slightly polluted.

In Figure 10.17 is shown an example of an optimization with respect to (a) BOD and (b) phosphorus. The drainage area, previously mentioned, is here supposed sewered with a separate system. The figure reveals that the optimization is seriously dependent on which component is to be removed.

The numerous unit operations necessary and the ever changing raw water quality and quantity will make a wastewater treatment plant a complex problem

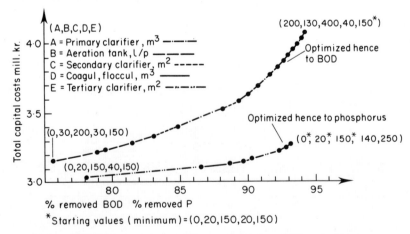

Figure 10.17 Sewage treatment plant optimization BOD and phosphorus

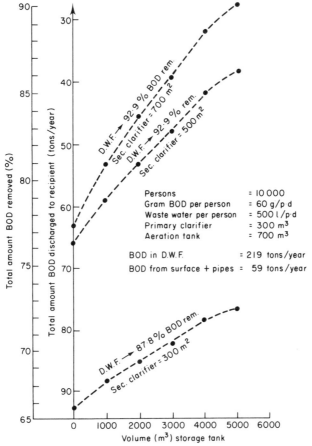

Figure 10.18 Performance of storage tank BOD in storm runoff—Curve A

to analyse. This is demonstrated in an example displayed in Figures 10.18 and 10.19.

In Figure 10.18 the storm water is heavily contaminated (curve A in Figure 10.13). The amount of BOD discharged to recipient per year decreases when the volume of the storage tank increases. This is in force for all reasonable sizes of the secondary clarifier. In Figure 10.19 there is a more moderate contamination in the storm water runoff (curve B in Figure 10.13). The important thing in Figure 10.19 is that the yearly discharge of BOD to the recipient increases with increasing size of storage tank when the secondary clarifier is 300 m². This seems very incorrect at first sight. The explanation is as follows. The storage tank will store rainwater in long periods. This rainwater will discharge into the plant at a constant rate of 2 DWF in long periods of the year. The secondary clarifier is dimensioned so scarcely that this overload causes great amounts of suspended solids in the clarifier's effluent in long periods of the year. The reason why this

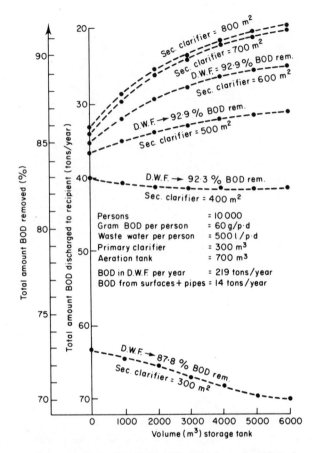

Figure 10.19 Performance of storage tank BOD in storm runoff—Curve B

effect did not show in Figure 10.18, is that the importance of a storage tank is greater when the degree of pollution is higher in the storm-water runoff. As we shall see in Table 10.2, the degree of pollution in storm runoff is very important concerning the amount of pollution discharged in overflows.

In Table 10.1 is shown discharge of BOD to recipient per year from a storm overflow combined with different volumes of a storage tank. See Figure 10.20.

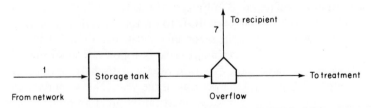

Figure 10.20 Storm overflow combined with storage tank

Table 10.1 Kilograms BOD per year passed a section. Pollution in stormwater runoff:
Curve B in Fig. 10.13

Section	Setting of storm overflow					
	2 DWF = 116 l/s Volume of storage tank m³			5 DWF = 290 l/s Volume of storage tank m³		
	0	2 000	4 000	0	2 000	4 000
Stream 1	233 750	233 750	233 750	233 750	233 750	233 750
Stream 7	23 490	10 270	3 570	8 910	5 200	1 460

Table 10.2 Kilograms BOD per year discharged in storm overflow

Setting of storm overflow	Without pollution in storm-water runoff Curve C	Moderate pollution in storm-water runoff Curve B	Heavy pollution in storm-water runoff Curve A
2 DWF = 116 l/s	8 760	23 490	69 510
5 DWF = 290 l/s	2 630	8 910	31 290

From Table 10.1 it is obvious that the use of storage tanks in combined sewerage networks is very important to control water pollution.

In Table 10.2 the overflow without storage tank is computed for three different degrees of pollution in the storm water runoff.

From Table 10.2 the conclusion must be that a knowledge of the amount of pollution in storm water runoff is very important in order to predict what happens in a sewerage system.

10.7 SUMMARY

The Norwegian Institute for Water Research (NIVA) has developed four computer programs for a total analysis and simulation of sewerage systems. The main objectives of these programs, written in FORTRAN, have been:

(a) to characterize the flow of waste waters through the various components of the sewerage system,
(b) to minimize the 'leakage' of waste water from the system, and
(c) to determine an economically optimal system.

Four specific models have been developed:

(1) A waste- and storm-water runoff model for sewer networks,
(2) A model of the wastewater treatment plant,
(3) A sludge treatment plant model, and
(4) An overall model for the economics of the system.

246

The models may be linked together, but they are mostly operated independently.

The time intervals for the simulation for a given input (rainfall) can be chosen by the operator for the models (usually 10 minutes). The one year rainfall activity is usually represented by 5 to 10 representative rainfalls. Because the second model is specially designed for interactive use, a sensitivity analysis can easily be made.

In the second model it is a subroutine which may automatically search for an optimal plant configuration with respect to BOD or phosphorus removal. The method is a combination of a trial-and-error method, and a gradient method. Several demonstrations and examples are used to show the applicability of models of sewerage systems.

10.8 REFERENCES

Deininger, R. A. (Ed.) (1973). *Models for Environmental Pollution Control*, Ann Arbor Science Publishers, Ann Arbor.

Eckenfelder, W. W., and O'Connor, D. I. (1961). *Biological Waste Treatment*. Pergamon Press, New York.

Eckenfelder, W. W. (1969). *Proc. 4th Int. Conf. Wat. Pollut. Res., Prague*. Pergamon Press, p. 592.

Eye, I. D. (1969). 'Extended Aeration Plant', *Jour. Water Pollution Control Federation*. **41**, 1313.

Hopwood, A. P. and Downing, A. L. (1965). 'Factors Affecting the Rate of Production and Properties of Activated Sludge in Plants Treating Domestic Sewage', *Journal and Proceedings*. The Institute of Sewage Purification, London. **64**, 435.

Munz, W. (1966). 'Die Wirkung verschiedener Gewässerschutzmassnahmen auf den Vorfluter', *Schweizerische Zeitschrift für Hydrologie*. **28** (2), 184.

Watkins, L. H. (1962). 'The Design of Urban Sewers System', *Road Research Technical Papers*. *No. 55*. Dept. of Scientific and Industrial Research, London.

11

Sedimentation

M. J. Hamlin and T. H. Y. Tebbutt

11.1 INTRODUCTION

Since sedimentation is a well-known physical process which occurs naturally in waters, as evidenced by the presence of alluvial plains, it is not surprising that it was readily adopted for the removal of suspended matter from wastewater. The date of the earliest sedimentation tank is unknown, but around 2000 BC the Minoans constructed small settling basins in their drainage systems so that it is clear that the process has a long history in wastewater treatment.

In spite of this long history, the prediction of the performance of sedimentation tanks, particularly when treating heterogeneous suspensions, is not an established practice. In many cases it would appear that variations in the nature of the suspension are likely to have greater effects on the sedimentation process than differences in tank design or operation.

11.2 SEDIMENTATION IN THEORY

The basic concept of the removal of suspended particles from a fluid by gravitational forces is one which is deceptively simple and which lends itself to a mathematical approach. Thus the formulation of Stokes's law,

$$v_s = \frac{g\,d^2(S_s - 1)}{18\,v}$$

where v_s = settling velocity
g = acceleration due to gravity
d = particle diameter
S_s = specific gravity of particle
v = kinematic viscosity of water

permits the calculation of theoretical settling velocities for spherical discrete particles under quiescent laminar flow conditions. The effect of particle shape on settling velocity is not likely to be significant at the velocities normally encountered in wastewater treatment. The limitations as to discrete particles and quiescent conditions are much more restrictive.

Discrete particles are those which have a fixed shape and do not tend to form agglomerates with other particles, hence they have a constant settling velocity.

Flocculent particles do not have fixed shapes and are likely to coalesce under the influence of flocculation forming larger particles so that settling velocities are not constant with time. The way in which agglomeration proceeds is dependent upon the nature of the particles, the degree of flocculation applied and the length of the settling path so that mathematical formulation would become somewhat complex. The Stokes's law velocity relates to a single particle in isolation, whereas the actual settling velocity of a suspension can be reduced at high solids contents due to hindered settling effects.

11.3 NATURE OF SUSPENSIONS

There are three basic types of suspension commonly encountered in wastewater treatment:

    (i) Large dense discrete inorganic particles, e.g. grit, sand
  (ii) Heterogeneous suspension of inorganic and organic particles with a wide size range and varying degrees of flocculent properties, e.g. crude sewage.
 (iii) Relatively homogeneous suspensions of flocculent inorganic or organic solids, e.g. activated-sludge mixed liquor, precipitation sludges.

Type (i) suspensions lend themselves to Stokes's law velocity determinations provided that the settling velocities do not exceed the limit for laminar flow conditions. With this type of suspension it is possible to determine the particle size distribution by sieve analysis and hence predict the settling characteristics of the suspension. Such techniques are not feasible for the other types of suspensions listed above and it is necessary to adopt other parameters for characterization of the suspensions. Thus the properties of most wastewater suspensions are assessed by determination of suspended solids and/or settleable solids contents.

The suspended solids determination using a gravimetric filtration technique gives only a gross measurement of the solids concentration with no indication of the size or density distribution of the particles in the suspension. Since the standard glass fibre filter paper used in the determination (Whatman GF/C) has a pore size of 1·2 μm, the suspended solids measurement includes particles much smaller than would be removed by sedimentation in a treatment plant. For example, a 10 μm diameter particle of organic matter with S. G. of 1·4 would have a Stokes's law settling velocity of about 0·14 m/d, a value well below the normal overflow velocity in sedimentation tanks for wastewater treatment which is usually 20–40 m/d. The specific gravity of secondary sludge particles is often very low, values of 1·02–1·04 being commonly obtained so that settling velocities can be very small. The settleable solids determination was originally based on a volumetric indication of the settleable matter in a suspension using an Imhoff cone—in effect a laboratory-scale settling tank. Because of poor reproducibility in the test when large particles are present a modified procedure is now detailed (Dept. of Environment, 1972) in which settleable solids by weight are determined by deducting the suspended solids in the supernatant after a period of settlement from the initial suspended solids content of the sample.

It will be clear that neither of these parameters is likely to give sufficient information about a suspension to enable its settlement behaviour in a particular situation to be predicted with any confidence. Data more specific to the settling behaviour of a suspension can be obtained from settling column tests, but if these are to be meaningful it is important that the experimental procedure be such as to simulate the full-scale conditions as closely as possible. For flocculent suspensions the depth of the column should be the same as the prototype tank and because of the low settling velocities normally encountered, convection currents in the column must be avoided by operation in a constant temperature environment. Results from settling column analyses give an indication of the settling characteristics of the suspension under test conditions, but these may not necessarily be appropriate to conditions in a full scale sedimentation tank.

11.4 SEDIMENTATION IN PRACTICE

Early sedimentation tanks both for primary settlement of crude sewage and for humus settlement following bacteria beds were operated on a fill and draw basis so that quiescent conditions were achieved after the tank had been filled. Retention times of 24h or more were common, and anaerobic decomposition of the sludge often resulted in poor quality effluents. Because of these troubles and the high cost of such large tanks almost all sedimentation tanks are now designed for continuous flow operation, the main exceptions to this practice being settling basins for some industrial wastes with intermittent discharge patterns. For continuous flow settlement tanks, the normal UK criteria are 6–12h retention at dry weather flow for primary sedimentation and 2–6h at dry weather flow for secondary settlement, the shorter periods being required for activated-sludge plants. Although it can be shown that, for discrete particles at least, the surface loading of a tank is of prime importance in controlling its performance, most designers consider retention time as the main parameter in sedimentation tanks. Because of the emphasis on retention time as being the important factor in sedimentation there has been a tendency in the past to think of the process as one governed simply by hydraulic considerations. The nature of the suspension and its effect on the performance of the tank are often not allowed for in design, partly no doubt because of the difficulty of quantifying these properties. Nevertheless it is known that many factors, e.g. heavy metal content, can affect the settleability of suspensions.

The fact that the inflow to a sedimentation tank contains suspended solids which are, in part at least, removed in the tank means that the inflow must be denser than the tank contents. Density currents are thus likely to be formed and these currents may themselves be modified by the density differences caused by temperature differentials. Attempts to introduce baffles into tanks to prevent the formation of density currents have not been noticeably successful and it could in any case be argued that the velocity gradients due to density currents are of some value in promoting flocculation.

In wastewater treatment, sedimentation tanks are usually of the horizontal flow type, either rectangular or circular in plan, although on small works hopper bottomed mixed flow tanks are often employed. There is no definite relationship known between tank configuration and settling performance, and decisions are usually taken on the basis of site requirements or sludge removal considerations. It should be appreciated that sedimentation has two functions in wastewater treatment viz:

(i) Solids–liquid separation
(ii) Solids thickening.

The latter function is rarely considered at the design stage but can in fact have a marked effect on the overall operation of a treatment plant. The solids thickening aspect of sedimentation tank design and operation has been discussed by Dick (1972) who suggested that in certain cases, e.g. final settling tanks for activated sludge, the sludge thickening criteria may be more restrictive than settlement criteria.

11.5 THE BIRMINGHAM PILOT PLANT

Studies on various aspects of primary settlement of sewage at Birmingham (Wills and Davis, 1964; March and Hamlin, 1966; Tebbutt, 1969; Hamlin and Wahab, 1970; Hamlin and Tebbutt, 1970) have underlined the effect that variables in the sewage quality have on settling behaviour. The result of these studies has been to confirm the belief that investigations into primary settlement are best carried out on a plant scale. Unfortunately, hydraulic restrictions and limitations on operating procedures often prevent examination of wide variations in hydraulic loading on a full-scale plant and thus pilot-scale studies may be more suitable. When dealing with flocculent suspensions such as crude sewage it is important that the depth of the experimental tanks should be of the same order as would be used on the full scale.

A pilot-scale sewage treatment plant has been constructed on the campus at Birmingham to permit experimental study of the interaction between the various processes and research on optimization of treatment processes is currently supported by an SRC grant. The plant the flow sheet of which is shown in Figure 11.1 is supplied with comminuted domestic sewage and includes primary sedimentation, activated-sludge treatment, and anaerobic sludge digestion. The design of the plant is such that the flow to the primary sedimentation stage can be varied over a wide range ($25-150$ m^3/m^2 d in the current investigations) and for each particular sedimentation tank loading activated-sludge treatment can be carried out with aeration times of approximately 2, 6, 10, and 21 h. The appropriate mixtures of primary and secondary sludges are fed to anaerobic digesters for each combination of loading conditions. It is hoped that the results of this study will provide more information about the overall performance of the treatment process which will aid in optimized design procedures.

Earlier work on the plant (Tebbutt, 1969) involved studies of the performance

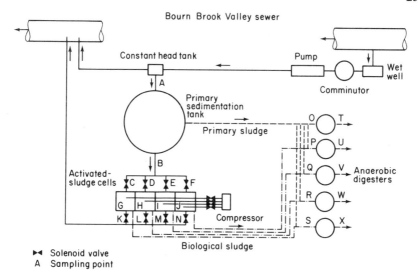

Figure 11.1 Schematic arrangement of the Birmingham pilot plant

of a conventional centre feed tank and a peripheral feed unit, but since no significant difference was apparent the current work has been restricted to the conventional tank.

11.6 MATHEMATICAL MODELLING OF THE SEDIMENTATION PROCESS

The description of any process in mathematical terms seems now to be accepted as mathematical modelling. By this definition the early work of Hazen (1904) and others (Camp, 1946; Rich, 1961) qualify for inclusion in this chapter. Before discussing mathematical models in particular, however, it is worth discussing their purpose. In the case of sedimentation tanks they may be devised to predict the removal of settleable suspended solids from the tank influent or to predict the load going forward to the following process stage.

The former type includes those models which seek to estimate the total volume of solids removed for sludge treatment, and must include allowance for sludge thickening within the tank, as well as those which in conjunction with settling column analysis are designed to predict the removal efficiency of the sedimentation basin. The latter type is concerned with the BOD load going forward to the biological stage and must make some estimate of the correlation between the suspended solids removal and the reduction of BOD which takes place within the tank. The use of hydraulic models is also of importance since these have often been employed to study the hydraulic performance of tanks so that the appropriate mathematical model is chosen for purposes of prediction.

11.7 DETERMINISTIC PREDICTION MODELS

The ideal sedimentation tank has been conceived on the basis of 'piston flow' but in reality this is never achieved. In any event piston flow is only optimum if the particles to be settled are discrete and non-flocculating. Hazen, in extending work from quiescent settling to continuous flow tanks, made the assumption of both piston flow and non-flocculation. He showed that if a particle with a settling velocity v commencing at the surface at the inlet of the tank is just settled at the outlet end of the tank then the relationship $v = Q/A$ holds and that if the particles are uniformly distributed throughout the depth of the tank at the inlet they will all be settled within the tank. In the more general case of a suspension having a settling velocity u (where $u < v$) the proportion of solids which will settle out in a tank for which the relationship $v = Q/A$ is valid, is u/v. Under these circumstances the effluent concentration C_e is related to the influent concentration C_0 by the equation

$$C_e = C_0 \left(1 - \frac{u}{v} \right)$$

This equation, if applied to a series of n basins all having the same volume as the original basins, leads to the equation

$$C_e = C_0 \left(1 - n\frac{u}{v} \right)^n$$

Hazen also showed that where complete mixing takes place the relationship is

$$C_e = \frac{C_0}{1 + \dfrac{u}{v}}$$

and if a series of n similar basins are arranged in series the relationship becomes

$$C_e = \frac{C_0}{\left(1 + n\dfrac{u}{v} \right)^n}$$

It should be noted that in the limiting case where n tends to ∞ both equations become

$$C_e = C_0 e^{-u/v}$$

A consideration of Hazen's theory leads to the conclusion that performance is dependent solely on surface area and is independent of detention period. In fact this is far from the truth and more recent work (Clements, (1966); Burgess and Green, 1958) has shown that the presence of density currents and circulation within the tank require a much more complicated approach. March (1969) in a theoretical consideration which allows for the presence of both density currents and recirculation shows that

$$C_e = \frac{C_0(1 - r)}{1 - \dfrac{\beta \left[1 - \left(\dfrac{\phi}{\gamma} \right)^n \right]}{\gamma - \phi} - r\left(\dfrac{\phi}{\gamma} \right)^n}$$

where

$$\beta = \frac{\dfrac{u}{v}}{\dfrac{u}{v} + \dfrac{1}{1-r}}$$

$$\gamma = 1 + \frac{u}{v} \cdot \frac{(1-r)}{rn}$$

$$\phi = \frac{\dfrac{1}{1-r}}{\dfrac{u}{nv} + \dfrac{1}{1-r}}$$

where r is $Q'/Q + Q'$ in which Q is the flow rate and Q' the flow in the reverse current.

As n tends to ∞ the expression becomes

$$C_e = \frac{C_0(1-r)\,e^{-u/v(1-r)}}{1 - r\,e^{-u/v((1-r^2)/r)}}$$

This relationship is indicated graphically in Figure 11.2.

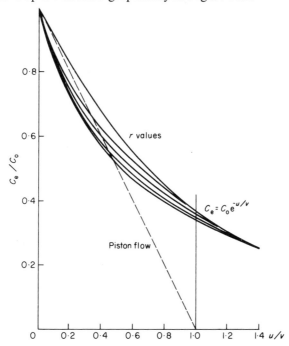

Figure 11.2 Tank performance based on the concept of hypothetical mixing chambers with various degrees of recirculation

254

11.8 HYDRAULIC MODELLING

The need for a clearer understanding of the behaviour of density currents and their effect on the removal efficiency of a tank led to a model study by Wahab (1969). In a physical simulation model density differences were obtained by

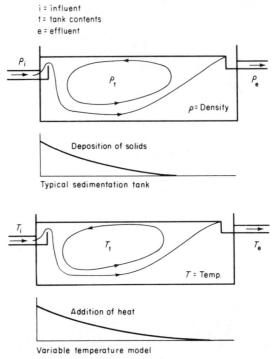

Figure 11.3 Temperature analogue of sedimentation

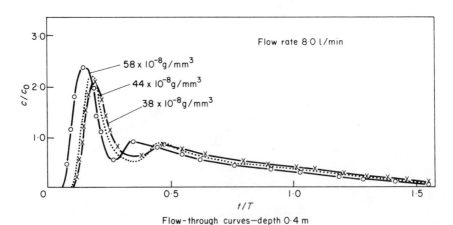

Figure 11.4 Typical results from temperature analogue model for various density differences

Table 11.1 Parameters for Varying Density Differences in Basin Depth of 0·2 m and Flow Rate of 8 l/min

Test no.	$\Delta\rho \times 10^{-s}$ (g mm^{-3})	t_i/T (%)	t_p/T (%)	t_A/T (%)	t_{50}/T (%)	t_{10}/t_{90}	N	$\dfrac{C_o C_e}{C_i}$ (%)
287	33	25·0	44·0	61·0	84·0	5·9	3·6	57
283	38	25·3	40·3	60·0	85·0	5·3	3·0	58
284	46	22·2	33·3	54·0	86·0	7·2	2·6	59
285	52	21·5	35·4	57·0	94·0	10·4	2·6	62
278	75	19·4	31·7	62·0	86·0	7·7	1·9	59
280	89	18·3	34·4	62·0	85·0	7·3	2·2	60
281	92	19·4	34·4	65·0	85·0	7·5	2·2	58
279	98	16·7	29·7	59·0	84·0	6·9	2·0	56
282	100	16·4	27·5	57·0	83·0	8·7	1·9	56
272	108	15·3	25·0	50·0	67·0	5·5	2·0	48

controlling the temperature differences between the tank influent and the tank effluent. This is shown diagrammatically in Figure 11.3. and typical results are given in Figure 11.4. If a settling column analysis is assumed then the efficiency of sedimentation can be inferred from the flow-through curves and these are shown in Table 11.1.

The work of March, Wahab and others indicates that the sedimentation process is a complex interaction of a large number of variables. There is an incomplete understanding of the process and it is unlikely that at the present time a definitive mathematical model can be specified. It is more likely that predictions of tank behaviour will be judged using performance relationships or the stochastic model referred to in a later section.

11.9 PERFORMANCE CURVES

Since most designers believe that retention time is the important factor in sedimentation it might be assumed that it would be easy to develop a relationship between sedimentation efficiency and retention time. Experimental work by Holmes and Gyatt (1929) suggested that sedimentation efficiency for a given suspension increased according to the fourth or fifth root of the retention time so that large increases in capacity would bring only relatively small increases in removal.

With the existence of a considerable amount of data relating to the performance of sedimentation tanks, it is not surprising that attempts have been made to produce performance curves for design purposes. Figure 11.5 shows performance curves for primary settlement of sewage derived from three sources. Escritt's curve (1956) was based on data from existing plants and relates the effluent suspended solids to retention time and feed suspended solids with the aid of constants depending upon the design of tank, the nature of the suspension and other, unspecified, factors. The CIRIA curve (1973) was based largely

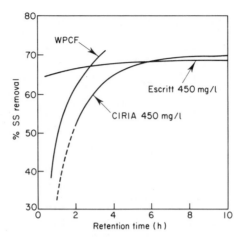

Figure 11.5 Performance curves for primary
settlement of sewage. Full lines show range
covered by data except for Escritt's curve
for which no limits were given

on the analysis of operating data from a number of large works and assumes
that retention time and the feed suspended solids concentration affect removal
efficiency. The WPCF curve (1959) is based on US data which implies that
surface overflow rate is the governing factor in tank performance and that the
initial suspended solids concentration does not affect the removal performance.
The WPCF curve shown in Figure 11.5 has been adapted from the original
curve relating removal to overflow rate by assuming a depth of 2·5 m. The
original curve was a median line for widely scattered data from full-scale plants.

There are clearly considerable differences between the performances predicted
by the three curves, particularly at the shorter retention times, but this is not
unexpected in view of the many factors which influence sedimentation efficiency.
All three curves have been derived from consideration of the performance of
tanks at different works designed to different criteria (retention time or overflow
rate) and from data relating to the performance of a particular tank or tanks
under different loading conditions. In both cases there is likely to be more than
one variable influencing performance. Comparison of performance at different
works must be affected by the characteristics of the individual sewages and
comparison of performances at one works under different flow conditions may
well be influenced by changes in the settling characteristics of the sewage with
variations in flow. When performance data are based on mean values over 24
hours or perhaps even a week the effect of plant size may be important since
small treatment plants are likely to be subject to much greater variations in
flow over a short period of time than large plants.

When considering tank performance the effect of return sludge must be
recognized and its presence as a solids load should be included in any removal
efficiency calculations. It will be noted that all three curves discussed express
efficiency in terms of suspended solids removal and as already pointed out this

parameter cannot be directly related to the settleability of the suspension.

Although the feed to secondary sedimentation tanks is in general much more homogeneous than crude sewage, little attempt appears to have been made to derive performance curves for this type of settlement in wastewater treatment. The recent CIRIA study was apparently unable to construct any performance curves for secondary settlement and could only recommend that conventional design criteria be used for this part of the process.

Ideally, performance curves for sedimentation tanks should be based on the following data:

(i) Settleable solids removal for fixed settleable solids feed concentration over a range of hydraulic loadings.
(ii) Settleable solids removal at a fixed hydraulic loading over a range of settleable solids in the feed.

Unfortunately, information is not usually available in this form from full-scale plants and thus published performance curves should not be expected to provide precise predictions of removal efficiency.

11.10 STOCHASTIC MODELS

The mathematical models typified by the work of Hazen and March indicate that the suspended solids removal which may be expected for a given situation, where the tank geometry, internal flow patterns and suspension characteristics are known, is fixed. Performance curves indicate that for a tank of a given volume the removal of suspended solids can be ascertained.

It is well known that neither of these approaches describes the actual performance of a sedimentation tank. In practice the inflow to a tank varies through a very wide range of flow quality and the effluent from the tank is also subject to considerable variation.

There are situations when the average performance of the tank is all that is required. If however the dynamics of the behaviour of the complete treatment plant are being considered the variation of load in the various units is of some significance. Where the overall performance of the treatment works is to be modelled it is essential to consider some alternative method of describing the sedimentation process if its dynamic nature is to be successfully described in mathematical terms.

The use of stochastic models is well known in other disciplines and in other branches of Civil Engineering, notably in hydrology, but it has only rarely been used to describe the unit processes of water or wastewater treatment. Essentially a stochastic model seeks to relate two variables in terms of their probability density function rather than by a simple least squares regression.

This may be best explained by a rather simple example before considering specific application to the sedimentation process. If a linear relationship exists between two sets of variables $x_1, x_2 \ldots x_n$ and $y_1, y_2 \ldots y_n$ the relationship can be defined by the equation

$$y = ax + c$$

where the coefficients a and c are evaluated by least squares regression. A measure of the accuracy of the equation is obtained from the correlation coefficient ρ which is defined as

$$\rho = \frac{\sum\limits_{i=1}^{n} x_n y_n - nx\bar{y}}{\left(\sum\limits_{i=1}^{n} x_i^2 - nx^2\right)^{1/2} \left(\sum\limits_{i=1}^{n} y_i^2 - ny^2\right)^{1/2}}$$

If ρ equals one all the points lie on the line and if ρ is equal to zero then there is no correlation between the x and y series and the process is totally random. Figure 11.6 indicates in graphical form a relationship in which $0 < \rho < 1$. From the graph it can be seen that for a value of x' the estimated value of y obtained from the equation $y = ax + c$ is y'. Nevertheless there exists in the parent series two values of y, y'', and y''' which correspond to observed values of x'. A measure of this variance which is not explained by the equation can be estimated from the magnitude of $(1 - \rho^2)$. Thus if ρ is 0·7, $(1 - \rho^2)$ is 0·51 and it can be said that 49% of the variance is explained by the equation and 51% is unexplained and arises from circumstances which have not been modelled. At this level of ρ value it is clear that there is very little correlation between x and y and some improved relationship must be sought. It may well be possible to achieve this through a transform of x to take account of an increasing number

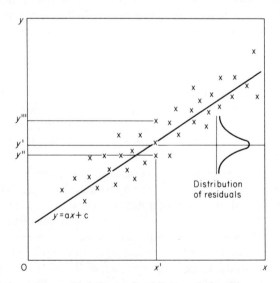

Figure 11.6 Generalized linear relationship

of variables. This may raise the value of ρ until for example the equation explains up to 80% of the variance leaving only 20% unexplained.

At this point the equation defining the process may be written in the form

$$y = af(x) + c + \eta\sigma_y(1 - \rho^2)^{1/2}$$

in which $f(x)$ is the transform of x.

The first two terms give an estimated value of y using the transformed value of x and the third term represents the stochastic component in the product of η, σ_y, which is the standard deviation of y, and $(1 - \rho^2)^{1/2}$ where ρ is as previously defined. η is defined as a set of standardized random numbers having a mean of zero, a variance of one, and a distribution similar to the distribution of the deviation of the observed points from the line defined by $y = af(x) + c$.

The complete equation will for a series of x values give a series of y values which have the same statistical properties as the observed series of y values. It will be possible to obtain, for a given value of x', a value of y which would include, with the appropriate probability, values such as y'' and y''' as well as y'. There is a considerable literature on stochastic processes and 'Measurement and Analysis of Random Data' (Bendat and Piersol, 1966) is a good starting point.

When applied to the sedimentation process, the first task would be to define the purpose of the model. If this was to predict the concentration of the suspended solids in the effluent, then the x series could be the calculated values of C_e using Hazen's equation and the y values would be the corresponding observed values of suspended solids in the effluent.

Thus

$$y = ax + c + \eta\sigma_y(1 - \rho^2)^{1/2}$$

where

$$x = C_0\left(1 - n\frac{u}{v}\right)^n$$

If this gave unsatisfactory values of ρ, x might be transformed using the March equation

where

$$x = \frac{C_0(1 - r)}{1 - \dfrac{\beta\left(1 - \left(\dfrac{\phi}{\gamma}\right)^n\right)}{\gamma - \phi} - r\left(\dfrac{\phi}{\gamma}\right)^n}$$

Alternatively an entirely new regression equation could be devised which might take the form of performance curves quoted earlier

where $\qquad x = f(\text{Detention period})$

As an example of this approach the mathematical model proposed for final clarifiers by the United States Environmental Protection Agency (1972) gives the suspended solids, x, in the final effluent as either

$$x = 18 \cdot 2 + 0 \cdot 0136m - 0 \cdot 003n$$

where m is the actual tank overflow range gal/ft^2 d and n is the mixed liquor suspended solids mg/l or alternatively for a different set of observations

$$\frac{382 . m^{0 \cdot 12} . \left(\dfrac{P}{n} \right)^{0 \cdot 27}}{n^{0 \cdot 35} . q^{1 \cdot 03}}$$

where P is the BOD loading in lb/d and q is the detention period in hours

The specification of tank performance by an equation which includes a stochastic term acknowledges mathematically that only part of the variance can be explained by a deterministic relationship and that part has to be explained by a stochastic variable.

The following example illustrates how this technique may be used to predict the suspended load passing from the sedimentation tank to the biological units of the pilot plant referred to earlier. Table 11.2 gives part of a typical set of daily results obtained from the pilot plant where the size of sedimentation tank is fixed. The flow rate was constant at 25 $\text{m}^3/\text{m}^2\text{d}$ and this corresponds to a nominal retention period of two hours. A plot of the data is given in Figure 11.7 and the equation relating C_e and C_0 for the particular set of flow conditions specified is

$$C_e = .15C_0 + 150 + \eta 67(1 - 0 \cdot 5^2)^{1/2}$$
$$= .15C_0 + 150 + 58\eta$$

If a series of input values of C_0 are known or can be estimated then the suspended solids in the effluent can be predicted and the load to the following units can be determined. The use of this equation for prediction purposes is illustrated by considering a series of input values of 500, 400, 550, 500, and 600. A set of random numbers having the appropriate distribution need to be generated and suppose the first five values of the random set are $+ 0 \cdot 04, - 0 \cdot 32, + 1 \cdot 08$,

Table 11.2 Suspended Solids Concentration in Influent and Effluent of Pilot Scale Sedimentation Tank

Suspended solids in mg/l					
C_0	C_e	C_0	C_e	C_0	C_e
554	486	388	228	355	139
406	168	1 046	274	369	144
380	248	250	111	456	233
584	206	342	189	577	201
509	215	424	149	1 133	343

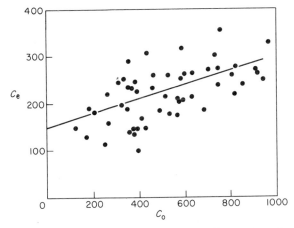

Figure 11.7 Plot of C_e against C_0 for pilot scale sedimentation tank

$+ 0.51$, and $- 0.73$. Substitution of these values in the equation gives estimates of C_e of 227, 159, 295, 255, and 198. Without the addition of a random component the values would be 225, 210, 232, 225, and 240, respectively.

Experimental work on the pilot plant has continued since the original presentation of this paper and has been reported by Tebbutt and Christoulas (1975). Analysis of the data obtained from the primary sedimentation unit for overflow rates of 25, 50, 100, and 150 $m^3/m^2 d$ has shown that the influence of overflow rate and initial suspended solids concentration on performance can be represented by a relationship of the form

$$\frac{C_0 - C_e}{C_0} E_0 e^{-(b/C_0 + cq^1)}$$

where E_0 = SS removal efficiency under quiescent conditions
q^1 = surface overflow rate
b and c are constants

For the particular sewage used in the studies regression analysis on the data in the range C_0 200–600 mg/l $E_0 = 0.955$, $b = 263$ and $c = 0.0021$ when C_0 and C_e are in mg/l and q^1 in $m^3/m^2 d$. The application of this form of performance relationship provides a more satisfactory prediction of solids removal than was previously possible, and the results demonstrate the relatively greater importance of initial suspended solids concentration on the removal efficiency.

11.11 CONCLUSION

There are serious problems in trying to describe the sedimentation process mathematically. Whilst many researchers have sought more and more complex models none of the models so far proposed is wholly successful in describing the performance of settlement tanks. Many are based on unrealistic assumptions of flow conditions within the tank or of the settling characteristics of the suspension. The use of performance curves based on a single parameter such as detention period can only reflect the average performance of average tanks and can never give more than a guide to possible performance figures. The concept of stochastic models is likely to be of more use in the overall modelling process not because it gives a better prediction of the average performance figures, which of course it does not, but because it recognizes mathematically that the actual performance of a tank varies about the average performance. Even where the predicted behaviour of a tank is written in complex terms obtained from regression analysis there is always some random variation about the predicted value. This random variation is accounted for by the interaction of complex variables which exist within the system. At present these variables are either not measured or alternatively where they are measured their interaction is imperfectly understood and therefore not adequately modelled. The importance of an accurate model arises from the need to design subsequent processes so that they can successfully treat the effluent from the unit being modelled. It would therefore be worthwhile to obtain as much information as possible about the performance of full-scale tanks operating under known conditions. Analysis of such performance data would be helpful in producing a more realistic mathematical model of the sedimentation process.

11.12 REFERENCES

Bendat, J. S., and Piersol, A. G. (1966). *Measurement and Analysis of Random Data*, Wiley, New York.

Burgess, S. G., and Green, A. F. (1958). 'The use of radioactive isotopes for the investigation of sewage treatment plant', *J. Inst. Civ. Engrs.*, **11**, 297.

Camp, T. R. (1946). 'Sedimentation and the design of settling tanks', *Trans. Amer. Soc. Civ. Engrs.*, **11**, 695.

Clements, M. S. (1966). 'Velocity variations in rectangular sedimentation tanks', *J. Inst. Civ. Engrs.*, **34**, 171.

Construction Industry Research and Information Association (1973). *Cost—Effective Sewage Treatment—The Creation of an Optimizing Model*, Report 46, 2, Ciria, London 13.

Department of the Environment (1972). *Analysis of Raw, Potable and Waste Water*, HMSO, London, 43.

Dick, R. I. (1972). 'Gravity thickening of sewage sludges', *Wat. Pollut. Control*, **71**, 368.

Escritt, L. B. (1956). *Sewerage and Sewage Disposal*, Contractors Record, London, p. 282.

Hamlin, M. J., and Tebbutt, T. H. Y. (1970). Sedimentation studies', *Surveyor*, **135** (4065), 42.

Hamlin, M. J., and Wahab, A. H. A. (1970). 'Settling characteristics of sewage in density currents', *Wat. Res.*, **4**, 609.

Hazen, A. (1904). 'On Sedimentation', *Trans. Amer. Soc. Civ. Engrs.*, **53**, 45.

Holmes, G. D., and Gyatt, W. P. (1929). 'Operation of Syracuse N. Y. Plant—Four Year Summary, *Sew. Wks. J.*, **1**, 318.

March, R. P. (1967). 'The performance of rectangular sedimentation tanks'. Birmingham University, *Ph.D. thesis*.

March, R. P., and Hamlin, M. J. (1966). 'An investigation into the performance of a full-scale sedimentation tank', *J. Proc. Inst. Sew. Purif.*, **2**, 118.

Rich, L. G. (1961). *Unit Operation of Sanitary Engineering*. Wiley, New York, p. 81.

Tebbutt, T. H. Y. (1969). 'The performance of circular sedimentation tanks', *Wat. Pollut. Control*, **68**, 467.

Tebbutt, T. H. Y., and Christoulas, D. G. (1975). 'Performance relationships for primary sedimentation', *Wat. Res.*, **9**, 347.

United States Environmental Protection Agency (1972). *A Mathematical Model of a Final Clarifier*, Washington, 75.

Wahab, A. H. A. (1969). 'An investigation into the hydraulic characteristics of model rectangular sedimentation tanks'. Brimingham University, *M.Sc. thesis*.

Water Pollution Control Federation (1959). *Manual of Practice No. 8 Sewage Treatment Plant Design*, WPCF, Washington, 93.

Wills, R. F., and Davis, C. (1964). 'Flow patterns in a rectangular sewage sedimentation tank'. Advances in Water Pollution Research, *Proc. 1st. Intl. Conf. Water Poll. Res. 2*, Pergamon, Oxford, p. 335.

12

A Mathematical Model for Bacterial Growth and Substrate Utilization in the Activated-Sludge Process

G. L. Jones

12.1 INTRODUCTION

Many models have been developed which attempt to describe in mathematical terms the behaviour of the activated-sludge process. Most of these fall into one of two main groups.

(i) *Pragmatic models* which consider the relationships between plant loading and performance, in respect of removal of carbonaceous matter and production of activated-sludge solids from domestic sewage, usually on the basis of first-order reaction kinetics. The model of Tuck, Chudoba, and Madera (1971) is of this type. The use of such models has been successful because domestic sewage contains a mixture of carbonaceous substrates metabolized by a variety of different bacteria with different growth rates, so that the overall kinetics of the system approximate to those of a first order reaction (Wilson, 1967; Jones in press). However, they are not so useful for describing the removal of a single substrate from the mixture normally applied to a treatment plant. For such purposes a closer examination of the kinetics of bacterial growth and substrate utilization is required.

(ii) *Fundamental models* which consider bacterial growth and substrate utilization, often on the basis of the equations

$$\mu = \mu_{max} \frac{S}{K_s + S} \tag{12.1}$$

$$ds/dt = -\frac{1}{Y}\mu_{max} X . \frac{S}{K_s + S} \tag{12.2}$$

developed by Monod (1942, 1950) for the growth of pure bacterial cultures. In these equations μ is the specific growth rate, μ_{max} the maximum specific growth rate, K_s the saturation constant, being the concentration of the growth limiting substrate S at which μ is $\mu_{max}/2$, Y is the yield coefficient (g cells produced/g substrate consumed) and X is the cell concentration. These equations were used successfully by Downing, Painter, and Knowles (1964) for the prediction of nitrifying conditions in activated sludge and by Garrett (1958) as a theoretical basis for his successful use of constant specific wastage rate to control the operation of a full-scale treatment plant.

While both types of model are useful, each having properties not found in the other, the greatest insight into a biological process should be obtained from a fundamental model. The model to be described has been based on our knowledge of bacterial physiology and applications of the model assume the conditions in the aeration stage to be those of complete mixing.

12.2 LIMITATIONS OF THE EQUATIONS

It can be shown that the Monod (1942, 1950) equations are unsuitable for predicting removal of carbonaceous matter from mixed substrates, concentrations of which are usually measured in terms of BOD and COD. Used for this purpose they give erroneous results (Jones, in press) often with very high values for the saturation constant (Tench and Morton, 1962). Their most successful application has been by Downing and colleagues (1964) to the development and decline of nitrifying activity in the activated-sludge process. It is thought that the success of this simple model is associated with the special nature of the organism ($Nitrosomonas$, a chemo-autotroph) and the substrate (ammonia) since a subsequent attempt to apply the same equations to the development of activity by an activated-sludge unit towards a carbonaceous substrate in the form of detergent (Dobane JNX) was less successful (Downing and Knowles, 1967). From the data available the authors had to postulate enzymic adaptation of the bacteria present in activated sludge leading to a thirteenfold increase in sludge activity over a period of thirty days, which, together with bacterial growth, resulted in a predicted fifteen- to sixteenfold increase in sludge activity towards detergent. Later work (Painter and Durrant, in preparation) provided further data which showed at most a one and a half-fold increase in sludge activity, together with a lower concentration of bacteria capable of growing on the detergent than calculated by Downing and Knowles. Using both sets of data the adaptation model has been modified (Jones, in preparation) to allow for cometabolism (Horvath, 1972) of the detergent by organisms unable to utilize it as a carbon source. The predictions of the two models are very similar (Figure 12.1) with the later model predicting a less than fourfold increase in sludge activity towards the detergent over the period of acclimatization.

This example shows that the Monod equations alone are unable to give a complete picture of the process. Although their use was satisfactory for predicting nitrification, considerable modification, involving an assumption of either enzymic adaptation of the micro-organisms or cometabolism of the detergent, was necessary to allow the removal of detergent by activated sludge to be quantitatively expressed.

Further problems arise from the prediction of too high a proportion of viable bacterial cells in the sludge. Herbert (1961) derived an equation which showed that the concentration of bacteria, X, in activated sludge could be

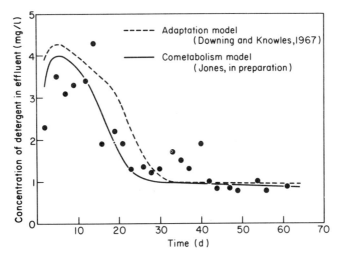

Figure 12.1 Observed and predicted concentrations of detergent (Dobane JNX) in effluent from a laboratory-scale activated-sludge plant

obtained from the expression

$$X = \frac{YD(S_0 - S)}{\mu} \qquad (12.3)$$

where D is the dilution rate of the aeration system (flow/volume), and S_0 and S are the initial and effluent concentrations respectively of the growth-limiting substrate. This equation has been used to calculate the predicted change in the proportion of cells in activated sludge which utilize carbohydrate (Curve A, Figure 12.2). The concentration of activated-sludge solids was calculated using the empirical equation given by Downing and Knowles (1967) assuming sewage with a BOD of 329 mg/l (the average value for sewage from a residential area in Stevenage from August to November, 1971, Pike and Carrington, 1972), retention time of sewage of 8 h ($D = 0.125\ \text{h}^{-1}$), recycle ratio 1, and temperature 15°C. The concentration of carbohydrate in the sewage was assumed to be 20% of that of organic carbon, as had been found by Painter, Viney, and Bywaters (1961), and it was further assumed that for each gram of carbohydrate consumed 0.31 g cells would be formed. This corresponds to the yield of cells quoted for a glucose-utilizing organism isolated from activated sludge by Washington, Rao, Hetling, and Martin (1965). This model predicts that the proportion of bacteria (≡ active mass) in the sludge increases with increasing sludge age, which is contrary to the experimental observations of Weddle and Jenkins (1971).

As pointed out by Wayman (1971), equation 12.1 cannot allow for death of bacteria unless S becomes negative. To overcome this many models have incorporated a term to allow for death of bacteria, for example Wayman (1971) and Lawrence and McCarty (1970) and some workers, for example Middle-

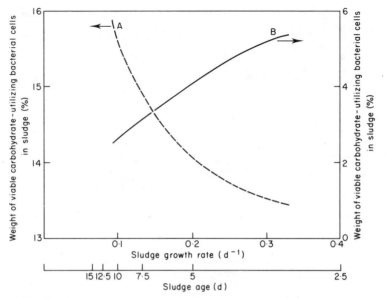

Figure 12.2 Comparison of concentrations of viable carbohydrate–utilizing bacteria in sludge calculated with (A) no maintenance requirement, and (B) maintenance requirement of 0·067 g/gh

brooks and Garland (1968), have attempted to measure this parameter in activated-sludge type systems. The technique used by Middlebrooks and Garland (1968) in fact measured maintenance requirements (Marr, Nilson, and Clark, 1963; Pirt, 1965; Washington and colleagues, 1965); this technique gave values very similar to those obtained by use of the equation for sludge growth given by Downing and Knowles (1967) (Jones, in press). The effect of maintenance energy requirements is to reduce the cell yield according to the equation

$$1/Y = 1/Y_G + m/\mu \qquad (12.4)$$

where Y_G is the 'true growth yield', being the yield that would be obtained in the absence of a maintenance requirement, m (g/g h). This equation was developed by Pirt (1965) in order to explain the observation that the concentration of bacteria in a continuous culture at low dilution rates ($\equiv$ specific growth rate) was lower than expected. Low specific growth rates are essential for successful waste treatment, and this equation may therefore be expected to be particularly relevant to the prediction of the concentration of bacteria in activated sludge, and to have a significant effect on the results. Combining equation 12.4 with equation 12.3 and using the values of 0·31 g/g and 0·067 g/g h for Y_G and m respectively obtained by Washington and colleagues (1965) for an organism growing on glucose, the proportion of sludge present as viable cells growing on glucose was calculated for the same conditions as previously (Curve B, Figure 12.2). The shape of this curve agrees well with that obtained experimentally by Weddle and Jenkins (1971), the proportion of viable cells increasing with decreasing sludge age; the proportion of cells present (2–5%) is also much closer to observation than was obtained with the simpler model.

12.3 BACTERIAL GROWTH

The classical bacterial growth curve (Figure 12.3) shows four phases. After an initial lag phase the rate of growth gradually increases to a maximum which is maintained as the logarithmic growth phase. As the substrate becomes exhausted the rate of growth decreases leading to the stationary phase, followed by a declining phase which occurs when the substrate present is no longer able either to support growth or to maintain the existing population, so that the rate at which cells die exceeds the rate of growth of new cells. At this point in a batch culture viability of the population begins to decrease, although as shown by Stephenson (1928), biochemical activity continues and the specific activity of some enzyme reactions may even increase.

It is this declining phase of growth which is most pertinent to waste treatment. The aim of waste treatment is to reduce the concentration of substrate and the endogenous oxygen demand of suspended cells to as low a level as possible and, no matter whether a complete-mixing or plug-flow system is used, the cultural conditions are very similar to the declining phase of a batch culture. Thus it is to be expected that the bacterial population in a sewage-treatment plant producing effluent of high quality will be almost moribund, and of low viability, and that some 'posthumous' biochemical activity may be retained by the non-viable part of the cell mass.

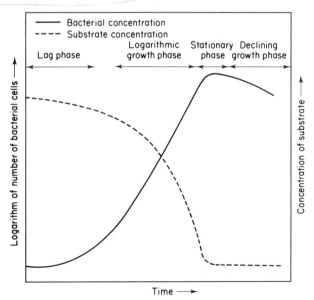

Figure 12.3 Phases of bacterial growth

Table 12.1 Bacterial content of sludge determined by alternative methods

Substrate	Specific rate of removal of substrate (mg/g h)		Bacterial content of sludge (% by weight)	
	Sludge	Pure culture	From activity	From count
Glucose	30	300	10	0·2
Acetate	50	500	10	0·1
Stearate	1	10	10	0·8
NH_4^+	3	300	1	0·01*
Anionic detergent	1	50	2	—
Total viable count (Pike and Carrington, 1972)				4·7

*Counts probably underestimated.

That significant BOD removal could be achieved by non-viable bacteria was shown by Wooldridge (1933) and Wooldridge and Standfast (1933, 1936). These workers demonstrated that removal of BOD from sewage was an enzymic process carried out by bacterial enzymes and could occur in the absence of viable bacteria. In addition to explaining the low counts of viable bacteria in treatment plants, this 'posthumous' activity could also explain the observation that the activity of a sludge towards a substrate is often greater than can be accounted for by the population of viable cells (Table 12.1).

12.4 MODEL FOR BACTERIAL GROWTH AND SUBSTRATE UTILIZATION IN WASTE TREATMENT

Three mechanisms, cometabolism of substrate, utilization of substrate for maintenance of cells, and 'posthumous' biochemical activity of cells, have been described, which can account for the small proportion of viable bacteria present in a typical activated sludge. Two of these, cometabolism and 'posthumous' activity, also allow for the activity of a sludge towards a substrate being greater than can be accounted for by the population of viable cells. In all three cases substrate is diverted from processes leading to the synthesis of cells to other enzymic processes. These reactions will have Michaelis–Menten kinetics of the form

$$r = \frac{RS}{K_m + S} \tag{12.5}$$

where r is the specific rate of reaction, R the maximum specific rate, and K_m the Michaelis constant, being that concentration, S, of the limiting substrate at which the rate is half the maximum. A better description of bacterial growth and substrate utilization in activated sludge would therefore be obtained if this Michaelis–Menten term were added to the Monod term of equation 12.2 so producing a hybrid equation of the form

$$ds/dt = -\left(\frac{\mu_{max}}{Y} \cdot X^v \cdot \frac{S}{K_s + S} + \frac{RS}{K_m + S} \cdot X^a\right) \tag{12.6}$$

in which X^v is the concentration of viable cells growing on the substrate and X^a the concentration of cells which metabolize the substrate without consequent growth.

12.5 APPLICATION OF THE MODEL

Table 12.1 shows for the more common substrates the rates at which pure cultures of bacteria isolated from sludge would metabolize these substrates. From these measurements and measurements of sludge activity towards the same substrates the proportion of viable cells in the sludge was calculated. Also shown is the concentration of total viable bacteria measured by Pike and Carrington (1972). Since this measured total concentration of bacteria is less than the calculated concentration for any of the individual substrates for which data are available, cometabolism probably has little influence in reducing the concentration of viable cells required for these more common substrates and it is likely that 'posthumous' activity will have the major role in reducing the apparent discrepancy between calculated and observed concentrations.

In order to make use of a model which includes a term for 'posthumous' activity of cells, it is necessary to provide some means of assessing the values of X^v and X^a (Equation 12.6) to be used. The method of assessment used depends on the observation that at low specific growth rates of bacteria their viability decreases with decreasing specific growth rate (Tempest, Herbert, and Phipps, 1967; Department of the Environment, in press), a loss of viability similar to that found in batch culture. Typically the growth rates of activated sludge are between 0·05 to 0·15 d^{-1} (sludge age 7–20 days Lawrence and McCarty, 1970) corresponding to doubling times in the range 4·6–14 days; for comparison, *Nitrosomonas*, one of the slowest growing bacteria, is capable of doubling in less than a day. At these low growth rates the maintenance requirement for substrate is the most important factor in relation to viability of the organism, and in the absence of a suitable energy source, the bacteria will lose viability.

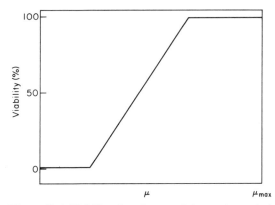

Figure 12.4 Viability function used in mathematical model

For the purposes of the model it has been assumed that viability is related to specific growth rate according to the simple function shown in Figure 12.4. Down to a given proportion of the maximum specific growth rate the viability was assumed to remain 100%, and then to fall in proportion to the specific growth rate until a minimum value of 1% was reached. The growth rate, and hence the proportion of viable cells in the biomass, was calculated taking the substrate concentration in the effluent from the aeration tank, adjusting the concentration of each component according to the assumed wastage rate of mixed liquor, and using the newly calculated concentrations of the two components as input parameters for a new cycle.

The model described has been developed on the basis of the behaviour of bacteria in both batch and continuous culture, when the concentration of substrate is low and, in the presence of 'posthumous' biochemical activity, the concentration of viable bacteria in the system needs to be sufficient only to maintain the total biomass at the required level and does not need to account for the total biochemical activity of the system.

12.6 PREDICTIONS OF THE MODEL

Figure 12.5 shows that the model predicts a difference between total activity and activity due to viable cells alone; such a difference has been observed in practice.

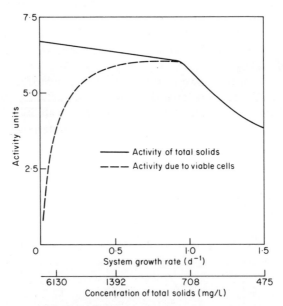

Figure 12.5 Biochemical activity of sludge and viable bacteria predicted by mathematical model. Concentration of total solids calculated from equation of Downing and Knowles (1966) with sewage BOD–329 mg/l

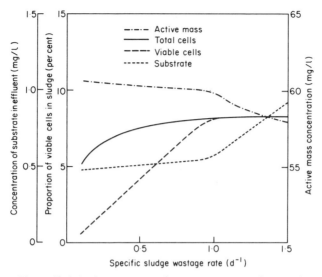

Figure 12.6 Active mass, substrate concentration and proportion of cells in sludge predicted by mathematical model

Figure 12.6 shows the changes in substrate concentration, active mass, and the proportion of cells in sludge predicted for a complete-mixing system of 8 h retention time. The proportion of viable cells in the solids is of the same order as found by observation. Total solids concentration was obtained from use of the equation given by Donning and Knowles (1967) (BOD = 329 mg/l, temperature = 15°C, $D = 0.125 h^{-1}$, $a = 1$); for the bacterial concentration it was assumed that the cells grew on carbohydrate present in the sewage at a concentration of 95 mg/l (20% of the organic carbon, Painter and colleagues 1961) with yield coefficient 0.31, maintenance coefficient 0.067 (Washington and colleagues, 1965), μ_{max} 0.4 h^{-1}, and $K_s = 5$ mg/l. The results show a marked discontinuity at the stage where the population becomes 100% viable; X^a is zero and equation 12.6 becomes identical to equation 12.2, and the concentration of the substrate in the effluent begins to increase much more rapidly. Up to this point changes in concentration of both substrate and biomass with change in growth rate are very slight. This is similar to the experimental results (Table 12.2) of Boon and Burgess (1972) where reducing the solids concentration to increase the system growth rate for the same aeration period had little effect on the BOD of the effluent, until at the lowest solids concentration performance suddenly deteriorated.

The model provides a basis for explaining this property of activated-sludge systems whereby effluent quality deteriorates abruptly when a critical value of sludge loading is exceeded. A reservoir of 'posthumous' activity due to non-viable bacteria brings a considerable inertia to the system so that any changes — for instance in specific wastage rate or retention time — generally have little effect on the concentration of substrate in the effluent. However, once this

274

Table 12.2 Variation with concentration of suspended-solids in mixed liquor of the BOD of effluent from an experimental activated-sludge plant—aeration period 0·62 h

Suspended solids (mg/1)	System growth rate (d⁻¹)	BOD (mg/1) of effluent	
		Range	Average
8200	0·36	8–13	10
6280	0·42	4–15	10
4490	0·59	5–19	10
1930	1·8	29–58	45

reservoir of 'posthumous' activity is removed or destroyed, the system becomes more sensitive to changes in conditions so that a slight increase in specific wastage rate, or loading, is accompanied by a larger increase in substrate concentration.

Inertia of this type has been shown in an experimental system by Clayfield (in press). In his experiments, doubling the loading on the plant resulted in a new steady state being achieved in about two days, whereas after halving the loading a much longer period elapsed before the new steady state was achieved. The model behaves in a similar fashion, and Figure 12.7 shows the results obtained by numerical integration of the model for similar changes in loading. The differences in response to the two situations can again be attributed to a reservoir of 'posthumous' biochemical activity. With increase in loading the concentration of substrate present also increases since the concentration of

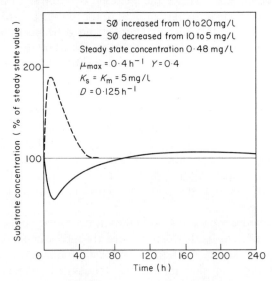

Figure 12.7 Response of effluent substrate concentration to changes in loading

active mass present is insufficient to cope with the new loading. The increase in substrate concentration allows the support of a higher concentration of viable bacteria so that viability increases and new biomass is built up at a rate, governed by the kinetic constants for growth of the micro-organisms concerned, which is considerably greater than the specific wastage rate. Changes in concentration of substrate with time are characterized by a rapid increase followed by a more gradual return to the steady state level (cf. experimental points for detergents in Figure 12.1).

When the loading is decreased the concentration of active mass present is now more than is required for the substrate being applied so that the concentration of substrate in the system is reduced, thereby reducing the viability of the bacteria present. In this situation the concentration of biomass can only be reduced by solids wastage together with some loss of biochemical activity by lysis, so that reduction of the biomass by the required 50% may take several days. In the model, after the initial fall in substrate concentration, the system over-corrected and the substrate concentration rose above the steady-state value before falling very gradually back to the steady-state value.

This same mechanism could explain the observations of Adams and Eckenfelder (1970) in which transient organic loads applied to activated sludge resulted in respiratory activity continuing to increase when the load was restored to the original level. The response of the model to this situation is shown in Figure 12.8. The slow decline in solids loading between 11·3 and 16·3 h is due to increase in biomass during this period. A similar observation was made by Jacquart, Lefort, and Rovel (1972) which they explained by postulating the biological storage of nutrients.

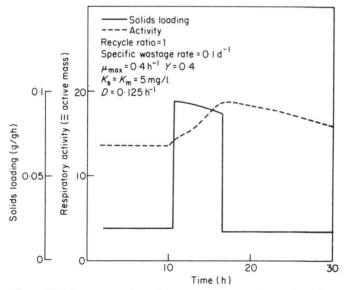

Figure 12.8 Response of respiratory activity to change in solids loading

Table 12.3 Proportion of maximum specific growth rate supportable by substrate
concentrations found in sewage effluents

Substrate	Concentration (μg/1)	K_s (mg/1)	Per cent μ_{max} supported
Ammonia	$1 - 2 \times 10^3$	1	50–70
Acetic acid	20 **†	6	0·3
Propionic acid	5 **†	2	0·25
Glucose	< 50 *†	1	< 4·8

*Data from Katz, Pitt, Scott, and Rosen (1972).
**Data from Murtaugh and Bunch (1965).
†Observed at the Water Research Centre.

12.7 NITRIFICATION

Straightforward application of the Monod equations to waste treatment
problems has been successful only in the case of nitrification (Downing and
colleagues, 1964). In order to account for removal of carbonaceous substrate
various modifications of the equations have been necessary allowing for activity
by non-viable material. For maximum usefulness the modified model should
still be able to satisfy the conditions of nitrification; the modified model in fact
gives results very similar to those of Downing and colleagues (1964). Table 12.3
shows the concentration of three constituents of sewage effluent together with
the proportion of the maximum specific growth rate which they can support.
Clearly, ammonia is a special case, and, at 50–70% of its maximum growth rate,
Nitrosomonas will be present at almost 100% viability. Thus the Michaelis–
Menten term in equation 12.6 is zero and the modified model becomes identical
to that of Downing and colleagues (1964). As a general rule organisms, such as
Nitrosomonas, having maximum specific growth rates not much greater than
the specific wastage rate of the systems will be present at 100% viability.

12.8 DISCUSSION

Of the two main types of mathematical model for waste treatment, the most
used, and probably the most successful, have been those devised to describe the
overall behaviour of treatment systems, i.e. the pragmatic models. Fundamental
models, with their basis in bacterial physiology, are more specialized, and
limited to single substrates. Attempts to apply these models to complex substrates have resulted inevitably in erroneous, and high, values for saturation
constants (Jones, in press) and use of parameters such as COD to measure
the removal of glucose (Storer and Gaudy, 1969) also distorts the values obtained
for kinetic constants since change in COD measures not only glucose removal
but also production of metabolites.

In developing fundamental biological models it is important to realize that
substrate utilization by bacteria is not necessarily followed by growth. When
cultural conditions are altered, the time until the next bacterial division may,

because of the fixed rate of gene replication, remain at its previous value (Donachie, Jones, and Teather, 1973) but the enzymic processes responsible for substrate removal can respond almost instantaneously. A case for separating bacterial growth and biochemical activity in waste treatment was first put forward by Wooldridge (1933) and Wooldridge and Standfast (1933, 1936), and the model described here shows how this concept can be applied. Three types of biochemical activity other than that due to growth have been postulated. These are a maintenance requirement, which is activity uncoupled from growth and exerted by viable bacteria, 'posthumous' biochemical activity exerted by non-viable bacteria, and cometabolism, being biochemical activity of cells towards a substrate on which they do not grow. The effects of these three types of activity are similar in that all divert substrate from bacterial growth. It is possible that maintenance and 'posthumous' activity are of significance with most substrates while the role of cometabolism could be most important for those compounds which are present only intermittently, and those which, being difficult to degrade biologically (Painter, in press), only support bacterial growth to a limited extent.

Any model of this type can only be applied to a single substrate, and a satisfactory description for all the substrates present, in for example, sewage, requires a summation of all the substrate/bacteria systems concerned. This will be very complex and much more information on the interaction (Department of the Environment, 1972) between bacteria in waste-treatment processes is needed. An all-embracing description of the waste treatment process will contain both pragmatic and fundamental models. Thus, the fundamental model described provides an explanation for the observed discrepancy between sludge activity and activity attributable to viable bacteria (Table 12.1), and, together with the empirical model for total solids production from BOD given by Downing and Knowles (1967) predicts viable cell concentrations of the same order as found in practice (Pike and Carrington, 1972).

Acknowledgement

Crown copyright. Reproduced by permission of the Controller, H.M.S.O.

12.9 REFERENCES

Adams, C. E., and Eckenfelder, W. W. (1970). 'Response of activated sludge to transient organic loadings', *Journal of the Sanitary Engineering Division. Proceedings of the American Society of Civil Engineers*, **96** (SA2), 333–352.

Boon, A. G., and Burgess, D. R. (1972). 'Effects of diurnal variations in flow of settled sewage on the performance of high rate activated-sludge plants', *Water Pollution Control*, **71** (5), 493–522.

Clayfield, G. W. (in press). 'The respiration of laboratory and works activated sludges', *Water Pollution Control*.

Department of the Environment (1972). *Water Pollution Research, 1971*. H. M. S. O., London, 70–72.

Department of the Environment (in press). *Water Pollution Research, 1972*. H. M. Stationery Office, London.

278

Donachie, W. D., Jones, N. C., and Teather, R. (1973). 'The bacterial cell cycle', *Microbial Differentiation*, Symposium 23, The Society for General Microbiology, 9–44.

Downing, A. L., Painter, H. A., and Knowles, G. (1964). 'Nitrification in the activated-sludge process', *Journal and Proceedings of the Institute of Sewage Purification*, 130–158.

Downing, A. L., and Knowles, G. (1967). 'Population dynamics in biological treatment plants', *Advances in Water Pollution Research*, Volume 2, Water Pollution Control Federation, Washington, 117–142.

Garrett, M. T. (1958). 'Hydraulic control of activated sludge growth rate', *Sewage and Industrial Wastes*, **30** (3), 253–261.

Herbert, D. (1961). 'A theoretical analysis of continuous culture systems', *SCI Symposium No. 12, Continuous culture of micro-organisms*, 25–53.

Horvath, R. S. (1972). 'Microbial co-metabolism and the degradation of organic compounds in nature', *Bacteriological Reviews*, **36** (2), 146–155.

Jacquart, J. C., Lefort, D., and Rovel, J. M. (1972). 'An attempt to take account of biological storage in the mathematical analysis of activated sludge behaviour', *6th International Conference Water Pollution Research*, Jerusalem.

Jones, G. L. (in press). 'Bacterial growth kinetics: measurement and significance in the activated sludge process', *Water Research*.

Jones, G. L. (in preparation). 'Acclimatization of activated sludge to detergent: a mathematical exploration of possible mechanisms'.

Katz, S., Pitt, W. W., Scott, C. D., and Rosen, A. A. (1972). 'The determination of stable organic compounds in waste effluents at microgram per litre levels by automatic high resolution ion exchange chromatography', *Water Research*, **6** (9), 1029–1037.

Lawrence, A. E., and McCarty, P. L. (1970). 'Unified basis for biological treatment design and operation', *Journal of the Sanitary Engineering Division, American Society of Civil Engineers*, **96** (SA3), 757–778.

Marr, A. G., Nilson, E. H., and Clark, D. J. (1963). 'The maintenance requirement of *Escheria coli*', *Annals of the New York Academy of Sciences*, **102** (3) 536–548.

Middlebrooks, E. J., and Garland, C. F. (1968). 'Kinetics of model and field extended aeration wastewater treatment units', *Journal Water Pollution Control Federation*, **40** (4), 586–612.

Monod, J. (1942). *Recherches sur la croissance des cultures bacteriennes*. Hermann et Cie, Paris.

Monod, J. (1950). 'La technique de culture continue: theorie et applications', *Annals Institute Pasteur, Paris*, **79**, 390–410.

Murtaugh, J. J., and Bunch, R. L. (1965). 'Acidic components of sewage effluents and river water', *Journal Water Pollution Control Federation*, **37** (3), 410–415.

Painter, H. A. (in press). 'Assessment of the environmental impact of chemicals', *Proceedings Royal Society, London B*.

Painter, H. A., and Durrant, K. (in preparation). 'A study of the acclimatization of activated sludge to an anionic detergent'.

Painter, H. A., Viney, M., and Bywaters, A. (1961). 'Composition of sewage and sewage effluents', *Journal Institute of Sewage Purification*, 302–310.

Pike, E. B., and Carrington, E. G. (1972). 'Recent developments in the study of bacteria in the activated-sludge process', *Journal of the Institute of Water Pollution Control*, **71** (5), 583–605.

Pirt, S. J. (1965). 'The maintenance energy of bacteria in growing cultures', *Proceedings of the Royal Society B*, **163**, 224–231.

Stephenson, M. (1928). 'On lactic dehydrogenase. A cell-free enzyme preparation obtained from bacteria', *Biochemical Journal*, **22**, 605–614.

Storer, F. F., and Gaudy, A. F. (1969). 'Computational analysis of transient response to quantitative shock loadings of heterogeneous populations in continuous culture', *Environmental Science and Technology*, **3** (2), 143–149.

Tempest, D. W., Herbert, D., and Phipps, P. J. (1967). 'Studies on the growth of *Aerobacter aerogenes* at low dilution rates in a chemostat', *Microbial Physiology and Continuous Culture*. Proceedings of the third International Symposium. E. O. Powell, C. G. T. Evans, R. E. Strange and D. W. Tempest (Eds), H. M. S. O., 240–254.

Tench, H. B., and Morton, A. Y. (1962). 'The application of enzyme kinetics to activated sludge research, *Journal and Proceedings of the Institute of Sewage Purification*, **5**, 478–486.

Tuček, F., Chudoba, J., and Madĕra, V. (1971). 'Unified basis for design of biological aerobic treatment processes', *Water Research*, **5**, 647–680.

Washington, D. R., Rao, S. S., Hetling, L. J., and Martin, E. J. (1965). 'Kinetics of the steady state bacterial culture with application to the activated-sludge process', *Final report, Project W-P 288*. Division of Water Supply and Pollution Control, Public Health Service, United States Department of Health, Education, and Welfare.

Wayman, C. H. (1971). 'Biodegradation of synthetic detergents', *Progress in Industrial Microbiology*, **10**, 219–271.

Weddle, C. L., and Jenkins, D. (1971). 'The viability and activity of activated sludge', *Water Research*, **5**, 621–640.

Wilson, I. S. (1967). 'Some problems in the treatment of effluents from the manufacture of organic chemicals', *Chemistry and Industry*, 1278–1288.

Wooldridge, W. R. (1933). 'The "stability test" of sewage and its relation to enzyme activity', *Biochemical Journal*, **27** (1), 193–201.

Wooldridge, W. R., and Standfast, A. F. B. (1933). 'The biochemical oxygen demand of sewage', *Biochemical Journal*, **27** (1), 183–192.

Wooldridge, W. R., and Standfast, A. F. B. (1936). 'The role of enzymes in activated sludge and sewage oxidations', *Biochemical Journal*, **30** (9), 1542–1553.

13

The Development of a Dynamic Model and Control Strategies for the Anaerobic Digestion Process

J. F. ANDREWS

13.1 INTRODUCTION

The systems engineering concepts and tools discussed earlier in Chapter 1 by the author can be made clearer and more meaningful by application to a specific system. The anaerobic digestion process has been chosen for this purpose since it was in attempts to understand and prevent digester failure that the author first started using some concepts and tools of systems engineering. A dynamic model is clearly called for in this instance since a failing digester is definitely not at steady state!

The anaerobic digestion process is widely used for the treatment of organic sludges for municipal wastewater treatment plants and has several significant advantages over other methods for the treatment of organic sludges. Among these are the formation of useful by-products such as methane gas and a humus like slurry which is well suited for land reclamation. Unfortunately, even with these advantages the process has in general not enjoyed a good reputation because of its poor record with respect to process stability as indicated through the year by the many reports of sour or failing digesters. The major problems with the process appear to lie in the area of process operation as evidenced by its more successful performance in large cities where skilled operation is more prevalent.

At the present time, operating practice consists only of sets of empirical rules and there is a great need for a more rational control strategy to put process operation on a quantitative basis. Dynamic modelling and computer simulation can be of considerable value in evaluating the effectiveness of different control strategies. The implementation of such strategies should result in a decrease in the frequency of process failure and ultimately permit the optimization of process performance. Dynamic modelling and simulating would also be of value in improving process design since they would allow evaluation and selection of design parameters which contribute to improved process stability. The incorporation of modern control systems would also improve process stability and decrease the need for oversizing.

13.2 MODEL DEVELOPMENT

The model presented herein has evolved over the past ten years and this evolution can be followed through the publications by Andrews and colleagues (1964), Andrews and Willimon (1967), Andrews (1968, 1969), Andrews and Graef (1971), and Graef and Andrews (1972, 1973). The first models (Andrews and colleagues, 1964; Andrews and Willimon, 1957) were more concerned with the series of reactions occurring in the process and were steady state models based on continuous culture theory using the Monod relationship (Monod, 1942) between limiting substrate concentration and growth rate. However, models using the Monod relationship can predict process failure only by hydraulic overloading and cannot be valid for those substrates, such as volatile acids, which limit growth at low concentrations and are inhibitory to the organisms at higher concentrations.

The model was extended to the prediction of dynamic behaviour by Andrews (1968). The inhibitory effects of the substrate were recognized by the incorporation of an inhibition function to express the relationship between substrate concentration and growth rate. The incorporation of this inhibition function permitted the model to predict process failure by both hydraulic and organic overloading as has been observed in the field. In a later version of the model, Andrews (1969) further modified the inhibition function to consider the unionized volatile acids as the substrate. This is an important modification since it assists in resolving the conflict in the literature as to whether inhibition is caused by high volatile acids concentration or low pH. Since the concentration of unionized acids is a function of both total volatile acids concentration and pH, both are therefore of importance.

Andrews and Graef (1971) further extended the model by considering the interactions which occur in and between the liquid, gas, and biological phases of the digester. Consideration of these interactions permitted the development of a model which predicted the dynamic response of the five variables most commonly used for process operation, these being; (1) volatile acids concentration, (2) alkalinity, (3) pH, (4) gas flow rate, and (5) gas composition. Graef and Andrews (1972, 1973) incorporated a relationship to express the effects of toxic materials on the process thus expanding the model to the point where it could predict process failure by hydraulic, organic, or toxic materials overloading. Using the hybrid computer for process simulation, they also explored the effects of changes in design and operational parameters on process stability and investigated a variety of control strategies for the process.

Biological reactions

The anaerobic digestion of complex organic wastes is normally considered to consist of reactions occurring in series as illustrated in Figure 13.1. It should be recognized that this is an approximation since many different microbial species participate in the process and their interactions are not always clearly defined. For example, recent evidence (Bryant and colleagues, 1967) indicates

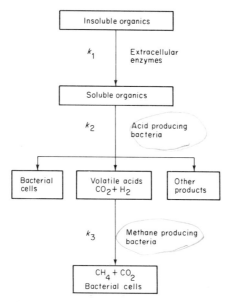

Figure 13.1 Biological reactions in the ana-erobic digestion process

that some non-methanogenic bacteria may be involved in the conversion of volatile acids to methane and carbon dioxide. However, since most species of methane bacteria have much lower growth rates (Andrews and colleagues 1964; Lawrence and McCarty, 1967) than the acid producing bacteria, the overall conversion of the intermediate products, volatile acids, to methane and carbon dioxide is usually considered to be the rate limiting step in the process and will be the only reaction considered herein. The methane bacteria also appear to be more sensitive than the acid producing bacteria to changes in environmental conditions such as pH, temperature, and inhibitory substances.

Stoichiometry

Buswell (1939) has studied the anaerobic decomposition of many organic materials and have presented a general formula (equation 13.1) for the conversion of complex organic materials to carbon dioxide

$$C_nH_aO_b + \left[n - \frac{a}{4} - \frac{b}{2}\right]H_2O \longrightarrow \left[\frac{n}{2} - \frac{a}{8} + \frac{b}{4}\right]CO_2 + \left[\frac{n}{2} + \frac{a}{8} - \frac{b}{4}\right]CH_4 \quad (13.1)$$

and methane. However, this formula does not include the fraction of substrate that is converted to microorganisms which, although small, is necessary for the development of a dynamic model of the process. A more general formula (equation 13.2) for the conversion of organic material to microorganisms, carbon dioxide, and methane is shown below.

$$\text{ORGANICS} \longrightarrow Y_{X/S}(C_6H_{12}O_6) + \frac{1}{Y_{CO_2/X}}(CO_2)_T + \frac{1}{Y_{CH_2/X}}(CH_4) \quad (13.2)$$

Where:

$Y_{X/S}$ = moles organisms produced/mole substrate consumed
$Y_{CO_2/X}$ = moles CO_2 produced/mole organisms produced
$Y_{CH_4/X}$ = moles CH_4 produced/mole organisms produced

Equation 13.3 is a more specific formula for the metabolism of acetic acid and is developed from an oxidation–reduction balance using the appropriate yield coefficients as defined.

$$CH_3COOH \longrightarrow 0.02\,C_6H_{12}O_6 + 0.94\,(CO_2)_T + 0.94\,CH_4 \quad (13.3)$$

The empirical composition of the microorganisms is assumed to be $C_6H_{12}O_6$. The value of $Y_{X/S}$ for acetic acid was estimated from the data of Lawrence and McCarty (1967).

McCarty (1964) has shown that acetic acid is the precursor for approximately 70% of the methane produced in anaerobic digestion and the simulations presented herein are for the conversion of acetic acid to microorganisms, methane, and carbon dioxide. Yields will be different for other volatile acids, especially noticeable being the increased ratio of methane to carbon dioxide produced as the length of the carbon chain increases. For field digesters utilizing complex substrates it would also be necessary to include the carbon dioxide generated by the acid producing bacteria.

The carbon dioxide produced is indicated as $(CO_2)_T$ since it can remain dissolved in the liquid, react chemically to form bicarbonate or carbonate, be transported into the gas phase, or precipitated as carbonate. The extent to which these reactions occur will be influenced by the interactions between volatile acids, alkalinity, pH, gas flow rate, and gas composition and are considered in the model with the exception of dissolution or precipitation as carbonate. The methane produced does not undergo any chemical reaction and being sparingly soluble is almost quantitatively transported into the gas phase.

Biological kinetics

One of the first quantitative expressions for microbial growth was the autocatalytic expression given in equation 12.4.

$$\frac{dX}{dt} = \mu X \quad (13.4)$$

where:

x = organism concentration, moles/litre
t = time, days
μ = specific growth rate, days^{-1}.

Organisms growing in accordance with this expression are said to be in the constant growth rate or logarithmic growth phase. Monod (1942) recognized that the specific growth rate was not a true constant but was instead dependent upon the concentration of some limiting nutrient. He described this relationship with a hyperbolic function (equation 13.5) similar

$$\mu = \hat{\mu}\left[\frac{S}{K_S + S}\right] \tag{13.5}$$

where:

$\hat{\mu}$ = maximum specific growth rate, days^{-1}
K_S = saturation coefficient, moles/litre

to the Michaelis–Menten equation used for describing enzyme–substrate interaction. The Monod function, as it is usually called, is still the most commonly used relationship between specific growth rate and substrate concentration although several others have been proposed.

However, the Monod function cannot be valid for those substrates, such as volatile acids, which limit growth at low concentrations and are inhibitory to the organism at higher concentrations. Andrews (1968) has proposed an inhibition function (equation 13.6) for this purpose.

$$\mu = \hat{\mu}\ \frac{1}{1 + \dfrac{K_S}{S} + \dfrac{S}{K_i}} \tag{13.6}$$

where:

K_i = inhibition coefficient, moles/litre

The use of this function in lieu of the Monod function permits development of a model which can predict process failure due to organic overloading as well as

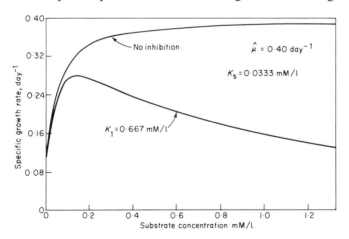

Figure 13.2 Monod and inhibition functions

hydraulic overloading. A graphical comparison of the Monod and inhibition functions is given in Figure 13.2. For the inhibition function there are two possible substrate concentrations for each value of the specific growth rate except at the maximum attainable specific growth rate. The higher of the two substrate concentrations represents an unstable situation for a continuous flow reactor and is the reason why a model using the inhibition function can predict failure by organic over-loading.

The form in which the substrate exists is also of importance and Andrews (1969) has further modified the inhibition function to consider the unionized volatile acids as the limiting substrate (equation 13.8).

$$HS \rightleftharpoons H^+ + S^- \tag{13.7}$$

$$\mu = \hat{\mu} \left[\frac{1}{1 + \dfrac{K_S}{HS} + \dfrac{HS}{K_i}} \right] \tag{13.8}$$

where:

HS = unionized substrate concentration, moles/litre
S^- = ionized substrate concentration, moles/litre
H^+ = hydrogen ion concentration, moles/litre

For pH values above 6, the total substrate concentration, S, is approximately equal to the ionized substrate concentration, S^- as stated in equation 13.9. Therefore, at a fixed pH and with a known total substrate concentration, the unionized substrate concentration can be calculated from the equilibrium relationship for the substrate (equation 13.10).

$$S = HS + S^- \text{ or } S \simeq S^- \tag{13.9}$$

$$HS = \frac{(H^+)(S)}{K_a} \tag{13.10}$$

where:

S = total substrate concentration, moles/litre
K_a = ionization constant, $10^{-4.5}$ for acetic acid at 38°C and an ionic strength of 0·2

The effect of pH on the inhibition function is shown in Figure 13.3. It will be noted that low pH values greatly enhance the inhibitory effect at high substrate concentrations because of the much higher unionized substrate concentrations at these pH values. However, at lower values of total substrate concentration, a low pH may have a beneficial effect because of the higher growth rates possible.

In addition to hydraulic and organic overloading, process failure can also be caused by toxic materials in the feedstream to a digester. As a first approximation, Graef and Andrews (1972) have assumed that the rate of organism killed can be assumed first order with respect to the concentration of the toxic agent as defined in equation 13.11.

$$r_K = K_T T_X \tag{13.11}$$

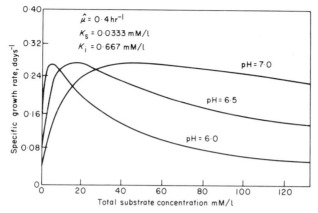

Figure 13.3 Effect of pH on substrate inhibition function

where:

r_K = rate of organism kill, moles/litre/day

K_T = toxicity rate coefficient, moles of organism killed/mole of toxic material/day

T_X = concentration of toxic material, moles/litre.

The rate of consumption of substrate or formation of products such as carbon dioxide and methane, can be related to the rate of organism growth through the stoichiometry (equation 13.2) and the growth rate expression (equation 13.4). These are expressed as follows:

$$\frac{dX}{dt} = -Y_{X/S}\frac{dS}{dt} \tag{13.12}$$

$$Q_{CH_4} = D V Y_{CH_4/X} \mu X \tag{13.13}$$

$$R_B = Y_{CO_2/X} \mu X \tag{13.14}$$

where:

Q_{CH_4} = rate of methane production, litres/day

D = litres of gas/mole of gas; 25·7 for digester at 38°C and one atmosphere of pressure

V = liquid volume of reactor, litres

Other relationships

In addition to the biological relationships which have been presented, it is also necessary to consider several physical and chemical relationships. Included among the liquid phase relationships which must be considered are the chemical reaction between dissolved carbon dioxide and bicarbonate (equation 13.15), the bicarbonate–carbon dioxide equilibrium (equation 13.16), and a charge balance on the ions in solution (equations 17, 18, and 19).

$$(CO_2)_D + H_2O \rightleftharpoons H^+ + HCO_3^- \qquad (13.15)$$

$$\frac{(H^+)(HCO_3^-)}{(CO_2)_D} = K_1 \qquad (13.16)$$

$$(H^+) + (C) = (HCO_3^-) + 2(CO_3^=) + (S^-) + (OH^-) + (A) \qquad (13.17)$$

$$(C) - (A) - (HCO_3^-) + (S) \qquad (13.18)$$

$$(HCO_3^-) = (Z) - (S) \qquad (13.19)$$

where:

$(CO_2)_D$ = dissolved carbon dioxide concentration, moles/litre

(HCO_3^-) = bicarbonate concentration, moles/litre

K_1 = ionization constant, $10^{-6\cdot0}$ at 38°C and an ionic strength of 0·2

C = concentration of cations other than the hydrogen ion, equivalents/litre

A = concentration of anions other than those shown in equation 13.17, equivalents/litre

Z = net cation concentration, $C - A$, equivalents/litre

The simplified equations presented above are based on a restriction of the model to a pH range between 6 and 8 and in this pH range (H^+), (CO_3^-), and (OH^-) can be considered negligible in the charge balance. The net cation concentration, Z, does have some limited chemical significance since it is approximately equal to the ammonium ion (NH_4^+) concentration. The rate of chemical reaction between dissolved carbon dioxide and bicarbonate is very fast compared to the other reactions involved and can be considered to go to completion.

The interactions between the gas and liquid phases must be considered in order to determine the partial pressure of carbon dioxide in the gas phase. The net rate of carbon dioxide transfer between the gas and liquid phases can be expressed by the 'two film' theory of gas transfer as given in equation 13.20.

$$T_G = K_L a \{(CO_2)_D^* - (CO_2)_D\} \qquad (13.20)$$

where:

T_G = gas transfer rate, moles/litre-day

$K_L a$ = overall gas transfer rate coefficient, moles/litre-day

$(CO_2)_D^*$ = concentration of dissolved CO_2 in the liquid phase when in equilibrium with the gas phase, moles/litre.

The equilibrium concentration of dissolved carbon dioxide is calculated using Henry's law.

Material balances

Most anaerobic digesters can be appropriated as continuous flow, stirred tank reactors (CFSTR) as illustrated in Figure 13.4. The final equations rep-

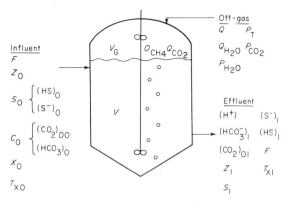

Figure 13.4 Schematic of the anaerobic digestion process

resenting the dynamic model for the process can now be developed for this type of reactor by making material balances on the appropriate components using the relationships which have been presented. The general form of a material balance is as follows.

Rate of Material Flow into Reactor	+	Rate of Appearance or Disappearance of Material due to Reaction	=	Rate of Material Flow Out of Reactor	+	Rate of Accumulation of Material in Reactor

It should be remembered that for a multiphase process, such as anaerobic digestion, it may be necessary to make material balances for specific components in each phase as well as considering transfer of the component between phases. Also, one or more of the terms may be zero or may have more than one component.

The material balance for the total substrate is:

$$V\frac{dS_1}{dt} = F(S_0 - S_1) - \frac{\mu X_1}{Y_{X/S}} V \qquad (13.22)$$

where:

F = liquid flow rate, litres/day

$_0$ = subscript indicating influent concentration

$_1$ = subscript indicating effluent concentration

The organism balance is:

$$V\frac{dX_1}{dt} = F(X_0 - X_1) + \mu X_1 V - K_T T_X V \qquad (13.23)$$

For net cations and conservative toxic chemicals

$$V \frac{dZ_1}{dt} = F(Z_0 - Z_1) \tag{13.24}$$

$$V \frac{dT_{X1}}{dt} F(T_{X0} - T_{X1}) \tag{13.25}$$

For carbon dioxide in the gas phase assuming the gas phase is completely mixed:

$$\frac{dP_{CO_2}}{dt} = P_T D \frac{V}{V_G} T_G - \frac{P_{CO_2}}{V_G} Q \tag{13.26}$$

where: P_{CO_2} = partial pressure of CO_2 in the gas phase, mm Hg
P_T = total pressure of CO_2 and CH_4 in the reactor gas volume, mm Hg
Q = total dry gas flow rate, $Q_{CH_4} + Q_{CO_2}$, litres/day
V_G = reactor gas volume, litres

Finally, the mass balance on carbon dioxide dissolved in the solution phase in given in equation 13.27.

$$V \frac{d(CO_2)_{D1}}{dt} = F\{(CO_2)_{DO} - (CO_2)_{D1}\} + T_G V + R_B V + R_C V \tag{13.27}$$

where:

$$R_C = \frac{F}{V} (HCO_3^-)_0 - \frac{F}{V}(HCO_3^-)_1 + \frac{dS_1}{dt} - \frac{dZ_1}{dt}$$

R_C, the rate of production of carbon dioxide from bicarbonate, results from the reaction of substrate (volatile acids) and bicarbonate to produce carbon dioxide as given in equation 13.28.

$$HS + HCO_3^- \rightleftarrows H_2O + CO_2 + S^- \tag{13.28}$$

A summary of the mathematical model showing the information flow between phases is given in Figure 13.5. The original references should be consulted for a more detailed development of the model.

13.3 MODEL VERIFICATION

At this stage of its development, the model must be considered only semi-quantitative in nature since, as previously stated, it has been necessary to make several simplifying assumptions in the development of the model and reliable values for many of the parameters are not available. However, qualitative evidence for the validity of the model can be obtained by simulating the process and comparing the simulation results with those which have been commonly observed in the field. In performing these simulations, estimates of $\mu = 0.4 \, \text{day}^{-1}$, $K_S = 0.0333$ mM/litre, and $Y_{X/S} = 0.02$ moles/mole were made from the data

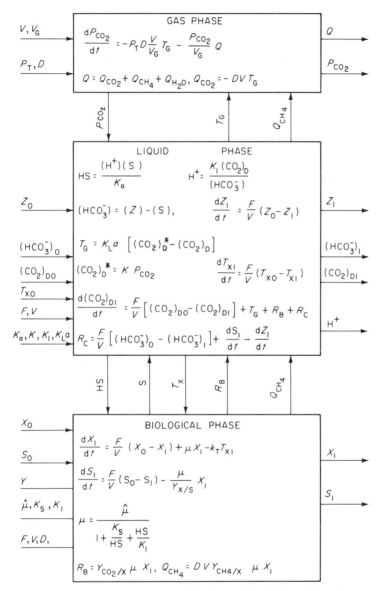

Figure 13.5 Summary of mathematical model and information flow

presented by Lawrence and McCarty (1967). For acetic acid, $Y_{CO_2/x}$ and $Y_{CH_4/x}$ are equal and were determined from the basic stoichiometry (equation 13.3) as 47·0 moles/mole. An order of magnitude estimate of $K_i = 0·667$ mM/litre was made using the 2 000 − 3 000 mg/litre of total volatile acids that Buswell (1939) considers to be inhibitory. All values are estimated for a temperature of 38° and the equilibrium constants used are for this temperature and an ionic strength of 0·2. A value of $K_L a = 100$ day^{-1} was selected to maintain the dissolved carbon dioxide concentration reasonably close to the equilibrium value since it was not the purpose of these simulations to investigate the effect of gas transfer rate on the process.

The model for a batch reactor is easily obtained from the CFSTR model by setting the liquid flow rate equal to zero wherever it occurs in the material balances. The results of simulation studies on a batch reactor indicate that increasing the initial organism concentration, decreasing the initial substrate concentration, or increasing the initial pH, all had the effect of decreasing the time required for batch digestion. These results are in accordance with those commonly observed in field studies on batch digestion.

Simulations of the start-up of a CFSTR indicated that: (1) increasing the quantity of seed sludge or increasing pH will decrease the time required for start-up, (2) digester failure may occur if insufficient seed sludge is present or the pH is too low when a CFSTR is started, and (3) digester failure during start-up may be avoided by slowly bringing the digester loading to its full value. These results are also in accord with field observations on anaerobic digestion.

The simulations were able to predict process failure by hydraulic, organic, and toxic material overloading. In addition, the course of failure, as evidenced by the behaviour of the operational variables, pH, alkalinity, volatile acids concentration, gas flow rate, and gas composition, is qualitatively the same as that observed in the field. Examples of the response of these operational var-ables to the simulated failure of the process by hydraulic and organic overloading are given in Figures 13.6 and 13.7. The simulations also indicated that process failure could be avoided by (1) stopping or reducing the flow to the reactor, (2) the addition of base, and (3) the recycle of sludge from a second stage reactor. The simulation of the latter two of these control actions are presented in Figure 13.8 and all three are commonly used in field operations.

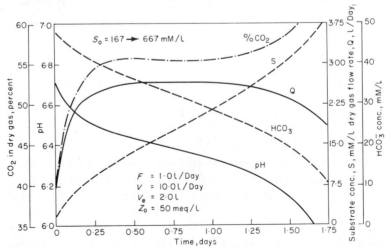

Figure 13.6 Process failure by a step change in influent substrate concentration

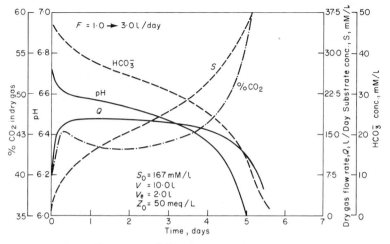

Figure 13.7 Process failure by a step change in flow rate

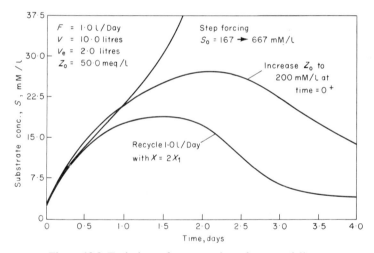

Figure 13.8 Techniques for prevention of process failure

13.4 PROCESS STABILITY

As previously mentioned, the anaerobic digester has a poor reputation with respect to stability. The model which has been presented can be used to predict instability and ultimate process failure due to hydraulic, organic, and toxic material overloading. More importantly, it can also be used to indicate those design and operating factors which can be used to improve process stability. Included among these are increases in residence time, alkalinity, influent substrate concentration, digested sludge recycle, loading frequency, and various control strategies.

294

Quantification of the effects of design and operational factors on digester stability is a tedious and time consuming task even with continuous system modelling programs for the digital computer. Fortunately, these problems can be overcome by using a hybrid computer for the simulations and the results reported herein were obtained by using an EAI 680 analog computer interfaced with a PDP 15 digital computer. Graef (1972) should be consulted for the details of the programming involved. The high speed of the analog computer permitted days of digester operation to be compressed into seconds and the 'hands on' aspects of the hybrid computer permitted the rapid evaluation of the effect of a wide range of the factors of interest.

The stability analysis procedure utilized involved simulating a change in a factor of interest, such as digester residence time, by changing a potentiometer (pot) setting on the analog computer. A step forcing in substrate loading rate was then simulated with the results being rapidly displayed on a cathode ray tube (CRT). If this forcing was insufficient to cause process failure, the loading rate was increased by changing the setting on a second pot and the step forcing again simulated. This trial and error procedure was repeated until the step increase in substrate loading rate which just caused process failure was determined. The pot representing digester residence time was then changed to a new value and the trial and error procedure repeated at this new residence time. By plotting the locus of points of the critical substrate loading rates vs residence time, or other factor of interest, it was possible to obtain a semiquantitative measure of digester stability.

Figure 13.9 illustrates the increased stability that can be obtained by increasing either the residence time or alkalinity. As an example, increasing the residence time from 10 to 15 days at an alkalinity of 2000 mg/litre, permits an

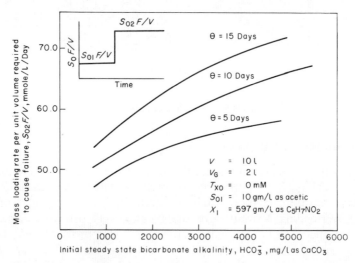

Figure 13.9 The influence of residence time and alkalinity on process stability

increase in the maximum step loading from 56 to 62 millimoles/litre-day. This increase in residence time can be obtained by increasing the effective reactor volume and/or by decreasing the flow rate to the reactor. For an existing installation, the effective volume can be increased by constructing additional digesters, improving mixing to eliminate dead zones in the reactor, or by removing grit and scum accumulations from the digester. Moreover, the flow rate can be reduced, while handling the same mass flow rate of solids, by concentrating or thickening the digester feed.

The simulation results presented in Figure 13.9 also show that process stability can be increased by increasing the alkalinity and one means of accomplishing this is by the addition of a base. However, another technique is to increase the concentration of solids in the feed stream by sludge thickening. The chief mechanism for the production of alkalinity in municipal digesters is the reaction of ammonia with carbon dioxide. The ammonia is derived from the degradation of proteinaceous material in the sludge and, since carbon dioxide is always in excess, the bicarbonate alkalinity can be increased by thickening the sludge which in effect concentrates more protein per unit volume of sludge.

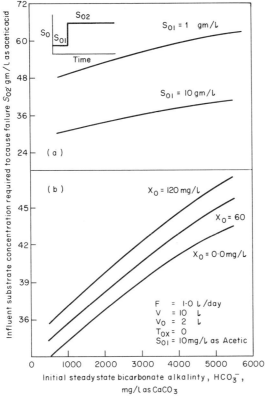

Figure 13.10 Influence of steady state influent substrate concentration and digested sludge recycle on process stability

Increasing the steady state influent substrate concentration and the recycle of digested sludge from a second stage can also improve stability as illustrated in Figure 13.10. A higher steady state loading rate ($S_{O1} = 18$ g/litre vs 10 g/litre in Figure (13.10a), which can be obtained by sludge thickening, results in a higher steady state concentration of microorganisms in the digester thus enabling it to sustain larger step increases in organic overloading. The organism concentration can also be increased by the recycle of digested sludge from a second stage. The lower curve in Figure 13.10b represents no return sludge, the middle curve 10% recycle of the digested sludge, and the upper curve 10% recycle of digested sludge which has been concentrated by a factor of two prior to recycle. The curves clearly depict the gain in stability that can be obtained by recycling sludge especially when it can be concentrated prior to recycle as is done in the activated sludge process.

Other factors which can improve stability are increases in the loading frequency and the provision of automatic control systems. However, the most significant results of the simulations are that three of the methods for improving stability, increased residence time, alkalinity, and influent substrate concentration, can all be obtained by sludge thickening.

13.5 CONTROL STRATEGIES

In selecting a control strategy for the anaerobic digester, a wide variety of output variables are available for initiation of the control action. The variables most commonly used are (1) volatile acid concentration, (2) pH, (3) alkalinity, (4) gas flow rate, and (5) gas composition. Combinations of these variables are also used and included among these are the volatile acids/alkalinity ratio, unionized acid concentration, and rate of methane production. There has been considerable speculation as to which variable, or combination of variables, is the best indicator of impending digester failure. Graef and Andrews (1972) have shown that this is dependent upon the type of overloading to which the digester has been subjected and their work should be consulted for a more detailed evaluation of possible process condition indicators. In selecting a variable for initiation of control action, the available analytical techniques must also be considered and, on this basis, preference would be given to gas phase measurements over liquid or solid phase measurements.

After selecting a variable for measurement, an appropriate control action must be initiated. Control actions available in anaerobic digestion include a temporary halt in organic loading, a decrease in the rate of organic loading, addition of a base such as lime or soda ash, dilution of the digester contents, or the addition of well digesting sludge from another digester. A new control action which has been proposed by Graef and Andrews (1972) includes the scrubbing of carbon dioxide from the digester gas with subsequent recycle of the gas. The most effective type of control action is dependent upon the type of over-loading to which the digester is subjected.

The proposed new control action, gas scrubbing and recycle, provides process control by adjusting digester pH through the removal of a weak acid, carbonic, instead of by the addition of a base as is the usual practice. The mechanics of this control strategy are portrayed schematically in Figure 13.11. The measured variable is pH and when this deviates from its set point, an on–off controller diverts a portion of the gas into a gas scrubber where carbon dioxide is essentially eliminated by absorption into an aqueous spray. The scrubbed gas is then recirculated to the reactor thus reducing the partial pressure of carbon dioxide in the gas phase by dilution which, in turn, causes dissolved carbon dioxide to be vented from the solution phase thus yielding an increase in pH because of the removal of carbonic acid.

The simulated digester response to this control strategy is shown in Figure 13.12. The solid lines depict the controlled and uncontrolled process behaviour when the digester is subjected to a step increase in organic loading sufficient to cause failure. The dashed lines are the controlled response to a step increase less than that required to cause failure. This particular strategy incorporates two fixed capacity compressors arranged in parallel; each of which is either on or off. Under routine loading conditions, one compressor is sufficient to maintain the pH within the designated range. The second compressor is actuated by severe loading increases and discharges an accelerated gas flow rate to the gas scrubber.

Simulations were also made to evaluate the effectiveness of this control strategy for the prevention of failure by hydraulic and toxic material overloading. The simulations indicated that although gas scrubbing and recycle was not detrimental to the process, failure could only be delayed and not prevented.

One of the best control strategies for the prevention of failure by an overload of toxic materials is the recycle of concentrated sludge from a second stage

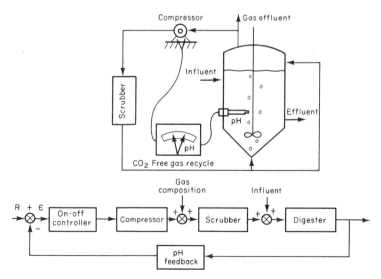

Figure 13.11 Schematic of digester control by gas scrubbing and recycle

298

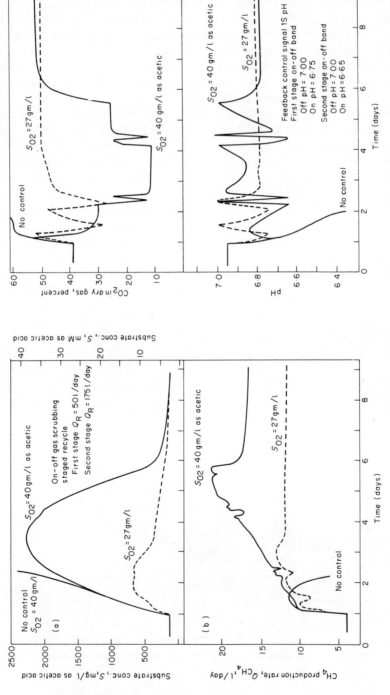

Figure 13.12 Digester response to control by gas scrubbing and recycle

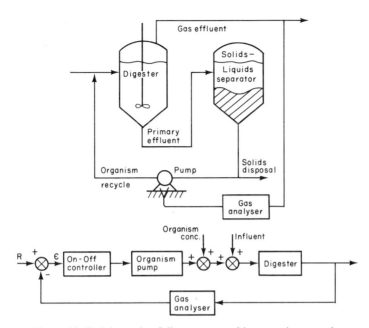

Figure 13.13 Schematic of digester control by organism recycle

digester using the rate of methane production as a feedback signal. A schematic
of this control strategy is given in Figure 13.13. Although the control action
proposed is not new, having first been proposed by Buswell and coworkers
(1939), the proposed control signal, rate of methane production, is new and
appears to be one of the best indicators of digester condition with respect to
overloading with toxic materials. The rate of methane production can be
easily calculated from the common measurements of flow rate and composition
for the gas phase and would be an excellent indicator of the activity of the
methane bacteria which are the most sensitive and critical organisms in the
digester. An analogue can be drawn between the use of the rate of methane
production as an activity indicator in the anaerobic digestion process and the
use of oxygen utilization rate as a measurement of microbial activity in the
activated sludge process.

The simulated digester response to this control strategy is shown in Figure
13.14 and represents the controlled and uncontrolled response of the process
when it is subjected to a pulse loading of a conservative toxic material. A two-
fold concentration of viable organisms was assumed in the recycled sludge
and the sludge return pump was started and stopped in accordance with the rate
of methane production.

The examples presented have illustrated the use of control systems for the
prevention of gross process failure. However, control can also be used to increase
the maximum organic and hydraulic loading rates which a digester can handle
or to improve the quality of the digester effluent. For example, pilot plant
studies by Torpey (1955) have demonstrated that stable operation can be

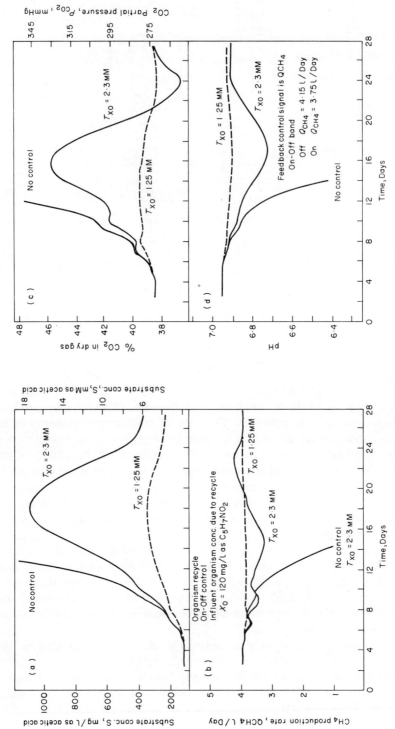

Figure 13.14 Digester response to control by organism recycle

attained at residence times as low as three days. However, current digester design criteria usually call for residence times in the range of 20 to 30 days. Improved operational strategies could result in substantial reductions in the required volumes of new digesters and permit existing digesters to meet increased demands without requiring plant expansion.

13.6 SUMMARY

A dynamic model for the anaerobic digestion process has been presented herein. The key features of the model are: (1) use of an inhibition function to relate volatile acids concentration and specific growth rate for the methane bacteria, (2) consideration of the unionized acids as the growth limiting substrate and inhibiting agent, and (3) consideration of the interactions which occur in and between the liquid, gas, and biological phases of the digester.

Simulation studies provide qualitative evidence for the validity of the model by predicting results which are commonly observed in the field. Some of these results are: (1) failure of the process can occur through hydraulic, organic, and toxic material overloading, (2) the course of failure, as evidence by the behaviour of the operational variables, pH, alkalinity, volatile acids concentration, gas flow rate, and gas composition, is qualitatively the same as that observed in the field, and (3) stopping or reducing the flow to the reactor, the addition of base, or recycle of sludge from a second stage digester are effective techniques for curing failing digesters.

Hybrid computer simulations were used to analyse process stability. Stability was improved by increasing residence time, alkalinity, and influent substrate concentration as well as by the recycle of digested sludge from a second stage and by the incorporation of suitable control strategies. It is significant that three of the measures for improving stability, increased residence time, alkalinity, and influent substrate concentration, can be attained by sludge thickening.

A variety of control strategies are available for the anaerobic digester and were investigated using the hybrid computer. The recycle of gas from which carbon dioxide has been scrubbed, a new control action proposed by Graef and Andrews (1972), was effective for the prevention of failure by toxic overloading. Failure by an overload of toxic materials was best prevented by the recycle of concentrated sludge from a second stage using the rate of methane production as a feedback signal.

The dynamic model, process stability characteristics, and control strategies presented herein have evolved over the past ten years and are discussed in more detail in other publications. They are by no means complete and still require considerable field verification. However, they should be of semiquantitative value for digester design and operation as well as serving as a guide for field experimentation and a valuable framework for future modifications.

Acknowledgement

This paper was published in *Water & Sewage Works*, March 1975, pp. 62–65 and April 1975, pp. 74–77, and is reproduced by permission of *Water & Sewage Works*.

13.7 REFERENCES

Andrews, J. F. (1969). 'Dynamic Model of the Anaerobic Digestion Process', *Journal Sanitary Engineering Division, American Society of Civil Engineers*, **95** (SA1), 95.

Andrews, J. F. (1968). 'A Mathematical Model for the Continuous Culture of Micro-organisms Utilizing Inhibitory Substrates', *Biotechnology and Bioengineering*, **10**, 707.

Andrews, J. F., and Graef, S. P. (1971). 'Dynamic Modelling and Simulation of the Anaerobic Digestion Process', *Anaerobic Biological Treatment Processes, Advances in Chemistry Series No. 105*, American Chemical Society, Washington, D. C., 126.

Andrews, J. F., and Willimon, E. P. (1967). 'Multi-Stage Biological Processes for Waste Treatment', *Proceedings 22nd Industrial Waste Conference*, Purdue University, Lafayette, Ind., 645.

Andrews, J. F., Cole, R. D., and Pearson, E. A. (1964). *Kinetics and Characteristics of Multi-Stage Methane Fermentations*, Sanitary Engineering Research Laboratory Report, 64–11, University of California, Berkeley.

Bryant, M. P., Wolin, G. A., Wolin, J. J., and Wolfe, R. S. (1967). '*Methanobacillus omelianskii*, a Symbiotic Association of Two Species of Bacteria', *Archiv fur Mikrobiologie*, **59**, 20.

Buswell, A. M. (1939). *Anaerobic Fermentations*, Bulletin No. 32, Illinois State Water Survey, Urbana, Illinois.

Graef, S. P., and Andrews, J. F. (1973). 'Mathematical Modelling and Control of Anaerobic Digestion', *Water—1973* (Bennett, C. F., Ed.), Chemical Engineering Progress Symposium Series, American Institute of Chemical Engineers, New York (In press).

Graef, S. P., and Andrews, J. F. (1972). 'Process Stability and Control Strategies for the Anaerobic Digester', *Journal Water Pollution Control Federation* (In press).

Lawrence, A. W., and McCarty, P. L. (1967). *Kinetics of Methane Fermentation in Anaerobic Waste Treatment*, Stanford University, Technical Report No. 75, Stanford University, California.

McCarty, P. L. (1964). 'The Methane Fermentation', *Principles and Applications in Aquatic Microbiology* (Heukelekian, H, and Dondero, N. C., Eds), John Wiley & Sons, New York.

Monod, J. (1942). *Recherches sur la Croissances des Cultures Bacteriennes*, Hermann et Cie, Paris.

Torpey, W. N. (1955). 'Loading to Failure of a Pilot High Rate Digester', *Sewage and Industrial Wastes*, **27**, 121.

14

An Ecological Model of Percolating Filters

A. JAMES

14.1 INTRODUCTION

Despite the widespread use of percolating filters over many years their design and operation are still on an empirical basis. This is due to the inadequacies of the attempts at design formulation. There have been two basic approaches which have been used. They may be summarized as follows:

(i) The empirical approach based on experience from existing installations. This is clearly limited in utility due to differences in climatic conditions, per capita water consumption, and so forth, between regions and countries. (e.g. NRC, 1946; UPB, 1951; Fairall, 1956; Escritt, 1971.)
(ii) The chemical engineering approach which describes the changes inside the filter bed in purely physical and chemical terms. This is a more rational basis but since it is not capable of handling a situation which depends also on biological phenomena it is also of limited application. (e.g. Velz, 1948; Stack, 1957; Howland and colleagues, 1963; Sinkoff and colleagues, 1959: Schulze, 1960; Ames and colleagues, 1962; Atkinson and colleagues, 1968; Monadjemi and Behn, 1970.)

This chapter describes an approach based on ecological concepts. The relationship between the environment and the filter community is analysed and then formulated into a simulation model. This model is then used to explore the effect varying design and operational parameters.

14.2 THE ENVIRONMENT

A trickling filter provides an artificial environment which can be colonized by aquatic microorganisms and moisture-loving invertebrates. The environment is considered below in terms of the more important ecological factors.

(1) Filter medium

The filter medium provides the support for the biological film, so it should be physically durable and chemically inert. The important characteristics of the medium are the amount of surface area per unit volume and the number and

Table 14.1 Difference in % void and number and size of voids for two sizes of clinker. All figures are in metres

Nominal Size (m)	% Voids	Average Volume per stone	Average Number of stones/m³	Average Number of voids/m³	Average Volume per void
0·0254	53·8	0·000069	7 000	10 000	0·000054
0·063	57·0	0·0011	400	600	0·001

size of the spaces between the stones. The former characteristic is a measure of the area available for biological growth; the latter represents the passages available for air and liquid flow. Since large biological growths would tend to restrict these passages, some measure of the space is often used as representing the safety factor to prevent ponding, see (2). Unfortunately, the engineering variable commonly used for this purpose—the void ratio or percentage voids—does not truly reflect number and size of voids. This is illustrated in Table 14.1.

As the number and size of voids can be calculated approximately from the nominal size, it is sufficient to specify the type of media required in terms of engineering variables, namely type of medium and nominal size.

(2) Hydraulic conditions

Hydraulically a trickling filter behaves like a saturated sponge. It holds the maximum quantity of liquid between doses and then each cycle provokes a chain reaction of flow, down through the filter. There are thus two types of conditions experienced in a filter bed:

(1) A period of quiescent conditions between dosing intervals, with a slow drainage occurring;
(2) A period of turbulent flow during dosing.

Any material passing through a trickling filter will experience regular cycles of (1) and (2). The transition between the two types of condition is quite abrupt, as shown in Figure 14.1 and it is thus possible to regard the overall retention time as being composed of a number of quiescent periods separated by abrupt dosing intervals which are responsible for both displacement and mixing. From an ecological viewpoint the hydraulic conditions in a filter may be defined in terms of two parameters, namely the retention time and the degree of mixing.

The degree of mixing is important because during each quiescent period the rate of transfer of organic matter will decrease due to the formation of concentration gradients in the liquid film. These gradients will be destroyed by mixing during the dosing interval. Also mixing will occur between adjacent bands of liquid passing through a filter bed. The degree of mixing is a function of the extent of mixing that each dosing causes and the number of such mixings that occur during the overall retention period.

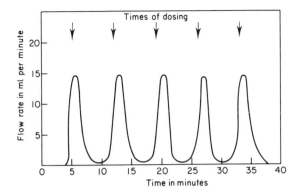

Figure 14.1 Flow pattern in the effluent of a segment of
an experimental filter bed

In terms of the variables normally used in design and operation of trickling filters, the hydraulic condition is specified by the hydraulic loading and the dosing frequency. The relationship between the engineering and ecological variables is shown below in Table 14.2.

In many plants, the frequency of dosing is a haphazard variable since it is controlled by variations of flow to the siphon chamber which doses the bed. As its importance becomes increasingly to be recognized, so more filters are being fitted with motorized distributors.

The hydraulic loading of a filter is often used to categorize the bed as standard-rate or high-rate. The two ranges differ somewhat in their ecological functioning. In standard-rate filters (hydraulic loading up to 1 000 $1/m^3$ day) the flow is not the primary mechanism responsible for control of film accumulation. High-rate filters (hydraulic loadings 2 000–20 000 $1/m^3$ day) use the short retention time and hydraulic flushing for film control. At any hydraulic loading

Table 14.2   Relationship between ecological and engineering variables

Engineering Variable	Ecological Variable	Relationship
Hydraulic Loading (Q)	Retention time (T)	$*T_\alpha \dfrac{1}{Q^{0.67}}$
	Degree of mixing (D)	Changes in Q alter the number of mixings and the extent
Frequency of Dosing (N)	Retention time	T is not related to N
	Degree of mixing	$D\ N$

*The nature of the relationship has been investigated by a number of workers (e.g. Bloodgood, and others (1959), Sheikh (1968), Eden and others (1964), with results that more or less agree on the theoretical $T_\alpha \dfrac{1}{Q^{2/3}}$

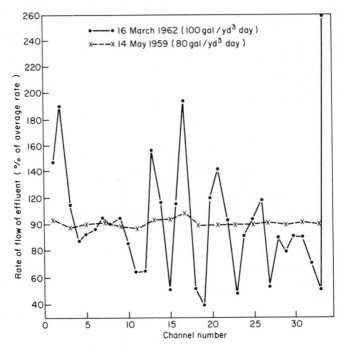

Figure 14.2 Comparison of rates of flow of effluent from individual channels in the base of a filter on two occasions

the instantaneous velocity on the surface will alter with frequency of dosing and the type of jets, but it is doubtful in standard rate filter if this is important in more than the surface 10 cm (Hawkes, 1964).

The effect of excessive slime growths on the flow pattern in filters has been investigated extensively because of the importance of ponding. It has been shown that the primary effect of the growths is to reduce the number of channels available for liquid flow and to cause short-circuiting of the liquid through the remaining channels (Truesdale and Eden, 1964). This is illustrated in Figure 14.2.

(3) Temperature

In the majority of waste treatment plants it is not economic to operate at above ambient temperatures so that the performance of trickling filters is to some extent a function of climatic conditions. The temperature within a trickling filter does not exactly correspond to the air or ground temperatures due to the interaction of a variety of factors. The various sources and sinks of heat energy in filters have been summarized by Bayley and Downing (1963) as follows:

   (i) Transfer by radiation
  (ii) Heat in the waste
 (iii) Liberation of heat from biological oxidation
 (iv) Conversion of potential energy

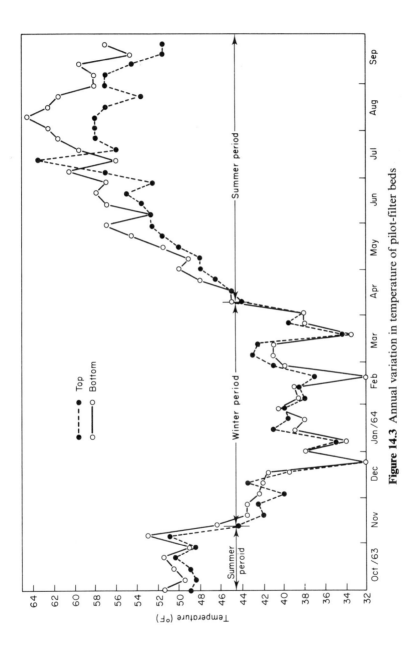

Figure 14.3 Annual variation in temperature of pilot-filter beds

(v) Exchange with atmosphere

(vi) Exchange with ground.

An annual cycle of temperature in a trickling filter resulting from the interaction of the above factors is shown in Figure 14.3. This figure shows that the temperatures at the top and bottom of the bed are often different, which may be important in helping aeration of the filter.

The relationship between temperature and filter efficiency is rather complex, Ware and Loveless (1958) suggest that temperature only plays a direct role in changing filter efficiency below 10°C and similarly Hawkes (1964) suggests that there is little difference in efficiency at the bacterial level between 10°C and 30°C. This is no doubt due to changes in species composition of the bacterial flora corresponding with the temperature changes. There are, however, marked seasonal fluctuations in the efficiency of trickling filters due to disturbance of the film-grazing fauna balance.

(4) Dissolved oxygen

Sewage as it reaches a treatment plant is normally devoid of oxygen; even if traces remain these are used up during primary sedimentation. The oxygen required for the aerobic breakdown of the waste enters partly during dosing and partly during the flow of liquid over the medium. The relative importance of these two processes depends upon a number of factors such as the type of dosing mechanism, height of filter arm above the bed and the nature of the flow pattern within the filter. It has been suggested that ventilation of filters may be an important factor, but as shown by Truesdale and Eden (1963) the ventilation rate is usually in excess of what is theoretically required. This was confirmed by examining the composition of air samples from the stream entering and leaving the filter.

(5) Energy and food supply

The effect of solar radiation as an energy source is limited to maintaining filter beds above ambient temperature; as a source of energy for the community it is insignificant compared with the chemical energy derived from the organic compounds in the waste. A filter surface in a temperate climate receives a maximum of 2500 $kcal/m^2/day$ from solar radiation compared with approximately 1500 $kcal/m^2/day$ of oxidizable organic compounds. The true comparison emerges when one considers the degree of utilization being respectively less than 0·02% and around 80%. Therefore in a trickling filter the primary source of energy is contained in the waste being treated and this is also the primary source of food. It is not possible to generalize about industrial wastes but domestic wastes contain, in addition to the organic carbon (energy source), sufficient proportions of Nitrogen, Phosphorus and the other elements required for bacterial growth.

The concentrations of both the energy and food sources are important and will obviously vary during the passage through the filter bed. The reduction in concentration of organic compounds with increasing depth and a consequent rise in the concentration of organic compounds has been noted by many workers (e.g. James, 1962).

14.3 THE COMMUNITY

(1) Composition

Although trickling filters differ somewhat from any natural environment, they provide an attractive habitat for a wide range of microorganisms and for some larger animals. The composition of the major groups in the filter community is described below, together with some observations on their feeding habits and metabolic activities.

(i) Bacteria

The Bacteria are the predominant group of organisms in the majority of trickling filters. The bacterial flora consists mainly of Gram-negative bacilli derived from the freshwater, which occur in numbers up to $2 \times 10^9/\text{ml}$. The species composition is extremely variable, but the variations do not appear to have any significance as a wide variety of species possess similar metabolic powers and have similar metabolic activities. The metabolic rates are usually in the range of 1×10^{-8} $\mu l/\text{cell}$ per hour at 15°C. They grow as a zoogleal slime on the filter medium, the volume of slime and the numbers of Bacteria diminishing in relation to the food and energy supply downwards through the filter. Biotic factors such as predation may alter this primary pattern.

Bacteria, sometimes together with Fungi, form the basis of filter slime. During the summer period the numbers of organisms are related to the food supply and are kept in check by predation but in winter their numbers increase dramatically as shown in Tables 14.3 and 14.4. The food supply in the case of the Bacteria is mainly organic material in suspension or colloidal solution that becomes adsorbed onto the surface of the zoogleal slime. There it is rendered soluble, if not already so, and passes by diffusion into the bacterial cells.

(ii) Fungi

These organisms are present in most filter beds and occasionally can dominate the community. Their metabolic requirements are slightly different as they can tolerate a higher Carbon: Nitrogen ratio and more acid conditions. Because they compete most successfully for complex organic compounds Fungi are mainly found in the upper layers of trickling filters.

Fungi utilize the same food and energy source as Bacteria and thus compete directly with them. Their metabolic rate is slightly lower and they synthesize a

Table 14.3 Biological results—average number per depth in pilot filters in summer period

Group	Depth I	Depth II	Depth III	Depth IV	Depth V	Depth VI
Bacteria	140 000 000	200 000 000	90 000 000	50 000 000	55 000 000	40 000 000
Sarcodina	26	28	19	20	7	7
Flagellata	30	100	130	100	80	115
Ciliata	350	360	620	480	480	330
Rotifera	0·5	1·7	9·2	1·7	2·2	2·8
Nematoda	66	67	62	71	55	50

Table 14.4 Biological results—average number per depth in pilot filters in winter period

Group	Depth I	Depth II	Depth III	Depth IV	Depth V	Depth VI
Bacteria	910 000 000	550 000 000	400 000 000	220 000 000	160 000 000	160 000 000
Sarcodina	32	34	23	30	31	24
Flagellata	200	150	160	180	190	200
Ciliata	530	410	400	420	410	410
Rotifera	1·2	1·8	2·6	1·8	1·9	1·9
Nematoda	100	88	110	110	140	120

All figures are expressed in millions per m³ of filter medium. (Results from some experimental filters.)

higher proportion of the substrate than do Bacteria. For the latter reason they produce growths with a more marked tendency to ponding, but this is not necessarily cause and effect since the conditions favouring fungal growths also predispose towards ponding.

(iii) Protozoa

Numerically the Protozoa are second only to the Bacteria and are clearly an important member of the filter community. The most important class of Protozoa is the Ciliate as shown in Tables 14.3 and 14.4. Amoeboid Protozoa and Flagellate Protozoa are common but their metabolic rates are so low that they make a small contribution to the overall activity.

Ciliate Protozoa are much larger than Bacteria but are still microscopic and live amongst the zoogleal clumps as part of the filter slime. They feed predominantly on Bacteria and Fungi as they are mainly particle feeders. The metabolic rate is lower than that of Bacteria.

As with Bacteria, the numbers of specific populations fluctuate widely but without any apparent effect on the effluent quality. The numbers of Protozoa increase in the Winter but not as much as the Bacteria.

(iv) Small Metazoa

These are chiefly Nematodes—roundworms, and Rotifers—wheel animals. The Nematodes are present in large numbers in filter slime and have a sufficiently high metabolic rate to make an important contribution to community activities. Nematodes feed mostly on particles and thus compete with the Protozoa, but they also absorb a certain amount of their food in solution.

Rotifers are common in filter beds especially in the lower layers. They do not contribute significantly to the breakdown process and their presence is merely symptomatic of an advanced degree of treatment.

(v) Large Metazoa

These are often referred to as the Grazing Fauna since they are macroscopic and feed on the smaller film organisms. The group consists of Annelid Worms and Insect Larvae. A reasonable amount of work has been carried out on this group since they are clearly very important in filter operation, but the factors determining whether flies or worms will predominate remain obscure. To some extent filter media with a large surface area and numerous crevices, e.g. slag, will tend to encourage the development of a worm population.

Whatever the composition of the Grazing Fauna, worms or Fly larvae are responsible for controlling the extent of film growths in trickling filters, at least during the summer period. During the winter period the prey–predator balance is destroyed.

312

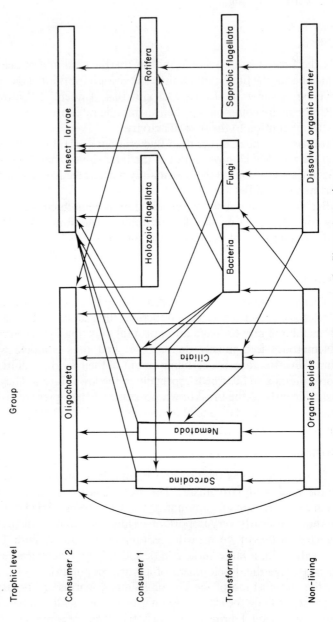

Figure 14.4 Trophic structure of a filter community

(2) Functional analysis

Various functional analyses have been produced of filter communities notably by Hawkes (1964). Unfortunately the approach is too generalized to show the main elements in the food web and has not been sufficiently developed to give quantitative predictions of population sizes or mass transfer. A more detailed approach is shown in Figure 14.4 where all the main populations and food chains are given. This has then been simplified to include only the significant nodes and elements in the energy flow network as shown in Figure 14.5. In Figure 14.5 each population is represented by its metabolic contribution as calculated from the number and the metabolic rates. It shows that almost all the organic matter which is removed from sewage is adsorbed by the Bacteria and transformed into inorganic matter and bacterial protoplasm. Some of the influent organic matter is taken up by the Ciliates and Nematodes, but their principal food is the Bacteria. The grazing fauna prey indiscriminately on the other groups and are undoubtedly the most important predators. Predation is extremely important not only because of the control of the bacterial population but also because the bacteria are transformed into larger units with better settling qualities.

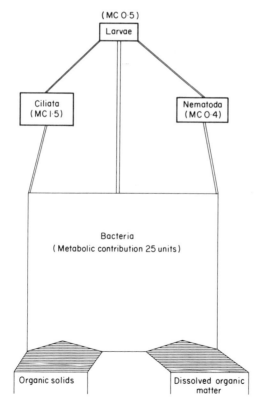

Figure 14.5 Food web in a filter community—simplified (see text for explanation of units)

It is possible to recognize an annual cycle in trickling filters, which is reflected in the community and in the performance. This annual cycle may be divided into summer and winter periods as shown in Figure 14.3. Transition between the two is not a step function, but the conditions are sufficiently distinct to be described separately. During the summer period there is a dynamic equilibrium between the numbers in all the major populations. The numbers are limited by the food supply and predation. For the bacterial population at any depth the available food can be expressed in terms of the average concentration of organic matter at the film-liquid interface. Any change in the initial concentration of food or in the hydraulic conditions would alter the bacterial numbers. More food is present than is required to just support the Bacteria and this would lead to an increase were it not for the predation by the Ciliates, Nematodes, and grazing fauna. The balance between prey and predators is not a steady equilibrium but is subject to violent fluctuations. However, over the 6–8 month summer period in some experimental filters, no long term trends were perceptible in any of the populations.

The onset of the winter conditions comes gradually in the autumn with lower temperatures and increasing weights of slime. As all the organisms concerned operate at the temperature of their surroundings, the lowering of the temperature causes a reduction in the rate of metabolism. The reduction is not the same for all groups but varies in the following manner

Group	Bacteria	Ciliate	Protozoa	Nematodes	Annelida	Insecta
Reduction factor for 10°C drop	1·5–2	3–2·5		2–2·5	2·5	1·5–2

thus disturbing the equilibrium which existed during the summer. As can be seen from Table 14.4, the numbers of bacteria increase tremendously because of the relaxation of predatory pressure and the lower food requirement per bacterium. The rate of bacterial build up is accentuated by the large numbers of bacteria in the influent and the lower death rate in winter. The numbers of Ciliates and Nematodes increase considerably but they do not compensate for a lack of activity on the part of the grazing fauna. In the winter period the Insecta are mainly present as eggs or pupae and Annelid worms, where present, require too long a period of development to take advantage of the increased food supply. The winter condition is, therefore, one of slime accumulation, the rate depending upon the degree of imbalance within the community. This, together with the duration of the winter period, will determine the extent of the slime accumulation and hence the deterioration in filter performance.

14.4 FILTER MODEL

The above description of a percolating filter has not done justice to the complexity of the biological and chemical interactions many of which remain to be elucidated. However, the understanding of the filter processes seems to have reached a stage where an attempt at simulation is possible.

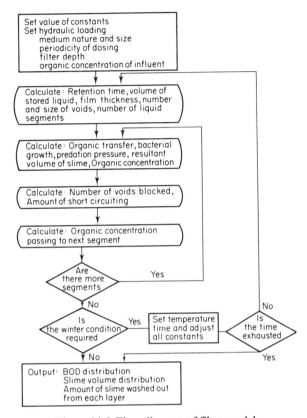

Figure 14.6 Flow diagram of filter model

(1) Model structure

The structure of the model, illustrated in Figure 14.6, is described below. The flow of liquid through a filter is regarded as pulsating so that the overall retention time may be divided into a number of quiescent intervals separated by very brief vertical displacements. The number of such intervals is, of course, given by Retention Time/Periodicity of Dosing. In the majority of cases this will not be an integer so that the last quiescent period is shorter than the others. The transfer of organic matter is represented as taking place from liquid bacterial film during the quiescent periods. The transfer is by molecular diffusion so the process may be represented by Ficks' Law as:

$$\frac{\partial^2 c}{\partial x^2} = \frac{1}{D}\frac{\partial c}{\partial t}$$

with the initial condition

$$c(x,0) = c_0 \qquad\qquad 0 \leqslant x \leqslant L$$

and the boundary conditions

$$c(0, t) = 0 \qquad\qquad 0 < t \leqslant T$$

$$\frac{\partial c}{\partial t}(L, t) = 0 \qquad\qquad 0 < t \leqslant T$$

where: c = concentration of organic matter
x = depth in the liquid layer with a maximum of L
D = diffusion coefficient
t = time interval up to a maximum of T which is the dosing interval

The mean concentration (c_m) at $t = T$ is given by

$$c_m(T) = \frac{1}{L}\int_0^L c(x, T)\,dx$$

which can be integrated to

$$c_m = c_0 \frac{8}{\pi^2}\sum_{n=0}^{\infty} \frac{e^{-DT(2n+1)^2\pi^2/4L^2}}{(2n+1)^2}$$

Hence the mean transfer per cycle (T) per unit area is:

$$(c_0 - c_m)L = c_0 L\left(1 - \frac{8}{\pi^2}\sum_{n=0}^{\infty}\frac{e^{-DT(2n+1)^2\pi^2/4L^2}}{(2n+1)^2}\right)$$

by substitution of

$$A = DT\pi^2/4L^2$$

a series solution can be obtained as

$$c_m = c_0\frac{8}{\pi^2}\left(e^{-A} + \frac{e^{-9A}}{9} + \frac{e^{-25A}}{25} + \ldots\right)$$

During these periods the rate of transfer decrease due to reduction in concentration and the formation of concentration gradients. In the displacement movements mixing occurs so that the concentration is once again uniform. Therefore more frequent dosing tends to maximize the organic transfer.

The organic matter transferred to the bacterial film provides a food source for these organisms. The relationship between film growth, food concentration, and number of predators may be summarized as follows:

(a) *Bacterial Growth*

$$\frac{S\,dN}{dt} = \frac{1}{N}\frac{Kc}{1+c}$$

where N = Bacterial Numbers
S = conversion factors for slime volume
(i.e. volume of slime associated with each bacterial cell)
K = Michaelis–Menten constant
c = concentration of organic matter

(b) *Washout*

$$\frac{dW}{dt} = SN^{K_2}(K_3 Q)$$

where K_2 = washout constant which takes into account of decreased stability
of large slime volume
K_3 = washout constant relating removal to hydraulic loading
Q = hydraulic loading

(c) *Predation*

$$\frac{dP}{dt} = K_4 SN$$

where P = numbers removed by predators
K_4 = conversion constant

The interaction of these processes is very dependent on temperature so that two
conditions have to be defined
(i) *Summer*

$$\frac{dP}{dt} = \frac{dN}{dt} - \frac{dW}{dt}$$

so that a balance slime volume exists
(ii) *Winter*

$$\frac{dN}{dt} < \frac{dP}{dt} + \frac{dW}{dt}$$

so that the slime volume steadily increases.
The effect of increased slime volume is felt to cause some short-circuiting of
liquid through that layer. The relationship between slime volume and short-
circuiting is not well defined but is of the following form:

$$B = (S - \Delta S) K_5$$

where ΔS = minimum volume at which by-passing occurs (this is a function of
the type and size of filter medium)
K_5 = by-pass constant
These calculations are carried out for each layer in the filter in succession as
shown in Figure 14.6 to give the BOD concentration at the top and bottom of
each layer. For the summer condition a steady-state solution is obtained. For
the winter condition the equations are solved on a daily basis and this is iterated
to give the solution at the required time.
The model described above is only at the development stage and cannot,
as yet, be used for design or operation.

318

14.5 REFERENCES

Ames, W. F., Behn, V. C., and Collings, W. Z. (1962). 'Transient Operation of Trickling Filters', *Proc. Amer. Soc. Civ. Eng.*, **88** (SA3), 3121, 21–38.

Atkinson, B., Daud, I. S., and Williams, D. A. (1968). 'A Theory for the Biological Film Reactor', *Trans. Int. Chem. Eng.*, **46**, 9, T245.

Bayley, R. W., and Downing, A. L. (1963). 'Temperature Relationships in Percolating Filters', *J. Inst. Pub. Hlth. Engrs.*, **62**, 55–78.

Bloodgood, D. E., Telezke, G. H., and Pohland, F. G. (1959). 'Fundamental Hydraulic Principles of Trickling Filters', *Sew. & Ind. Wastes*, **31**, 3, 243–53.

Eden, G. E., Brendish, K., and Harvey, B. R. (1964). 'Measurement and Significance of Retention in Percolating Filters', *J. Inst. Sew. Purif.*, pt 6, 3–15.

Escritt, L. B. (1971). *Sewers and Sewage Works*. Allen and Unwin.

Fairall, J. M. (1956). 'Correlation of Trickling Filter Data', *Sew. Ind. Wastes*, **28**, 9, 1069–74.

Galler, W. S. and Gotaas, H. B. (1964). 'Analysis of Biological Filter Variables', *Proc. Amer. Soc. Civ. Eng.* **90** (SA6), 4174, 59–79.

Hawkes, H. A. (1964). *The Ecology of Waste Water Treatment*. Pergamon Press.

Howland, W. E., Pohland, F. G., and Bloodgood, D. E. (1963). 'Kinetics in Trickling Filters', *Int. J. Air Water Polln.*, **5**, 233.

James, A. (1962). 'Ecological and Metabolic Studies on Bacteria in Percolating Filters', *PhD Thesis*, Univ. of Durham.

Monadjemi, D., and Behn, V. C. (1970). 'Oxygen Uptake and Mechanism of Substrate Purification in a Model Trickling Filter', *5th Int. Water Polln. Research Conf.*, Pergamon Press.

National Research Council (1946). 'Sewage Treatment at Military Installations', *Sew. Wks. J.*, **18**, 5, 787–996.

Schulze, K. L. (1960). 'Load and Efficiency of Trickling Filters', *J. Wat. Polln. Cont. Fed.*, **32**, 3, 245–61.

Sheikh M. I. (1968). 'Organic and Liquid Retention Times in Percolating Filters', *Ph.D. thesis*, University of Newcastle.

Sinkoff, M. D., Porges, R., and McDermott, J. H. (1959). 'Mean Residence Time of a Liquid in a Trickling Filter', *Proc. Amer. Soc. Civ. Eng.*, **85** (SA5), 2251, 51–77.

Stack, V. T. (1957). 'Theoretical Performance of Trickling Filter Process', *Sew. Ind. Wastes*, **29**, 987–1001.

Truesdale, G. A., and Eden, G. E. (1964). 'Comparison of Media for Percolating Filters', *J. Inst. Pub. Hlth. Engrs.*, **62**, 283–302.

Upper Mississippi Board of Public Health Engineering and Great Lakes Board of PHE (1951). Report of Joint Committee.

Velz, C. J. (1948). 'A Basic Law for the Performance of Biological Filters', *Sew. Wks. J.*, **20**, 607.

Ware, G. C., and Loveless, J. E. (1958). 'The Construction of Biological Film in a Percolating Filter', *J. appl. Bact.*, **21**, 2, 308–12.

15

CIRIA Model for Cost-effective Wastewater Treatment

K. Bowden and D. E. Wright

15.1 INTRODUCTION

This chapter describes a feasibility study carried out under contract to the Construction Industry Research and Information Association (CIRIA) by the Local Government Operational Research Unit (LGORU) between June 1971 and December 1972.

The aim of the feasibility study was to examine whether a mathematical model could be built that would enable the optimal (i.e. most cost-effective) design of a sewage treatment works to be found for a prescribed standard of effluent.

The chapter outlines the underlying principles of a simple prototype model that was constructed during the study. The chapter is a condensed version of a full report on the study issued by CIRIA.

Every day in Britain an average of eighteen million cubic metres (4000 million gallons) of sewage pass through sewage treatment work of some kind. In fact, sewage treatment outranks many industrial processes of the first order. It is bigger than liquid fuel consumption, for example, by around fifty to one, and brewing by a hundred to one.

Sewage treatment is an area to which the country will be devoting an increasing proportion of its resources. The Government has announced that it intends to permit local authorities substantially to increase spending on new sewage treatment works.

It is appropriate therefore that at this juncture renewed attention should be paid to the techniques by which works are designed, for improvements to technique could result in large savings in capital resources and operating expenditure.

It is the job of the civil engineer, with those specialists who advise him, to build sewage treatment works that achieve the required degree of purification with the greatest economy of resources. At the present time this is taken to be when the present value of capital and operating costs is minimized, although it should be noted that the time is coming when an improvement to the standards of effluent discharged to natural waters will be habitually regarded as conferring tangible and measurable cash benefits. In this latter case the scale of works desirable will be assessed on the normal industrial investment criterion of maximum net present value.

This minimum cost doctrine has particular significance for sewage treatment. A complete works consists of a sequence of *Process stages* in which the performance of one stage influences the effectiveness of the next. The designer of a works cannot ignore this interaction if he wishes to select the particular combination of process stages (and the relative capacities and capabilities of each stage in that combination) that represents optimal *cost effectiveness.*

The principal purpose of this project has been to examine the feasibility of building a mathematical model of the sewage treatment process suitable for optimizing cost effectiveness.

15.2 STUDY OBJECTIVES

Given that mathematical techniques are available on which to base the model (and this was by no means certain at the start of the work), the quality and quantity of available data determine completely the relevance of the model to the real treatment works. Thus one object of the study was to see whether the best data available were sufficient for the construction and effective operation of a mathematical model of a sewage treatment works. This could not be decided from an abstract and theoretical point of view, and it was agreed that the only realistic way to appreciate the many problems involved was to construct a simple prototype model.

The second point is that the provision of a working optimization method is intended as an *aid* to a designer in his selection of the most economic scheme of treatment. Although it may reduce the area in which the designer's intuition plays a part, it cannot eliminate the need for his clear and considered judgement. An aim that was kept clearly in view was to work in terms familiar to the practising engineer, so that the model when developed would find ready application. It was not intended to produce a research tool as such, although the model will have major ramifications in the research field.

Finally, it was agreed that the prototype model should be built so as to form a sound basis for future development.

15.3 LIMITS AND ASSUMPTIONS OF THE PROTOTYPE MODEL

The limits and assumptions described below were adopted in order to give realistic boundaries to the study. The potential breadth of the project was enormous, and unless limits were set there was the danger that too much would be attempted and too little achieved. Every limit and assumption has been discussed in detail and made, bearing in mind, the need to retain a strong link with reality.

It was decided that:

(1) the model should begin at the point following preliminary treatment, and should end with the final disposal of sludge residues and with the

discharge of treated liquor immediately prior to its being released into a watercourse.

(2) the capital and operating costs to the local authority only should be considered; thus, the evaluation of benefits to amenity resulting from improved treatment would be excluded.

(3) Only commonly used processes would be incorporated in the model.

(4) The processes of purification would be described in terms of generally accepted indicators of sewage quality.

(5) The model should not be built with the ability to optimize the design of the plan layout of the works site, since this degree of sophistication is not warranted in a feasibility study.

(6) The model would optimize for a given flow and site plan and slope.

Two simplifying assumptions had to be made that reflect the present state of knowledge of sewage treatment processes. These were:

(1) That the works is operating under steady-state conditions at the 'design flow (e.g. some arbitrarily chosen multiple of dry weather flow). We realize that many will think that by ignoring the variability of flow and load we have departed too far from reality. We accept that variability is of great importance, but being faced with the task of expressing its effect on performance had to conclude that at the present time there are no data on which to assess its effects. It should be noted that a massive volume of additional data will be required, as well as more research, into the effects of transient phenomena in order to extend the model to cater for these variations.

(2) That different incoming sewages are 'equally treatable'. It is assumed that the constituents of sewage not specifically included in the list defining the influent, but that can affect performance, do not vary from sewage to sewage. If it were required, for example, to assess the effects of the presence of an industrial discharge on a works, further constituents would have to be added and traced through the model.

It was necessary to limit the number of process stages to be included and three groups were considered:

(1) Group 1 comprised a chosen set of process stages representing a 'practical and frequently used' system of sewage treatment. This was intended to be the basic minimum requirement for a feasible model.

(2) Group 2 included alternative and commonly used stages. It was intended that the study would embrace as many of these as time permitted.

(3) Group 3 included other processes which, although of interest and potential future importance, were not in sufficiently common usage to justify inclusion in the feasibility study. However, if data on them were found, these data were to be stored for future use.

The list of process stages drawn up at the start of the study is given in Table 15.1.

Table 15.1 Process stages incorporated as modules in the model

Treatment stage	Module
Liquor stream	
Primary treatment:	1. Rectangular settlement tanks (horizontal flow)
	2. Circular settlement tanks (radial flow)
Secondary treatment:	3. Aeration tanks (activated sludge—diffused air system)
	4. Aeration tanks (activated sludge—surface aerators)
	5. Biological filters (single and recirculation)
	6. Biological filters (alternating double filtration)
	7. Circular settlement tanks (radial flow)
	(a) Following aeration tanks
	(b) Following biological filters
Tertiary treatment:	8. Grass plots
	9. Microstrainers
	10. Upward flow sand filters
	11. Downward flow sand filters
Disposal of final effluent:	To natural waters (direct)
Sludge stream	
Raw sludge thickening:	12. Thickening tanks (gravity consolidation)
Digestion:	13. Digesters (heated anaerobic)
Digested sludge thickening:	14. Thickening tanks (gravity consolidation)
Conditioning:	15. Chemical conditioning tanks
Dewatering:	16. Drying beds
	17. Filter presses
Sludge disposal:	18. Dump on land
	19. Spread on agricultural land
	20. Dispose at sea

The final list of process modules (the model's representation of each process stage) included in the prototype model is given in the first column of Table 15.4. It includes all of Group 1, with the exception of storm and balancing tanks (which were excluded following the decision to limit the study to steady state conditions), nearly all Group 2, and one process module from Group 3.

15.4 GENERAL CONCEPTS

Optimization

The aim of the model is to seek the combination of process stages and the relative sizes of stages within that combination that produce a stipulated degree of purification at minimum cost, taking into account capital and operating costs.

This type of objective occurs commonly in other process design fields (e.g. the petrochemical industry and iron and steel manufacturing). In these cases the quality of the final product often depends on the cumulative effects of several operations applied in sequence. The same is true in a sewage works; the crude liquor passes through one process stage after another until the final effluent emerges. Sewage treatment is complicated by two factors; not all works use the same combination of process stages; and some of the liquor or sludge may be fed back into the process at an 'upstream' point. These add to the complexity of the model, but the problems can be overcome.

We have devised a means of optimizing the cost effectiveness of a whole sewage treatment works that recognizes that each process stage has to conform to the conditions created by the preceding stage and to the constraints imposed by the following stage. Each process stage clearly has a cost that depends on the flow to be treated and the reduction of polluting load achieved by that stage, and this gives rise to the need to formulate *performance relationships* and *cost relationships*.

The advantage of being able to divide up the works into process stages is that each can then be described separately. The representation of each process stage must correctly express the two characteristics of performance and cost, in addition, the relationship to the next stage and to a previous stage if feedback is involved. The *process module* is the model's representation of a process stage.

A brief look at what goes on in a process module is essential in order to understand how the overall optimization works. All modules are alike in that they receive inputs and produce outputs. The description of the flow of sewage and sludge through each stage of the works and its nature at that stage is accomplished by the use of indicators of quantity and quality. These are called '*process stream parameters*'.

Inside the module an alteration occurs in one or more of the process stream parameters, resulting in the production of the output process streams. This alteration is a consequence of the biological, chemical, or physical mechanism inherent in the stage represented by the module. It is described algebraically through the medium of the performance relationships which are expressions describing how the degree of alteration in the process stream is linked to a specified physical and controlling characteristic of the stage. In practice the controlling characteristics are very numerous. For simplicity and because of the general lack of detailed analysis of performance relationships, the controlling characteristics are usually confined in the model to a single factor on which it is judged the performance of a stage principally depends. This specified characteristic is called the *design variable* and a list of those used for each process stage in the prototype model is given in Table 15.2.

The design variable controls the relationships between the input process stream(s) and the output process stream(s) expressed by changes in the composition of the process stream. It also has another function of great importance. Because of its role in describing a characteristic feature of the stage whose performance it controls, the design variable must, through appropriate relation-

Table 15.2 Design characteristics

Process module	Design variable(s)	Structural factors
Liquor stream		
1. Primary settlement tanks (rectangular—horizontal flow)	Retention time (h)	Length/breadth ratio; average tank depth (m)
2. Primary settlement tanks (circular—radial flow)	Retention time (h)	Average tank depth (m)
3. Aeration tanks (activated sludge—diffused air system)	Retention time (h); power* (kW); proportion of activated sludge entrained	Average tank depth (m)
4. Aeration tanks (activated sludge—surface aerators)	Retention time (h); power* (kW); proportion of activated sludge returned	Average tank depth (m)
5. Biological filters (single and recirculation)	Hydraulic loading (m^3/m^3d), recirculation ratio	Average filter depth (m); length/breadth ratio; specific surface area of media[†] (m^2/m^3)
6. Biological filters (alternating double filtration)	Hydraulic loading (m^3d).	Average filter depth (m); length/breadth ratio; specific surface area of media[†] (m^2/m^3)
7. Secondary settlement tanks (circular—radial flow)		
(a) Following aeration tanks	Solids loading rate[‡] (kg/m^2d). Sludge pumping rate (m^3s)	Average tank depth (m)
(b) Following biological filters	Upward velocity[‡] (mm/s)	
8. Grass plots	Loading rate[‡] (m^3/m^2d)	—
9. Microstrainers	Fabric grade[§]; loading rate* (m^3/m^2d)	—
10. Upward flow sand filters	Loading rate (m^3/m^2d)	—
11. Downward flow sand filters	Loading rate (m^3/m^2d)	—
Sludge stream		
12. Raw sludge thickening tanks (gravity consolidation)	Retention time (d)	Average tank depth (m)
13. Digesters (heated anaerobic)	Retention time (d); temperature[‡] (°C)	Average tank depth (m)
14. Digested sludge thickening tanks (gravity consolidation)	Retention time (d)	Average tank depth (m)
15. Chemical conditioning tanks	Dose rate (expressed as a percentage of the weight of sludge dry solids)	Storage required (d); chemical density (kg/m^3); storage tank height (m)

Table 15.2 (*Contd.*)

Process module	Design variable(s)	Structural factors
16. Drying beds 17. Filter presses	Loading rate (kg of dry sludge/m²); final solids content Pressing time* (h)	—
18. Dump on land 19. Disposal at sea 20. Spread on agricultural land	—	No. of drops/day, no. of days/week worked. Round trip journey time (h), no. of work periods in a week, duration of each work period.

*The values of these parameters are determined within the model and depend on the current values of the process stream parameters. The user, therefore, has no direct control of these values.
†Strictly, the specific surface of the media can affect both performance and cost and hence should be listed as a design variable. However, in practice, the choice of media may well be limited and the user would be advised to work only with those values that he knew to be available.
‡Performance relationships within the model only exist for specific values of these parameters. They are listed as design variables because it is accepted that they do influence both performance and cost.
§At present the model contains relationships relating only to Mk I grade micro-fabric.
¶The user is required to specify a loading rate and final dry solids content that are consistent with one month's drying time, during the drying season.

ships, be involved in the sequence of calculations that determine the cost of the stage.

When individual modules are linked together to form a model of a sewage treatment works, the interaction of the process stages enables us to examine alternative arrangements of stages that give the same overall performance. The opportunity to exploit alternative combinations is limited by the operational range of the design variables for the stages being used and by the constraints on final effluent quality (i.e. the pre-determined set of process stage parameters defining the output of the last module).

Effluent quality is specified by the parameters of the liquor stream, three of which (ammoniacal nitrogen, BOD, and suspended solids) can be set to the desired standards. The operation of the model ensures that in the final effluent each of the three fixed parameters falls on or within the required standard. One parameter may be precisely at the level of the design standard, and the model will bring the others exactly into line only if it can gain economic advantage.

The overall optimization of the works model is thus reduced to a search through all feasible combinations of design variables to find a single combination that minimizes cost. Carried out at random, this search would be a massive task. The function of the search procedure incorporated into the computer program is to direct the search according to strict mathematical criteria. Given a starting set of design variables, the search procedure embarks on a cycle of calculations the result of which is the selection of a design variable that can

326

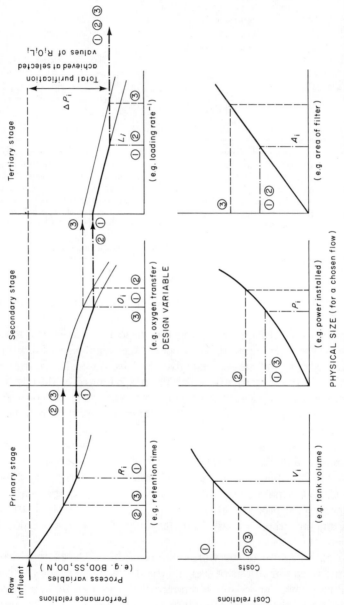

Figure 15.1 Principle of optimization. For selected values of the design variables R_i, O_i and L_i a pollutant reduction of ΔP is achieved at a total cost $C_t (= C_1 + C_2 + C_3)$. Various combinations of R, O and L can be tried (e.g. routes 1, 2 and 3) to find the combination that gives the minimum total cost (capital and operating). Note: Liquor stream only illustrated. Curves shown have no technical or economic significance

profitably be changed. This change is made and a new cycle initiated. This procedure is repeated until no further changes can usefully be made.

The principle of optimization as applied to sewage treatment is illustrated in Figure 15.1, which will also help the reader to visualize the significance of the various elements of the process module. In so far as the diagram is specific, it relates to sewage liquor, but the prototype model includes all the important sludge treatment processes.

The three columns in the diagram correspond to the main stages of liquor treatment. The upper row shows the performance relationships that link the reduction in pollutant concentration to the most significant design variable. The lower row shows the cost relationships that link the total costs (capital and operating) to the physical size of the stage. The physical size of a particular stage depends directly on the value of the design variable selected by the model and the flow. The diagram shows how purification proceeds from one stage to the next for three selected values of design variables, R, O, and L. Each value of R, O, and L has an associated value of the physical size, V, P, and A and hence total cost $C_T = C_I + C_{II} + C_{III}$.

Using its optimization routine the model examines a whole set of alternative combinations of the design variables (in the diagram represented by routes 1, 2, and 3), and searches for the particular combination that has minimum total cost.

Process stream parameters and structural factors

It was necessary to keep the number of process stream parameters down to a manageable level and within the limits of available data and knowledge. For this reason, many parameters of research interest have had to be excluded. On the other hand, those process stream parameters that have been selected include all the major ones currently used in design practice. pH is a parameter often measured at works, but has been excluded from this model because there is little quantitative knowledge of its effects on performance. Furthermore, it is difficult to calculate the pH of a combined liquor flow from those of the component liquor streams.

Table 15.3 lists the ten parameters eventually selected for use in the model, divided into the two subgroups. In the model it was sometimes necessary to consider the dissolved and suspended components of certain parameters. Rather than consider these components as parameters in their own right, and hence have to calculate each component for every process stream, it was decided to make the distinction only when required by a particular process stage.

For each of the process modules included in the model, it was necessary to establish those design characteristics that were sufficient to determine performance on the one hand, and cost on the other. There may be more than one cost relationship in a process module. Because the design variable may not carry enough information about size to permit a complete evaluation of cost, additio-

Table 15.3 Process stream parameters

Liquor Stream
Volumetric flow (m³/s)
Temperature (°C)
Suspended solids (mg/l)
Biochemical oxygen demand (5 day with A.T.U.*) (mg/l)
Ammoniacal nitrogen expressed as N (mg/l)
Total oxidized nitrogen expressed as N (mg/l)
Dissolved oxygen (mg/l)

Sludge Stream
Volumetric flow (m³/s)
Mass flow (tonnes/day)
Dry solids content (% by weight)
Volatile matter (% by weight of dry solids)

---

*Allylthiourea. BOD_s with ATU measures carbonaceous demand only. ATU suppresses nitrification.

nal items of information about the physical characteristics of the unit may be required. These are called *structural factors*. They are distinguished from the design variable because, although they affect the cost of the stage, they do not exert a prime influence on its performance (at least within the limits of present knowledge).

For example, for a given level of influent suspended solids, retention time was considered sufficient for the purposes of the study to determine the performance of a primary settlement tank. In conjunction with the sewage flow it can be used, through tank volume, to estimate the capital cost of the tank. Because the cost of the scraper may also be required (which can be considered to be related to the span of the tank), a knowledge of the tank geometry is needed. The model therefore requires two additional structural factors in the case of a rectangular tank (i.e. length-to-breadth ratio, and depth) in order to determine the total capital cost.

Because the relationship between structural factors and performance is not sufficiently well known to be included in the model, the user must fix their values at conventional levels. The model will not vary them as it will design variables; they exist merely to provide additional (but nonetheless important) information in order to complete the cost picture.

15.5 PERFORMANCE RELATIONSHIPS

It was initially decided that all the performance relationships to be used in the model would be derived either from the results of published research or from data extracted from the existing records of selected sewage treatment works. Five Process Stage Groups [These were Primary settlement, Activated sludge treatment, Biological filtration, Tertiary treatment, Sludge treatment and disposal. Each comprised practising engineer(s) expert in a particular process and research worker(s).] directed the search for these relationships. It was agreed that in the first instance requests for data should be restricted to a small number of major authorities (Greater London Council; Upper Tame Main Drainage Authority; Middle Lee Regional Drainage Scheme).

In only one case were useful results obtained from this part of the exercise; these related to the performance of rectangular primary settlement tanks. For the other processes the analytical approach was abandoned and performance relationships had to be obtained by other means.

After much discussion in the Steering Group and Process Stage Groups (see Acknowledgements), it was decided that it would be valuable for the purposes of the feasibility study for those involved to postulate what they considered to represent reasonable performance relationships. This technique of postulating figures and curves has always resulted in vigorous discussion and comment, and we are convinced that this approach has produced the best knowledge available at the present time. Where appropriate, plant manufacturers were also consulted and their views on the performance relationships were taken into account.

Thus, using a variety of approaches it was possible to establish performance relationships for all the processes given in Table 15.1. Some of these relationships may be imperfect, but at least they distill, combine and make available in numerical form the knowledge and experience of several public health engineers and specialists. As such they are an advance in themselves, although clearly one of the most urgent tasks following this work will be to obtain corroborative data. These performance relations are given in CIRIA (1973).

15.6 COSTS

The second function of each process module is to determine the cost of the process. The costs associated with each process can generally be considered to come under the two main categories.

(1) Capital costs, taken to mean those costs that are incurred in the construction or purchase of a piece of plant or equipment.
(2) Operating costs, those costs that are incurred during every year of the life of the installation.

The latter costs include items such as labour, insurance, maintenance and repair, power, consumable materials, and tools, rates, rents and taxes, administration and general support and transport, and may be of the order of 40% of capital (on a present value basis). They will clearly have a vital role in determining the optimum combination.

In a model concerned with minimizing the cost of a complete installation it is essential to bring these two fundamentally different types of cost on to a common footing. This can readily be done on discounted cash flow principles by using annual cash flow. To obtain this, the capital cost is converted to a sequence of uniform annual payments, which for a specified life and at a given interest rate are equivalent to the capital sum. The sum of these annual costs and the annual operating costs is called the annual cash flow.

The real problem is to establish a suitable method of determining the capital and operating costs of those process stages involved in the optimization. Because of the way these costs are known to vary from one area of the country to another, and often even from site to site, it was decided not to build into the model a standard set of costs. Instead, a framework was produced into which the user can insert a set of his own cost data and from which the model builds up the costs it requires. These gives the user control and ensures that the model can be made applicable to any situation. The Costs Group agreed the terms in which the user should specify his cost data. The result was that the user is asked to specify unit cost curves based on suitable physical parameters of the process stage required. Typically the user will have to provide a curve relating cost per unit volume with volume. An example of the cost calculation procedure relating to primary settlement is given in Table 15.4 and the complete procedure is described in detail in CIRIA (1973).

The Costs Group considered the format prepared by the IWPC Technical Sub-Committee on Costs (1973) but decided that a somewhat simpler format should be adopted for the prototype model. As cost information on the lines recommended by the IWPC is collected, so the cost calculation procedures used in the model can be made more sophisticated.

Because the user has such an important role to play in the provision of cost data, the detailed points that were considered when drawing up the cost calculation framework are given below for reference:

(1) Absolute numerical costs are not required for the optimization procedure provided the model is given accurate information on relative costs. However, the calculation of relative magnitude required knowledge of absolute figures, and realism is better served if actual cash figures are used.

(2) Cost information is to be presented at a level of detail sufficient for a designer wanting to make a reasonable first estimate of the cost of sewage treatment works.

(3) The user will have to specify his own numerical costs (or relative costs) to suit his own particular local conditions.

(4) The geometrical proportions of the unit(s) required for each process stage must be fixed by the user as the prototype model will not sub-optimize the geometry of individual units.

(5) To optimize for a particular configuration of structures, the user will have to insert costs (or relative costs) that suit the proposed excavated depths of the various structures. Thus, if the user wishes to examine the alternatives of pumping at the inlet to the works, or pumping downstream of the works, he will need to use the model twice, each time varying his cost figures (or relative costs) to match the differing degrees of excavation required.

(6) In the developed versions of the model, the costs of each unit can be built

Table 15.4 Cost calculation procedure

1	2	3	4	5	6	7	8
Process module	Design variable(s)	Process stage input	Structural factors	Total size of stage	Name of process unit (P.U.)	Max. size of P.U.	Min. no. of P.U.'s
	Model/user	Model	User	Model	—	User	User
1. Primary settlement tanks (rectangular—horizontal flow) 2. Primary settlement tanks (circular—radial flow)	Retention time T_j (hours)	Flow Q (m³/s)	Average depth: D (m) Length to breadth ratio for rectangular tanks $= \alpha$	Volume $V_j = Q.T_j \times 3600$ (m³)	Tank	Volume $\hat{v}$ (m³)	$\hat{n}$

Table 15.4 (*Contd.*)

9	10	11	12	13	14	15
Derived no. of P.U.'s	Derived size of P.U.	Standby capacity %	Total no. of P.U.'s	Cost curve Parameters (x:y)	Capital cost of each P.U.	Capital cost replication factor
Model	Model	User	Model	User's curves	Model	User

Civils

$$n_j = \text{max. of } \hat{n} \text{ and } \left[\frac{V_j}{\hat{v}}\right]^{\text{Int}} + 1$$

$$v_j = \frac{V_j}{n_j}$$

$$(\text{m}^3)$$

$$e$$

$$n_j^{\text{T}} = \left[\left(1 + \frac{e}{100}\right)n_j\right]^{\text{Int}} + 1$$

Volume v: $\left(\dfrac{\text{cost}}{\text{m}^3}\right)$

Mechanicals
Span s: $\left(\dfrac{\text{cost}}{\text{m}}\right)$

$$C_j^{\text{c}} = \left(\frac{\text{cost}}{\text{m}^3}\right)_{v_j} \times v_j$$

$$C_j^{\text{m}} = \left(\frac{\text{cost}}{\text{m}}\right)_{s_j} \times s_j$$

$$a$$

$$b$$

Table 15.4 (Contd.)

16	17	18	19	20	21	22	23
Total capital costs	Annual operating cost function	Annual operating cost	Interest rate	Economic life	Capital recovery factors	Total annual cash flow	Comments
Model	User's curve	Model	User	User	Model	Model	
$S_j^c =$ $C_j^c \times (n_j^T)^e$	x = total capital cost	$s_j^o =$ $g(S_j^c + S_j^m)$	1% p.a.	Civils N^c years	$f(\cdot, N^c)$	$S_j = S_j^o +$ $S_j^c \times f(i, N^c) +$	$s_j = \sqrt{\dfrac{v_j}{\alpha.D}}$
$S_j^m =$ $C_j^m \times (n_j^T)^b$	y (annual operating cost) $= g(x)$			Mechanical N^m (years)	$f(i, N^m)$	$S_j^m \times f(i, N^m)$	For circular tanks $\alpha = \pi$

Notes 1. The model is assumed to be evaluating the 'jth' set of design variable values.
2. The notation...$[Z]^{Int}$ should be read as..."the largest integer, less than Z". eg. $[4 \cdot 6]^{Int} = 4$ and $[3]^{Int} = 2$.
3. In appropriate cases the model will use the depth measurement to calculate the site area required.

up from the principal components (e.g. excavation, shuttering, concrete and steel). These data will be needed in any case if sub-optimization is to be done in later models to achieve the most economic size and shape of unit.

(7) Land costs are important, but it will be sufficient for the prototype model to calculate the land required for each alternative the designer wishes to examine on the basis of the plan area of the units.

(8) The model does not take account of the cost of pipework, culverts, and pumps *between* process stages, as for a given flow, site slope or plan layout these works do not affect the point of optimization. (Pipes and pumps integral to a process stage are costed in that stage.)

The effect of these points is that the model will provide the user with costs that are valid for comparison with other alternatives it gives, but these costs will *not* be the total costs of a complete sewage treatment works, which must be calculated by conventional methods. All costs used in the model must refer to a common year, to avoid distortion as a result of inflation.

15.7 CONCLUSION

The model, even in prototype form, confers the five major benefits outlined below.

(1) Design

The model produced by this feasibility study will only broadly resemble all the complexities of a real works. However, for the first time it will provide an overall framework for the economic evaluation of many combinations of sewage treatment processes, and in responsible hands should give significant results. It is envisaged that for some time the model will have to be used in conjunction with those who have been involved in its development.

(2) Sensitivity analysis

Once a framework is available that brings together all the main elements involved in the design of treatment works it becomes possible to assess which of these many elements bear most significantly on the economics of sewage treatment. The model will thus provide the means of highlighting 'sensitive' areas and will be of great assistance to those responsible for directing attention and resources to the improvement of sewage treatment.

(3) Research and the collection of performance data

One very important aspect of the sensitivity analysis referred to above is the role of the performance curves built into the model. Many of these are postulated relationships and the effect they have on the overall economics of sewage treatment can be tested in the model. It is anticipated that the performance relationships that have the greatest impact on cost will become the objects of closest scrutiny, and that where data are lacking effort will be directed to obtaining reliable information. Thus, the model will have an important part to play in the effective direction of applied research.

(4) New Attitudes

The existence of a framework into which performance relationships and costs can be inserted, and that enables an overall view of sewage treatment to be taken, should have a stimulating effect throughout the sewage treatment world. The publication of postulated performance relationships will provoke comment and, with the additional information that will result, the curves can be steadily improved.

(5) New treatment methods

The existence of the model will mean that those advocating new processes of treatment will have to present their data in terms of the relevant process stream parameters and show how these are affected over a range of design variables. In this way the economic impact of new methods can be properly assessed.

The conclusion that most clearly emerges from the work so far is that no value would be gained by proceeding immediately to build a full model for general use. There is no doubt that a full model is mathematically feasible. Its construction is unlikely to pose any mathematical problems that have not been met and overcome in building the prototype, and should any arise they are, as far as can be judged, likely to be amenable to solution by standard methods. Ultimately, however, the value and quality of the full model depends on the quality of the knowledge of individual sewage treatment processes, and this is still in its very early stages. A more gradual process of development in which modifications to the model keep pace with advances in understanding is, we believe, more appropriate in this case.

Moreover, it must be stressed that there is no intention of developing the prototype model to the point where the user is required merely to specify the required flow through the works and 'press the button' to have the computer produce a fully optimized works design. This may be technically feasible, but it has been a basic tenet of this study that the resulting computer model should be an *aid* to design, albeit a sophisticated one. The aim of the model is to evaluate alternative designs proposed by the sewage works designed. To this end it is considered essential that he should be able to make his own assumptions freely and to introduce his own design specifications.

Acknowledgements

The study exemplifies CIRIA's declared policy of involving its members and other specialists in the selection and management of its projects, and could not

336

have been completed without the efforts and cooperation of a great many people from a wide diversity of backgrounds. The successful outcome of the work owes much to their willingness to subordinate their specialized interests to the needs of overall appraisal, and their collective ability to cut through several levels of detail to get at the core of the problem.

The authors must express their very real gratitude for the work done by their colleagues on the CIRIA Project Steering Group which managed and directed the study.

R. Best	Directorate General Water Engineering, Department of the Environment
P. Coackley	University of Strathclyde
R. A. R. Drake	Greater London Council
R. S. Gale	Water Pollution Research Laboratory, Department of the Environment
J. D. Swanwick	Water Pollution Research Laboratory, Department of the Environment
G. A. Truesdale	D. Balfour & Sons
J. A. Green	Local Government Operational Research Unit.

The Steering Group reported every six months to the Working Party listed below:

J. T. Calvert (Chairman)	John Taylor & Sons
G.K. Anderson (from April 1972)	University of Newcastle upon Tyne
R. Best	DGWE
K. Bowden	LBORU
P. Coackley	University of Strathclyde
R. A. R. Drake	Greater London Council
E. A. Drew	Middle Lee Regional Drainage Scheme
E. B. Fletcher	Ipswich County Borough
R. S. Gale	WPRL
J. A. Green	LGORU
M. J. Hamlin	University of Birmingham
R. Hibberd (until April 1972)	University of Newcastle upon Tyne (Now with Satec Limited)
A. E. Naylor	Leeds City and County Borough
D. H. Newsome	Water Resources Board
E. H. Nicoll	Scottish Development Department
H. R. Oakley	J. D. and D. M. Watson
R. F. Pearson	Formerly, Greater London Council
D. A. D. Reeve	Upper Tame Main Drainage Authority

J. R. Simpson (until November 1971)	D. Balfour & Sons.
T. H. Y. Tebbutt	University of Birmingham
J. D. Swanwick (from February 1972)	WPRL
G. A. Truesdale (from November 1971)	D. Balfour & Sons.
D. E. Wright (Secretary)	CIRIA

The Process Stage Groups and Costs Group were formed to be representative of the designer, operator, and research worker and were chaired by individual members of the Steering Group as follows:

Primary settlement	R. A. R. Drake
Activated sludge treatment	P. Coackley
Biological filtration	R. Best
Tertiary treatment	G. A. Truesdale
Sludge treatment and disposal	R. S. Gale and J. D. Swanwick
Costs Group	D. E. Wright

The names of all who served on these groups are given in CIRIA (1973).

General responsibility was borne by CIRIA's Hydraulic and Public Health Engineering Committee chaired by Mr. H. R. Oakley (J. D. and D. M. Watson).

The study was funded by CIRIA and special contributions from the Department of the Environment and local authorities, new town development corporations, main drainage boards, and county councils.

The permission of the Director, Construction Industry Research and Information Association to publish and present this paper is gratefully acknowledged.

15.8 REFERENCES

CIRIA (1973). 'Cost-Effective Sewage Treatment: 'The Creation of an Optimising Model', *Report No. 46* (in two volumes), Construction Industry Research and Information Association, London.

IWPC (1973). 'Capital and Operating Costs of Sewage Treatment', *Water Pollution Control*, **72**, No. 1, 103.

N.B. A bibliography is given in the first reference.

16

Optimization Model for Tertiary Treatment Rapid Filtration

K. J. Ives

16.1 INTRODUCTION

Although there are alternative methods of tertiary treatment to produce effluents of a better quality than the Royal Commission standard, rapid filtration is probably the most widely used. This chapter offers no rationale for the choice of tertiary treatment method, but assumes that rapid filtration has been chosen.

From this starting point the engineer has to consider the design and operating variables, and should choose values for these to produce the required effluent filtrate quality at least cost. This means that he should try to satisfy two important criteria:

(1) The treatment plant must accomplish the required level of purification for the desired water output.
(2) The design should be the most economically feasible one: an optimum design as far as product and its cost are concerned.

Satisfaction of these two criteria poses difficult questions to the engineer concerned with design of rapid filters. Is it better to have a deeper filter of coarse media, or a shallow filter of fine media? Is it better to have a deeper filter at high flow rate, or a shallow filter at low flow rate? What are the methods of obtaining filtered effluent at least cost?

In order to answer these questions it is necessary either to have extensive data for the filtration of effluents through different thickness of media of different grain sizes at many different rates, with respect to filtrate quality, head loss, and cost per cubic metre of filtrate, or to have reliable mathematical models of the filtration process. Recent developments in filter theory allow the mathematical model method to be used.

16.2 MATHEMATICAL MODELS

For the purposes of optimization, mathematical models of the filtration process must describe the relationship between the quality of the filtrate, and the total head loss and the independent variables of filter depth and time during the filter operation. Assuming that the nature of the inflowing suspension and the water temperature remain reasonably constant, then the influence of filtration velocity, grain size, and grain size distribution with depth must be known in a mathematical form, as they affect filtrate quality and head loss.

It is generally agreed among research workers on filtration theory that the first-order relation between concentration (C) and depth (L) proposed by Iwasaki (1937) is valid for the initial condition of filtration through a uniform bed of media.

$$-\frac{\partial C}{\partial L} = \lambda C \qquad (16.1)$$

where λ is the filter coefficient.

Also a continuity equation relates the quantity of suspension particles removed from the flowing liquid to the quantity of deposit accumulating in the filter pores.

$$-\frac{\partial C}{\partial L} = \frac{1}{v}\frac{\partial \sigma}{\partial t} \qquad (16.2)$$

where σ is the specific deposit (volume of deposit per unit filter volume). t is filter run time, and v is the approach velocity of filtration.

During the filter run the efficiency of the filter changes due to the accumulation of deposits in the pores. There are two principal theories describing these changes: one due to Mints (1966) which attributes changes to scour of deposits by the water flowing through the pores, and one due to Ives (1969) which attributes changes to geometric and velocity changes in the pores. Mints described optimization procedures based on his theory, and these have been extended in a very detailed manner by Gur (1969).

The relationship derived by Ives is

$$\frac{\lambda}{\lambda_0} = \left(1 + \frac{\beta\sigma}{f}\right)^y \left(1 - \frac{\sigma}{f}\right)^z \left(1 - \frac{-\sigma}{\sigma_u}\right)^x \qquad (16.3)$$

where λ_0 is the filter coefficient at $t = 0$ (clean filter), f is the initial filter porosity, σ_u is the ultimate saturation value of specific deposit, and β, x, y, z, are empirically derived factors.

It has been shown that equation (16.3) is general in that all other theories, except Mints', can be described by appropriate choices of x, y, and z.

Equations (16.1), (16.2), and (16.3) describe clarification of a suspension by filtration at constant rate (v), in a uniform filter of constant grain size (d). If the filter is operated at a different rate, or with a different grain size, then the factors λ_0, β, σ_u, x, y, z will change. Although some theoretical investigations have been made of the effects of changing flow rate and grain size (Ives and Sholji, 1965; Ison and Ives, 1969), it is necessary to obtain experimental information because the relationships depend on the nature of the suspension (Mohanka, 1969).

These relationships may be written

$$\lambda_0 = \Lambda/d^{m_1} v^{n_1} \qquad (16.4.1)$$

$$\beta = Bd^{m_2} v^{n_2} \qquad (16.4.2)$$

$$\sigma_u = S/d^{m_3} (1 + v)^{n_3} \qquad (16.4.3)$$

$$x = X/d^{m_4} v^{n_4} \qquad (16.4.4)$$

$$y = Y/d^{m_5} v^{n_5} \qquad (16.4.5)$$

$$z = Z/d^{m_6} v^{n_6} \qquad (16.4.6)$$

Taking granular material with a porosity (f) of Q.44, filtering ferric hydroxide floc suspension in water, Mohanka obtained the following results, in cm g min units:

$$\Lambda = 5 \cdot 9 \times 10^{-3} \quad m_1 = 1 \cdot 35 \quad n_1 = 0 \cdot 25$$

$$B = 11 \cdot 8 \qquad\qquad m_2 = 0 \cdot 75 \quad n_2 = 0$$

$$S = 0 \cdot 44 \qquad\quad m_3 = 0 \qquad\; n_3 = 0 \cdot 75$$

$$X = 0 \cdot 95 \qquad\quad m_4 = 0 \cdot 61 \quad n_4 = 0 \cdot 24$$

$$Y = 1 \cdot 5 \qquad\qquad m_5 = 0 \qquad\; n_5 = 0$$

$$Z = 0 \cdot 75 \qquad\quad m_6 = 0 \qquad\; n_6 = 0$$

Note that λ_0 has dimensions of cm^{-1}, all other factors (β, σ_u, x, y, z) are dimensionless.

In addition to the action of clarifying the suspension, the filter shows a loss of head. The Kozeny–Carman equation

$$\left(\frac{dH}{dL_0}\right) = 5\frac{\mu v}{pg}\left(\frac{1-f}{f^3}\right)^2\left(\frac{6}{d}\right)^2 \qquad (16.5)$$

where μ/p is the kinematic viscosity, g is the gravitational acceleration, and $(dH/dL)_0$ signifies the hydraulic gradient at $t = 0$. During the course of filtration the specific deposit occupies part of the pore space, causing an increase in head loss.

$$\frac{\partial H}{\partial L} = \left(\frac{dH}{dL_0}\right) + k\sigma \qquad (16.6)$$

where k is a dimensionless head loss factor, dependent upon grain size (d) and flow rate (v).

$$k = Kv^{n_7}/d^{m_7} \qquad (16.7)$$

Mohanka's experiments gave, in cm g min units:

$$K = 0 \cdot 8 \quad m_7 = 0 \cdot 9 \quad n_7 = 0 \cdot 4$$

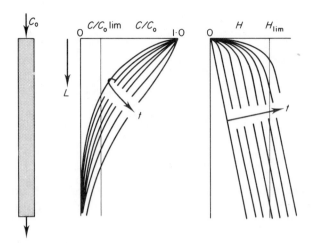

Figure 16.1 Curves of concentration ratio (C/C_0) and head loss (H) varying with depth in the filter (L) and with time of filter run (t)

Equation (16.6) has been reported empirically by several research workers (Mints, 1966), but it was given a theoretical basis by Horner (1968).

Integration of equations (16.1) to (16.7) can be carried out by finite difference techniques on a digital computer.

For a uniform filter, with given values of grain size (d) and flow rate (v) kept constant, the resulting curves of concentration and head loss are as shown on Figure 16.1. Alteration of the values of d or v will lead to different, but similar, curves.

16.3 OPERATIONAL OPTIMUM FOR UNIFORM FILTERS

As can be seen on Figure 16.1 the filtrate concentration changes during the filter run, rising in value as time proceeds. Also the total head loss rises with time.

Although the filter could be designed to produce satisfactory filtrate quality during the early stages of the filter run, there will come a time when the filtrate quality will become unsatisfactory, and the filter run will have to be terminated (time $= t_0$). Similarly, the head loss will rise until it reaches a limit set by the hydraulic conditions of the design; when this limit head loss is reached the filter run must be stopped (time $= t_H$).

A filter will be in an operational optimum condition if these two times are equal; that is, when the clarification capacity of the filter has been exhausted simultaneously with the hydraulic capacity.

$$t_{OPT} = t_c = t_H \tag{16.8}$$

This operational optimum can be determined simply by drawing on Figure 16.1. the criteria for the limiting filtrate concentration $(C/C_0$ limit) and for the limiting head loss $(H$ limit), as shown in dashed lines. These dashed lines

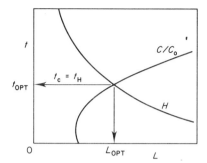

Figure 16.2 Curves of values of depths (L) and time (t) which meet the criteria C/C_0 limit and H limit, for a uniform filter

intercept the concentration and head loss curves at depths and times where the limit values are reached. The values of time, with corresponding values for depth can be plotted as curves, as shown on Figure 16.2. Where these two curves cross, equation (16.8) is satisfied and the optimum filter run time, and the optimum media depth, are defined.

If the curves of Figure 16.1 are recalculated for a different filtration velocity, then a different pair of depth–time curves will result, with a different intersection for t_{opt}. This can be repeated for several filtration velocities, with the result shown on Figure 16.3. Any point on the locus of optima will define the filtration velocity (v), filter depth (L), and run time (t) for an operational optimum filter of uniform grains of a given size.

Again, if the curves of Figure 16.1 are recalculated for a different grain size, then a different pair of depth–time curves will intersect at another value for t_{opt}. Repeating this for several different grain sizes will produce several different loci of optima as shown on Figure 16.4.

The result of the interaction of these four variables: depth, time, velocity,

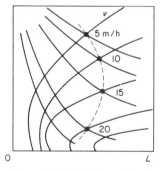

Figure 16.3 Variation of optimum run time (t_{OPT}) with filtration velocity (v) for a uniform filter

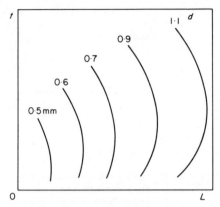

Figure 16.4 Loci of optima for different grain sizes of uniform filter medium

and grain size, is to produce a response surface as shown on Figure 16.5. Any point on this surface produces an operational optimum design, according to the constraints of limiting filtrate concentration (C/C_0 limit) and limiting head loss (H limit). If either or both of these constraints are changed, then a new response surface will be formed.

It can be seen that there are an infinite number of design solutions which conform to an operational optimum.

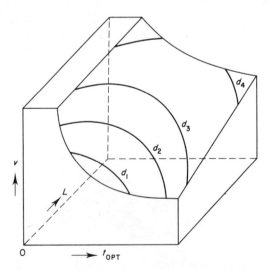

Figure 16.5 Response surface for operational optimum, depending on depth, time, velocity, and uniform grain size

16.4 OPERATIONAL OPTIMUM FOR GRADED-MEDIA FILTERS

Filters used in practice rarely contain media of uniform grain size. The media are either graded from fine to coarse in the direction of flow due to backwashing, or form a graded sequence of coarse to fine as in upflow and multilayer filters. Consequently, an additional variable is introduced into the design: the sequence of size in the depth of the filter.

In considering alternative designs for size-graded filters, the following variations are possible, either singly or in combination:

1. The total depth of the media can be changed;
2. the inlet face grain size can be changed;
3. the outlet face grain size can be changed;
4. the size distribution within the depth can either be changed
   (a) as a continuous gradation
   or (b) as a series of discrete size jumps
       (i) monotonically increasing or decreasing,
       (ii) non-monotonically changing.

It is this range of possibilities which make formal approaches to optimization very difficult, so that a computer is indispensable.

The programming of optimization procedures for size-graded filters depends on the principle that if t_c is not equal to t_H, then a step change in design is applied, and t_c and t_H are again compared, and this process is reiterated until t_c and t_H converge.

For example, a sequence of grain sizes can be read into the programme, covering a very long range of depth 0 to L_{max} (e.g. 3 m depth). Within this range a starting value L_{initl} and end value L_{end} can be chosen, with associated grain sizes representing the inlet face and outlet face grain sizes, respectively. The distance L_{initl} to L_{end} represents the depth of the filter. This may be represented diagrammatically as in Figure 16.6.

The values of t_c and t_H are calculated for the first choice of filter depth. If $t_c > t_H$ the bed must be made shorter by removing a layer of fine grains at the

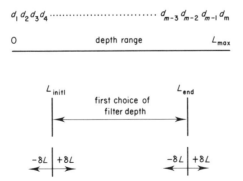

Figure 16.6 Diagram of array of grain sizes through filter depth

top ($L'_{initl} = L_{initl} + \delta L$) and a layer of coarse grains at the bottom ($L'_{end} = L_{end} - \delta L$). Conversely, if $t_c < t_H$ the bed must be made longer by adding a layer of grains at top and bottom ($L'_{initl} = L_{initl} - \delta L : L'_{end} = L_{end} + \delta L$). The process is repeated until $t_c = t_H$. It is necessary to have a sufficient depth range (0 to L_{max}) for the optimum to be selected. This can be readily achieved by digital computation with a deck of data cards, each of which represents a depth interval δL containing grains of size d_L.

By not varying either L_{initl} or L_{end}, it is possible to change the design by varying either only the outlet face grain size, or only the inlet face grain size.

With the data cards representing depth intervals containing given grain sizes, it is possible to calculate t_c and t_H for a full range 0 to L_{max}. If $t_c > t_H$, the filter can be made shorter by multiplying L_{max} by a factor less than 1·0. For example, $L'_{max} = 0·8 L_{max}$; the filter has been shortened by 20%. The depth interval value assigned to each data card is correspondingly shortened, $\delta L' = 0·8 \, \delta L;$ but the grain size assigned to each card (depth interval) remains unchanged. This effectively contracts the bed depth, but does not alter the grain size distribution. Conversely, if $t_c < t_H$, the bed depth can be stretched by multiplying L_{max} and δL by a factor greater than 1·0, which still preserves the grain size distribution. This is shown as a diagram on Figure 16.7.

The procedures described so far, deal with cases 1, 2, and 3 or the variations in design of size-graded filters, that is, changes in total depth, and inlet, and outlet grain sizes. For case 4 where the size distribution can be changed, a different method has been used. As the number of size distributions which may be used with either continuous or stepped size changes is infinite, some

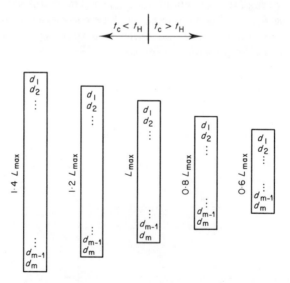

Figure 16.7 Contraction or extension of filter depth without changing the grain size distribution, to achieve an operational optimum

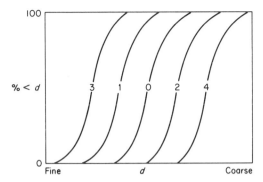

Figure 16.8 Grain size distribution, as continuous functions, available for a size-graded operationally optimum filter

choice of possibilities has to be established before computation is started. It is assumed, therefore, that a number of practical size distributions (depending on availability of media) are possible, as shown on Figure 16.8. One of these, marked 0 on Figure 16.8 is selected as the original grading, and with this providing the grain size data, the values of t_c and t_H are calculated, in a manner similar to that for Figure 16.6, but with a fixed depth 0 to L_{max}. However, in this case if $t_c - t_H$ is negative, then finer gradations are required, represented by the odd numbered distributions on Figure 16.8. If $t_c - t_H$ is positive, then coarser gradations are required, represented by the even numbered distributions on Figure 16.8. Size distributions in an appropriate direction will be selected sequentially until $t_c = t_H$, that is, an operational optimum is achieved.

The process can also be used for stepped size distribution as in dual media (anthracite–sand) or multilayer filters.

It is obvious from the number of variations of size-graded media, that very many operationally optimum designs are possible.

16.5 ECONOMIC OPTIMUM DESIGN

It has been shown that to achieve an operational optimum very many alternative designs are possible. But of these designs only one will produce filtered water at the least cost per cubic metre. The designs for operationally optimum filters considered only the media (depth, grain size), flow rate, and the hydraulic conditions limiting the head loss. In an economic optimum design the effects of the filter structure, hydraulic appurtenances, energy, and maintenance requirements, and size of the total filter plant also have to be considered.

The filter will be considered to be constructed as shown in Figure 16.9, and it is assumed that all filters in the treatment plant are identical, with no shared walls.

The following items contribute to the cost of filtration, established as a cost per month (30 days):

348

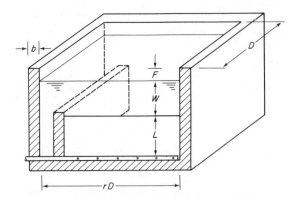

Figure 16.9 Sketch of filter giving principal dimensions

(1) Cost of walls

$$T_1 = C_1 ab[(2r+3)LD + 2(r+1)(WD+FD)] \qquad (16.9.1)$$

(2) Cost of floor including underdrains and piping

$$T_2 = C_2 arD^2\left[\frac{v}{v_s}+u\right] \qquad (16.9.2)$$

(3) Cost of media assuming wash gully width $\ll rD$

$$T_3 = C_3 aLrD^2 \qquad (16.9.3)$$

(4) Cost of energy due to head loss

$$T_4 = C_4 H_{\lim} rD^2 vt\left(\frac{30\times24}{t+t_w}\right)\frac{10^3\,\text{kgf/m}^3}{3\text{·}68\times10^5\,\text{kgf}-\text{m/kWh}} \qquad (16.9.4)$$

(5) Cost of backwashing

$$T_5 = C_5 v_w t_w rD^2\left(\frac{30\times34}{t+t_w}\right) \qquad (16.9.5)$$

Values for equation (16.12)

a	= capital recovery factor per month (amortization)	0·0075
b	= wall thickness (m)	0·3
D	= width of filter unit (m)	8·0
L	= media depth (m)	
r	= ratio of length to width of filter unit	1·5
W	= water depth over media (m)	
H_0	= initial head loss (m)	
$H_{\lim}$	= terminal head loss (m)	
t	= filter run period (h)	
t_w	= washing period (h)	0·12
v	= filtration velocity (m/h)	
v_s	= standard filtration velocity (m/h)	10

v_w	= washwater velocity (m/h)	50
u	= proportionality factor for underdrains	0·5
C_1	= unit cost of finished concrete (£/m³)	43
C_2	= unit cost of floor including underdrains (£/m²)	270
C_3	= unit cost of filter media (£/m³)	28·5
C_4	= unit cost of energy (£/kWh)	0·006
C_5	= unit cost of treated water (£/m³)	0·0085

It is assumed that

$$0·8(W + L) = H_{lim} \qquad (16.10)$$

which is reasonable considering the shape of pressure profiles through filters (see Ives, 1969), and allows 25% head loss in underdrains, piping, and so forth. Also it is assumed that $H_0 \ll H_{lim}$ and, therefore, that H_0 can be neglected. Some simplification can be made by assuming $t_w \ll t$, therefore, $t + t_w \simeq t$; also draining and recharge times have been neglected.

The volume of filtrate produced per month is $vrD^2 \times 24 \times 30 \mathrm{m}^3$. Therefore, the total cost per cubic metre filtered is

$$T = (T_1 + T_2 + T_3 + T_4 + T_5)/720 vr D^2 \qquad (16.11)$$

If a choice is made for r and F, it appears from inspection of equations (16.9.1) to (16.9.5) that the variables are D, W, L, v, and t. In an analysis of the effect of width D on monthly cost, Gur (1969) found that for different flow rates v, the cost was not sensitive (varying only $\pm 7\%$) for a range of D from 3 m to 11 m. If an extra filter is provided for standby, then the monthly cost increases markedly with filter size, as would be expected. Assuming no standby, but that load sharing takes place during washing of a filter, the choice of D can be arbitrary, not affecting cost optimization.

The variables W, L, v, and t (and grain size d of a uniform filter which is implicitly included) are related for operational optimum conditions as given in Figure 16.5, and other graphs similar to Figure 16.5 for different H_{lim} values. The water depth W and media depth L, taken together, specify the head loss limit H_{lim} by equation (16.10). The value $C/C_0 = 0·05$ is assumed constant in these calculations, but another value could be taken depending on input and output effluent quality.

So the procedure for calculating equation (16.11) is to specify values of W, L, and v. This fixes H_{lim}. From Figure 16.5 or analogues of it (see Gur, 1969), the value of t (operational optimum run time) is read. This, with fixed data for costs, amortization, and so on, enables T to be calculated.

For example, taking the cost and other values assigned in the symbol list, equations (16.9.1) to (16.9.5) and (16.11) simplify to:

$$T = \left[\frac{1}{v}(36·35L + 5·6W + 144 + 5100/t) + 1·3L + 1·3W + 28·3 \right]$$
$$\times 10^{-5} \text{£/m}^3 \qquad (16.12)$$

If $W = 2$ m, $L = 1$ m, then $H_{lim} = 2·4$ and $t = 17$ h,

Table 16.1 Table of values of cost of filtered water
in 10^{-3} £/m³

W = water depth over media, m
L = depth of media, m
v = filtration velocity, m/h
Calculated from equation (16.12).

$W = 1$

v \ L	1·0	1·8	2·5
10	0·84	0·74	0·72
20	0·92	0·74	0·68
30	1·23	0·87	0·76

$W = 2$

v \ L	1·0	1·8	2·5
10	0·81	0·74	0·71
20	0·88	0·83	0·68
30	1·07	1·02	0·70

$W = 3$

v \ L	1·0	1·8	2·5
10	0·80	0·75	0·71
20	0·83	0·71	0·68
30	1·01	0·84	0·71

$T = 0.81 \times 10^{-3}$ £/m³ or 0·081 pence/m³

Equation (16.12) has been calculated for an array of values $W = 1, 2, 3; L = 1.0$, 1·8, 2·5; $v = 10, 20, 30$. The resulting cost data are presented on Table 16.1. It can be seen that these do not include the optimum conditions (minimum cost). Up to a value of $W = 3$ m the optimum is at $L = 5.5$ m, $v = 20$ m/h, with corresponding values of $t = 30$h, $H_{lim} = 6.8$m and $d_3 = 1.5$mm, giving a cost of 0.66×10^{-3} £/m³.

As can be seen from Table 16.1, the cost is not very sensitive to variation in W, near the optimum. Extending W to 4m, produces a slight advantage to $T = 0.65 \times 10^{-3}$£/m⁵, at $L = 4$ m, $v = 20$ m/h, $t = 25$ h $H_{lim} = 5.6$ m, $d = 1.3$ mm.

The response of cost, with respect to L and v, has been calculated by Gur (1969) using the theory of Mints (1966), but with the same basic cost data. In this case, because of the nature of the cost function, the optimum (minimum cost) could be obtained by differentiation. The result was very similar to the present calculated optimum, with a depth $L = 3.8$ m, velocity $v = 20$ m/h, and cost 0.67×10^{-3} £/m³. It may be argued that, therefore, it does not matter very much which mathematical model of the filtration process is used.

Based on a simplified mathematical model of the filtration process, Herzig and Legoff (1971) have studied the theory of optimization, including an initial comment on the economic optimum. Taking standard conditions of uniform 0·45 mm sand, with a head loss limit of 5·2 m, they calculated a maximum production of filtrate (optimal solution) with a filter 1·35 m depth at a filtration rate of 10·8 m/h, giving a filter run of 2 h. This is not necessarily the optimum solution as they preset their grain size and head loss limit.

Using a standard form of filter, Huang and Baumann (1971) calculated the cost of water per unit filtrate for a range of sand sizes, filtration rates, sand depths, and limiting head losses. As the sand size increased, unit cost of filtrate diminished, with increasing filtration rate, and sand depth, but with lower head losses. Unfortunately, they did not continue their calculations to reach a minimum in unit cost, but stopped at 1·3 mm sand size, at 14 m/h, 0·9 m depth, and 2·5 m head loss. Their minimum lies at a coarser, deeper filter, operating at higher rate.

It is interesting to note that the standard conditions for rapid filtration in Europe and North America are $W = 2$ m, $L = 1$ m, $v = 5$ m/h. This gives $H_{lim} = 2·4$ m and $t = 40$ h (if operational optimum conditions are achieved, which is rare). The corresponding cost is $0·96 \times 10^{-3} £/m^3$, and doubling the rate of filtration reduces the cost to $0·81 \times 10^{-3} £/m^3$, a saving of 15·6% in the cost of the filtrate. This is similar to the figure given by Miller (1971) who estimated a saving of 17% on the cost of filtering water by doubling the rate.

16.6 REFERENCES

Gur, A. (1969). 'Theory and optimization of water filtration', *Ph.D. thesis*, Univ. London.

Herzig, J. P., and Le Goff, P. (1971). 'Le calcul previsionnel de la filtration a travers un lit epais. 2 Partie: Optimisatio d'un filtre homogene a sectio constante', *Chimie et Industre—Genie Chimique*, **104**, 2477.

Horner, R. M. W. (1968). 'Water clarification and aquifer recharge', *Ph.D. thesis*, Univ. London.

Huang, J. Y. C., and Baumann, E. R. (1971). 'Least cost filter design for iron removal', *J. San. Eng. Div., Proc. Am. Soc. Civ. Engrs.*, **97** (SA2), 171.

Ison, C. R., and Ives, K. J. (1969). 'Removal mechanisms in deep bed filtration', *Chem. Eng. Sci.*, **24**, 717.

Ives, K. J., and Sholji, I. (1965). 'Research on variables affecting filtration', *J. San. Eng. Div., Proc. Am. Soc. Civ. Engrs.*, **91** (SA4), 1.

Ives, K. J. (1969). 'Theory of filtration', Special Subject No. 7. *International Water Supply Congress Vienna*. Published by I. W. S. A., 34 Park St., London, W.1.

Iwasaki, T. (1937). 'Some notes on sand filtration', *J. Am. Wat. Wks. Ass.*, **29**, 1591.

Miller, D. G. (1971). 'Filtration: experimental developments', *J. Inst. Wat. Engrs.*, **25**, 21.

Mints, D. M. (1966). 'Modern theory of filtration'. Special Subject No. 10. *International Water Supply Congress Barcelona*. Published by I. W. S. A., 34 Park Street, London, W.1.

Mohanka, S. S. (1969). 'Theory of multilayer filtration', *J. San. Eng. Div., Proc. Am. Soc. Civ. Engrs.*, **95** (SA6), 1079.

IV

Application to Water Resources

17

The Trent Mathematical Model

A. E. WARN

17.1 INTRODUCTION

The River Trent rises 11 km to the north of Stoke-on-Trent and flows from west to east some 274 km, discharging ultimately to the Humber Estuary. The river is heavily polluted, largely through its tributary, the Tame, which drains the area of Birmingham and the industrial West Midlands. So far as the development of water resources is concerned, a major problem for this area is whether the River Trent can be reclaimed as a source of potable water and if so, how the costs would compare with those of other methods of obtaining water, for example by building large storage reservoirs to utilize relatively clean resources within the Trent basin, or imports from other catchments.

The Trent Mathematical Model was envisaged as an aid to the planning of the future use of the water resources in the Trent system. The development of the model was undertaken by the Local Government Operational Research Unit (LGORU) (WRB 1972k; Bowden, Green, and Newsome, 1971; Newsome, Bowden, and Green, 1972) under contract to the Water Resources Board (WRB) as one of the projects which comprised the Trent Research Programme (TRP) (WRB, 1972a).

Subsequently, the model has been operated by a small team drawn from the Trent River Authority (TRA), The Water Research Centre (WRC), and the WRB, with a view to its use in the solution of specific problems (Brewin, Chang, Porter, and Warn, 1972; WRB, 1972l). This work led to several improvements to the structure of the model and this paper describes the present structure, and outlines some of the results produced.

17.2 THE STRUCTURE OF THE MODEL

The model comprises three interacting sections. These are:

(1) the river model,
(2) the allocation model,
(3) the investment model.

The river model is concerned with forecasting river water quality and with the relation between river quality and fisheries and the relation between river

356

quality and the type of water treatment plant which would be required to produce a potable supply. The river model also selects an optimal allocation of costs between effluent treatment and water treatment for specified river abstractions and discharges. The allocation model incorporates the river model and chooses the most economic way of meeting a set of demands for water corresponding to one specific year. Such results are obtained for a number of specified years and are processed into a plan of development for the basin using the investment model.

17.3 THE RIVER MODEL

The structure of the water quality section of the river model

Figure 17.1 illustrates the division of the Trent system into reaches. The reaches, which are labelled Trent 1, Tame 2, and so on, define lengths of river roughly uniform in flow and quality. Table 17.1 is a list of the types of users of the system. Users are defined as any agency which affects or which is affected by river water quality. Thus the list includes brooks and streams (i.e. rivers which are not defined as reaches), water treatment plants, sewage treatment works, and fisheries. Agencies which affect river quality are called affective users and those which are influenced by water quality are recipient users.

For some users, the model is permitted to exercise some choice of policy as to the extent or type of use to be made. These alternative choices are called decisions. Thus several possible effluent quality standards may be considered for effluent treatment works, or several alternative types of plant may be stipulated for a water treatment installation. Users associated with decisions may be called active users to distinguish them from agencies for which the model has no decisions, i.e. passive users such as brooks and streams.

Individual users are arranged in geographical sequence along the reaches in the manner illustrated in Figure 17.2. The hatched lines indicate transfers

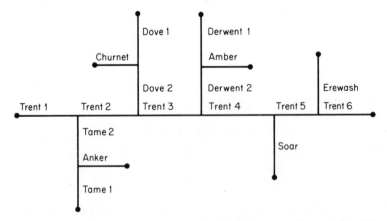

Figure 17.1 Diagrammatic map of reaches in the Trent mathematical model

Table 17.1 River users defined for the Trent mathematical model

Type of User	Nature of Influence
Sewage Treatment Works*	Adds flow and influences quality
Industrial Effluent Treatment Works*	Adds flow and influences quality
Stream or tributary	Adds flow and influences quality
Power Stations	Removes and returns flow
	Influences quality
TRA Routine Sampling Point	Null influence
Fishery*†	Influenced by quality
River Treatment Lake*	Influences quality
Weir	Influences quality
Water Treatment Works*†	Removes flow
	Indirect influence on quality
	Influenced by quality
Industrial Grade Water Treatment Works*†	Removes flow
	Indirect influence on quality
	Influenced by quality
Artificial Recharge Works*†	Removes flow
	Indirect influence on quality
	Influenced by quality
River Standard	(Mechanism for applying river quality constraints)

*Active users: decisions specified for the model. Other users are passive
†Recipient users: affected by water quality.

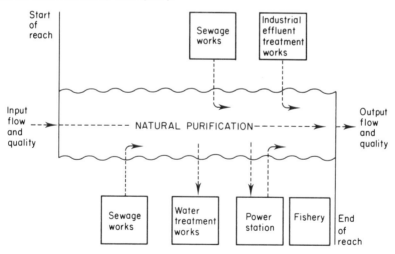

Figure 17.2 Diagrammatic map of a single reach

of water quantity between the river and the users. The action of the model starts at the furthest upstream user and proceeds downstream, sequentially from user to user, adding up the effects on flow and quality specified for each user. The calculation proceeds from reach to reach from the upstream limit to the downstream boundary of the model.

Table 17.2 Results of calibration of the river model for 1969

| Location River | Water Quality Characteristics Units—mg/l except Temp (°C) | | | | | | | | | | | | | | | |
	BOD	Am N	Ox N	DO	SS	Temp	TDS	PhOH	Cr	Cu	Ni	Zn	Pb	Cd	TOC	Cl
Great Haywood	*12.9	3.3	6.1	6.7	63	11.5	647	0.003	0.00	0.03	0.01	0.20	0.00	0.00	12	127
(Trent)	15.0	3.0	7.3	6.5	43	12.4	848	0.004	0.00	0.04	0.01	0.32	0.00	0.00	18	128
Wychnor	5.7	0.6	7.0	8.3	26	11.0	802	0.000	0.00	0.02	0.00	0.10	0.00	0.00	12	167
(Trent)	7.3	1.0	7.1	7.7	27	11.1	883	0.004	0.00	0.02	0.00	0.12	0.00	0.00	11	182
Lea Marston	27.4	6.2	8.5	5.3	135	11.8	571	0.089	0.00	0.06	0.23	0.28	0.01	0.00	30	87
(Tame)	29.6	7.2	9.7	5.3	61	12.1	693	0.171	0.08	0.18	0.27	0.47	0.00	0.01	35	92
Chetwynd Bridge	18.9	5.9	9.6	4.1	68	11.8	625	0.026	0.00	0.05	0.20	0.23	0.01	0.00	19	91
(Tame)	23.5	4.8	11.1	4.1	52	11.7	688	0.028	0.06	0.14	0.20	0.33	0.00	0.01	28	91
Burton	15.5	2.5	8.4	5.2	35	15.2	705	0.005	0.00	0.03	0.05	0.14	0.00	0.00	19	121
(Trent)	15.8	2.4	10.1	5.3	42	14.2	774	0.007	0.04	0.09	0.12	0.23	0.00	0.00	19	129
Monks Bridge	3.6	0.4	3.3	10.7	36	10.0	354	0.001	0.00	0.02	0.00	0.05	0.00	0.00	7	31
(Dove)	3.5	0.5	3.4	9.4	15	10.4	409	0.000	0.00	0.02	0.00	0.04	0.00	0.00	8	29
Shardlow	10.5	1.3	8.1	5.7	44	16.6	646	0.002	0.00	0.03	0.06	0.12	0.02	0.00	10	104
(Trent)	11.0	1.2	8.7	6.0	33	16.7	667	0.001	0.02	0.07	0.08	0.15	0.00	0.00	13	99
Wilne	5.4	1.0	3.5	7.9	24	13.6	437	0.003	0.00	0.02	0.00	0.06	0.01	0.00	10	56
(Derwent)	6.3	0.9	3.3	8.9	19	12.1	451	0.004	0.00	0.02	0.00	0.05	0.00	0.00	11	54
Trent Bridge	8.0	0.9	7.8	8.2	28	16.2	653	0.003	0.00	0.04	0.05	0.12	0.00	0.00	11	100
(Trent)	8.6	1.1	7.6	8.8	30	15.6	666	0.003	0.01	0.05	0.05	0.11	0.00	0.00	11	104

*For each location the top row of figures is the data from TRA. The lower row contains the estimate of the model.

River chemistry

The Trent Model operates on annual mean values of water quality characteristics. Where some indication of the variation in quality about the mean values is required, statistical relationships developed by the WRC are used (Garland and Hart, 1971; WRB, 1972d).

Before the river model was used to make forecasts, it was first calibrated by estimating the quality observed in the river in past years. Table 17.2 illustrates results of this exercise for 16 water quality characteristics for the year 1969. In this Table, the calculated results from the model are set alongside the values measured in the river sampling programme of the TRA.

The results of calibration have indicated that chloride (Cl), total dissolved solids (TDS), suspended solids (SS), chromium (Cr), copper (Cu), nickel (Ni), lead (Pb), and cadmium (Cd) behave conservatively in the river. The total inorganic nitrogen (i.e. the sum of the concentration of ammoniacal nitrogen, AmN, and oxidized nitrogen, OxN) appeared also to behave conservatively. For other parameters (biochemical oxygen demand, BOD; total organic carbon, TOC; monohydric phenols, PhOH; ammoniacal nitrogen, AmN, and, surprisingly, zinc, Zn), it was necessary to include mathematical functions which represented the effects of natural purification. These functions, generally first-order equations, were established on the assumption that they should be chemically reasonable, and that the associated empirically determined rate constants should produce agreement between the observed and calculated river quality for 1969. Similarly a cooling law was adopted for temperature, and the dissolved oxygen concentration was assumed to depend on the balance between processes tending to consume and to replenish oxygen in the river (Brewin, Chang, Porter, and Warn, 1972; WRB, 1972 l).

The relation between river quality and fisheries

The quality of the fisheries in the rivers of the Trent system was assumed to depend solely on river water quality. Any set of levels of water quality characteristics was classified into one of five fish states. These corresponded to the classifications:

(1) no fish,
(2) poor coarse fishing,
(3) fair coarse fishing,
(4) good coarse fishing,
(5) game and coarse fishing.

The method of predicting the type of fishery was based on work of the WRC in which an empirical relation was found between the calculated toxicity of mixtures of poisons to rainbow trout and the presence or absence of fish in the rivers of the Trent basin (Brown, 1968; Alabaster, Garland, Hart, and Solbe, 1972). Concentrations of certain chemical substances (AmN, PhOH,

Cu, Cr, Ni, Zn, Pb, and Cd) which would, in a laboratory test, kill half a batch of rainbow trout in two days (known as the 48 hr median lethal concentrations, 48-hr LC_{50}) were calculated as functions of other water quality characteristics (Hart and Warn, 1972; WRB, 1972 l). The inverse ratio of each of the values of the 48-h LC_{50} to the corresponding in-river annual mean concentration was calculated and all ratios summed. Ranges of values of this sum were set to produce agreement between the observed and calculated types of fishery for the year of 1969.

The relation between water treatment and river quality

In order to relate river quality to the type of water treatment plant which would be required to produce a potable supply, the model uses the results of research by the Water Research Centre (Miller and Short, 1972; WRB, 1972e). This work involved the specification of critical levels of river water quality for each of the following types of water treatment plants:

  (1) no treatment necessary,
  (2) storage, softening, coagulation, sedimentation, filtration, pH correction, and chlorination,
  (3) as (2) plus reduction of ammoniacal nitrogen,
  (4) as (2) plus granular carbon adsorption,
  (5) as (4) plus reduction of ammoniacal nitrogen,
  (6) as (2) plus desalination,
  (7) as (3) plus desalination,
  (8) as (4) plus desalination,
  (9) as (5) plus desalination,
 (10) no treatment possible.

Use of the water quality section of the river model

The river model may be used to forecast the future annual mean quality of the rivers in the Trent system as determined by changing use of the river, in particular, by increasing volumes of effluent. The fisheries and the types of water treatment plant associated with the forecast quality may also be indicated.

17.4 OPTIMIZATION IN THE RIVER MODEL

There is a very large number of ways in which a river system may be managed in order to achieve certain objectives. Such objectives could include:

  (1) the disposal of specified volumes of effluent,
  (2) the provision of specified volumes of potable water,
  (3) the maintenance of fisheries or river quality standards.

An extreme solution of such a problem is to spend as little money as possible on the treatment of effluents and use the river as a carrier of effluent. Subsequently, should the river be required to supply drinking water, the attendant costs of water treatment would be very large. In addition, the value of the river as a facility for recreation would be minimal, and the river might be a hazard to health as well as an aesthetic nuisance. A second extreme is to invest heavily in the treatment of effluent and to reap the benefits of a clean river, namely, good fisheries and perhaps a plentiful supply of potable water available at low treatment cost (in as much as the cost of water treatment may be considered as isolated from the cost of producing the clean river). Between these two extremes are overall optimal management policies which may be defined as those which achieve the specified objectives at minimum total costs. The river model attempts to seek out such optimal solutions. For this purpose, the costs are calculated on an annual basis. Annual charges on capital (based on a specified rate of interest) and annual costs of operation and maintenance are lumped together to form a total annual cash flow (TACF) which form the criterion by which optimal solutions are isolated. The TACF includes costs associated with each type of river user for which decisions have been specified. The sources of these costs are summarized below.

(1) *Costs of effluent treatment*
The costs of achieving three specified qualities of effluent discharge have been derived for sewage treatment works and industrial discharges by the WRC. Standards for effluents were specified and treatment plants costed so that the levels of water quality characteristics would be exceeded for not more than 5% of the annual flow. The standards were then restated in terms of the annual mean quality (Porter and Boon, 1971; Garland and Hart, 1971; WRB, 1972d). The effects and costs of river purification lakes located on the Tame have also been studied (Lester, Woodward, and Raven, 1971; WRB, 1972f).

(2) *Costs of water treatment*
Costs of producing potable water from raw water of specified annual mean quality were determined by the WRC (Miller and Short, 1972; WRB, 1972e). The production of potable water by the artificial recharge of the Nottinghamshire Bunter sandstone aquifer with water of Trent quality has also been considered (Satchell and Edworthy, 1972; WRB, 1972g).

(3) *Benefits of recreation*
The assessment of the economic value of the current fisheries in the Trent basins has been studied by Kavanagh and Gibson (1971) (WRB 1972j). This work has been extended to establish a general relation between the type of fishery and economic benefit, and these values are then adjusted to take account of recreation activities other than angling (Gibson and Porter, 1972; WRB, 1972l).

In the Trent system there are over 80 discharges of effluent in excess of 500 m^3/d, each of which may be considered to be treated at a cost to one of

362

several standards. Taken together these lead to about 10^{40} permutations or management choices for the basin. To scrutinize this range of alternatives and to seek out optimal policies, a method based on the mathematical technique known as dynamic programming (Bellman, 1957) has been derived. The implementation of this technique requires the introduction of two concepts: the river state, and the river stage.

The river state

For every point in the river system there is an amount of money (TACF) which, within the decisions open to the model, is the maximum which could be spent on the treatment of effluents discharged upstream. Moving downstream this cost increases because more effluents are discharged. Similarly for every point there is a minimum cash flow resulting from the treatment of each upstream effluent discharge to the standard associated with lowest cost. The range of costs between this minimum and the maximum represents the area within which the model is allowed to exercise choice. To define the river state, this range of costs is divided into a number of equal segments; the number of segments defining the number of river states. The segments are numbered in order from minimum to maximum costs, and management policies whose expenditure on effluent treatment falls into a particular segment are said to produce the river state whose number is that of the segment.

The ranges which define the boundaries between river states shift as the river is processed in the downstream direction, because more effluent treatment works have to be considered and the difference between the minimum and maximum cash flows attributable to effluent treatment expands.

The river stage

River stages are points in the river system at which the river state of the management policies is calculated. Stages are therefore the locations of each effluent treatment works for which costed decisions are specified. At each stage, the approximation afforded by dynamic programming is introduced. This is described below.

The dynamic program

The operation of the dynamic program of the river model is illustrated in Figure 17.3. In this diagram, only 10 river states have been defined. There are 30 in the current model.

At the upstream limit of the river system, the river has received no significant discharges of effluent and accordingly no costs attributable to effluent treatment. Up to the first stage, the river is labelled state 1. At the first stage, the river water quality associated with this state is mixed with the quality of the discharged effluent. In Figure 17.3 there are three effluent standards (labelled

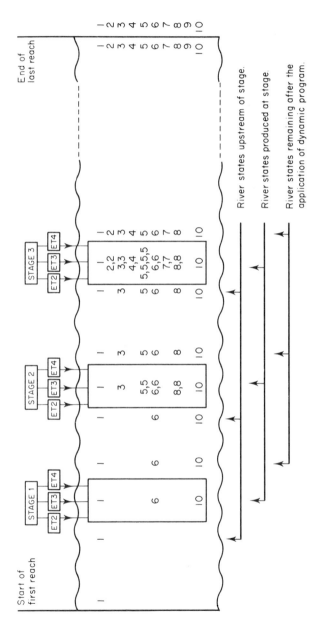

The designation 5,5 etc, indicates that two separate sequences of decisions (policies) produce river state 5.

Figure 17.3 Operation of the dynamic program of the river model

ET2, ET3, and ET4 using the nomenclature of Porter and Boon, 1971), each with an associated cost, and each producing a different river water quality after mixing with the river. The costs of effluent treatment define the river states: 1, 6, and 10. Downstream of the first stage, there may be recipient users such as fisheries and water treatment works. From the river quality associated with each state, the type of fishery or type of water treatment plant required may be established. The type of fishery is associated with a financial benefit to recreation, and the water treatment plant with a cost depending on the volume of the abstraction and the type of treatment. These benefits and costs are summed for each available river state and thus each state is characterized by:

(1) an annual cash flow attributable to effluent treatment (defining the state),
(2) a policy of decisions at effluent treatment works (i.e. active effective users),
(3) a resultant quality of rivers in the system,
(4) consequential decisions at active recipient users,
(5) a total annual cash flow comprising the sum of the costs of all upstream river users (affective and recipient).

At the second stage, there are three possible qualities of river water and each effluent standard for the second discharge must be considered with each possible river quality in the mixing process. This produces nine possible policies covering the first two stages and with these, nine resulting sets of levels of river water quality parameters and nine total costs (Figure 17.3). However, from the costs of effluent treatment, and according to the definition of the river state, these nine policies turn out to be distributed between only six of the ten possible river states. Whenever more than one policy leads to the same river state, the action of the model is to retain only that policy associated with a minimum total cash (as defined in note 5 immediately above). Thus only six policies with their attendant river states, qualities of river water, and total costs are passed on to the third stage.

It is important that the dynamic program tends to delete only those policies which would not feature in the absolute optimal solution were it determinable. It will be noted that the policies are deleted before the downstream recipient users are considered. This fact has not led to problems associated with the deletion of the wrong policies because:

(1) When choosing between two policies characterized by identical costs to upstream recipient users and the same river state, the model tends to retain high effluent standards at the larger effluent treatment works because of economies of scale. In the model, this leads invariably to a downstream river water quality better than that from the deleted policy, even when the deleted policy can take advantage of more extensive natural purification. Such an advantage would occur if most of the effluent load in the deleted policy were discharged further upstream than in the retained policy.

(2) For two policies of different costs to upstream recipient users and the same river state, the model selects that policy in which the total cost to these

users is a minimum. Where this occurs the river quality associated with the retained state is always better downstream of the stage than that associated with the deleted state.

The river system is processed stage by stage until after the last stage, there are only a limited number of policies for the entire basin, each policy corresponding to the retained method of achieving each river state. Each policy has achieved the objectives listed above and the state associated with the lowest total annual cash flow is taken as providing the optimal policy.

In reducing a set of policies from 10^{40} to 30, it is likely that the solitary absolute optimal solution will be eliminated by the dynamic program because alternative policies are deleted before the downstream lengths of river are considered. However, the isolation of the absolute optimal solution is unimportant because its selection involves the scrutiny of cost differences which are minute in relation to uncertainties in data. What is important is to be able to select the appropriate type of policy between the extremes of minimum and maximum expenditure on effluent treatment.

17.5 THE ALLOCATION MODEL

Given projections of the future population, per capita water consumption, and industrial requirements, the expected demand for water and the consequent returns of effluent have been estimated for centres of demand in the Trent catchment (TRA, 1968, 1971; WRB, 1972b). The allocation model selects those abstraction points in the river model which meet demands at minimum total annual cash flow. In addition, imports of water from neighbouring catchments are considered as alternatives to the indigenous sources. The allocation model uses the river model as a subroutine and, in addition to those costs which feature in the river model, the TACF in the allocation model includes the cost of the following items.

Reservoir storage

For the cleaner rivers in the Trent system, the natural low flows are inadequate to support further abstractions. Consequently to utilize these sources it will be necessary to construct pumped storage or regulating reservoirs (WRB, 1972b). Reservoir storage will be required also for the imports considered by the model (WRB, 1971; WRB, 1972l).

Transmission

The model includes the means of calculating the cost of transporting water from supply points to centres of demand. These costs include elements for pumping and for the construction of intakes, pumphouses, and pipelines (WRB, 1970; WRB, 1972l).

366

The use of the allocation model

The allocation model is operated for the set of demands associated with a single year in the future. The results for a number of such years are juxtaposed to form the basis of a development plan extending to the planning horizon of 2001.

In any one simulation, a list of potential supply points is made for each centre of demand. Such a set of possible links between supplies points and demand centres is shown in Figure 17.4. The allocation model operates by assuming initially that all the specified possible links are constructed and the river model is run to establish the optimum distribution of costs between affective and recipient river users. The sum of these costs with those of transmission and any reservoirs provide a total annual cash flow for the network of

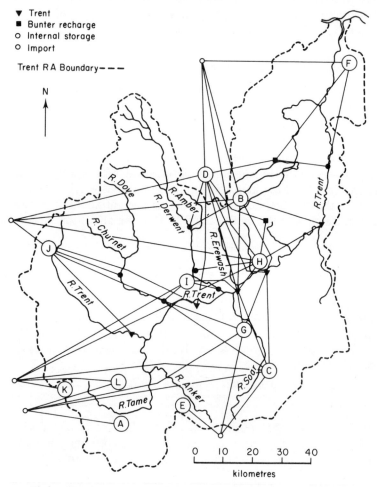

Figure 17.4 Starting set of links for allocation model simulations

supply-demand links. This cost is grossly unrealistic because each demand is being supplied several times over.

For each supply-demand link in the network, a total unit cost (£/tcmd) may be defined to contain elements associated with:

(1) development of regulating or pumped storage reservoirs (where necessary),
(2) water treatment,
(3) transmission from the treatment plant to the centre of demand.

Item (2), water treatment, covers the cost of the treatment plant selected as part of overall optimum distribution of costs between affective and recipient users as determined by the river model. It does not include the costs of effluent treatment of amenity benefits which may have influenced the type of water treatment plant selected and costed.

The allocation model operates by isolating that link with the largest total unit cost which may be removed from the network and yet still leave its associated demand centre with a supply. A new TACF is calculated for the revised network and the model compares the actual saving in cash flow brought about by deleting the link, i.e.:

$$\text{Actual Saving} = \text{TACF (with link)} - \text{TACF (without link)}$$

with the saving anticipated on the basis of the unit cost:

$$\text{Expected Saving} = \text{Total Unit Cost} \times \text{Demand}$$

These savings will differ because the deletion of a link and the removal of its water treatment plant modifies the balance of costs within the river model. The link is permanently deleted from the network if the actual saving exceeds the expected saving. If the converse is true, the model investigates first whether it might be preferable to delete the link of second highest unit cost in the initial ranking. If the actual unit cost of the first link exceeds the expected unit cost of the second link the model still deletes the first link. Otherwise the first link is restored to the network and the search procedure repeated, this time comparing the second link with the third in the original list. The search procedure is repeated until a link is finally and permanently deleted.

After the permanent deletion of a link, the list of expected unit costs is revised following a simulation of the new network, and the most expensive of the remaining removable links considered. The entire link deletion procedure is repeated until no redundant links remain in the network. This final network is taken as the optimal solution. Such solutions are determined for each of several demand years in the planning period.

This method of selecting optimum solutions has functioned economically and rapidly despite the fact that, in theory, it is only an approximate method. As in the case of the dynamic program of the river model, the nature of the data has been exploited to determine methods of optimization adequate for the problems to be solved. The allocation model has been checked by carefully planned programs of test simulations and, to date, it has never been used to produce solutions from a single computer run. Instead, the initial solutions have been checked to repeat simulations in which the choices are progressively limited in the light of the results obtained thus far. Nonetheless, it has been necessary only rarely to revise the indications of the initial simulations (WRB, 1972l).

The allocation model has been used to produce plans of development for the Trent basin under a variety of constraints. In some of these, certain types of resource have been used to their practical limit; in others, modes of supply have been excluded altogether. In all, some twenty development plans have been assembled (WRB, 1972l).

17.6 THE INVESTMENT MODEL

The investment model was designed to consider the evolution of schemes over the planning period, or, more precisely, the resolution of the TACF values produced by the allocation model into discounted or net present costs. In practice, it has proved convenient to perform this calculation by hand (WRB, 1972l).

17.7 RESULTS

Figure 17.5 shows the final network of links between supply points and demand centres selected from the initial set features in Figure 17.4. The network uses supplies from the River Trent to satisfy several demands. The network of links produced for the constraint that supplies from the River Trent are not available for selection is given in Figure 17.6. The values of TACF for both networks

Table 17.3 Summary of total costs* associated with two strategies determined by the Trent mathematical model

Description of Strategy	Total Annual Cash Flow** from Allocation Model				Net Present cost at 1971[†]	Total Expenditure to 2001
	(£ millions/annum)				(£ millions)	
	1971	1981	1991	2001		
Selected optimum solution (network as Figure 17.5)	− 0·7	2·76	7·2	13·7	60·0	155
Selected solution allowing no abstractions for potable water from the Trent (Figure 17.6)	− 0·7	2·7	8·0	14·5	62·4	161

*All costs are in 1967 prices.
**Interest rate: 8·5%.
[†]Discount rate: 8·0%.

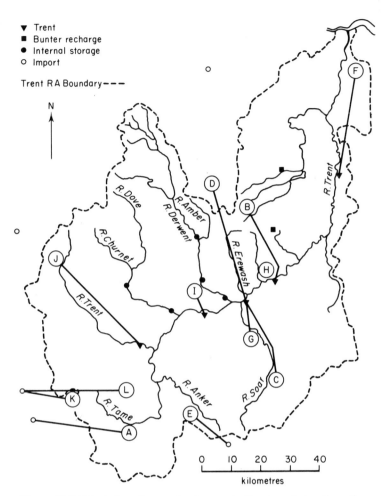

Figure 17.5 Selected optimum development allowing abstractions from
the River Trent

are given in Table 17.3 together with the discounted costs (NPC) and total
outlays obtained from the investment model. It will be noted that the costs
associated with these solutions are of similar magnitude. In fact the numerical
differences are small compared with uncertainties in the data. The costs of
both solutions are broken down into elements associated with various types of
installation in Table 17.4.

One feature of the results is that under the operating rules of the model
and with current data, the use of the River Trent for supplies of potable water is
calculated to be economically competitive with the development of other
resources. This result is achieved by including a river treatment lake on the
River Tame and by treating 55% of the total volume of effluent entering the
system by 2001 to advanced standard (i.e. ET4 as defined by Porter and Boon,

Table 17.4 Summary of elements of total net present costs produced for two strategies determined by the Trent mathematical model

| Item | Net Present Costs* at 1971 (£ millions at 1967 prices) | |
	Selected Optimum Solution (network as Figure 17.5)	Selected solution allowing no abstractions for potable water from the Trent (as Figure 17.6)
Transmission	25·0	23·8
Reservoir storage for imports	5·8	9·6
Reservoir storage with the Trent catchment	0·0	8·0
Basic water treatment	29·0	29·0
Advanced water treatment	4·7	0·0
Effluent treatment	6·4	1·3
Amenity benefits	− 10·9	− 8·3
Total	60·0	62·4

*Discount rate: 8·0%.

1971; WRB, 1972c); the selection of these standards being attributable to realizable savings in the costs of water treatment (Miller and Short, 1972; WRB 1972e). As a consequence the fisheries of the river system are, by 2001, far superior to those of today as is illustrated in Figure 17.7. A similar diagram (Figure 17.8) shows the fisheries for the same year under the constraint that 'supplies from the Trent are not available for selection'. In this situation, the model has no incentive to improve effluent standards, since the river water is not used for potable supplies. In addition, the realizable benefits to recreation are in themselves too small to offset costs of effluent treatment which would be incurred for the network which used the Trent for potable supplies.

The model output includes development plans and quality standards for each effluent treatment installation and these are described in WRB, 1972l.

These results outline just two of several programmes of development investigated by the model (WRB, 1972a; WRB, 1972l). The difference between the costs of such programmes have usually been found to be small relative to the effects of uncertainties concerning data, with the exception of schemes which rely heavily on water imported from outside the catchment area of the Trent.

This fact leads to the conclusion that so far as cost alone is concerned, the opportunities for water resource development within the Trent basin are equally attractive. However, in this situation, it must be noted that the option to use the Trent for potable supplies is untried on a large scale. In addition the results of the model have been based on water treatment processes recom-

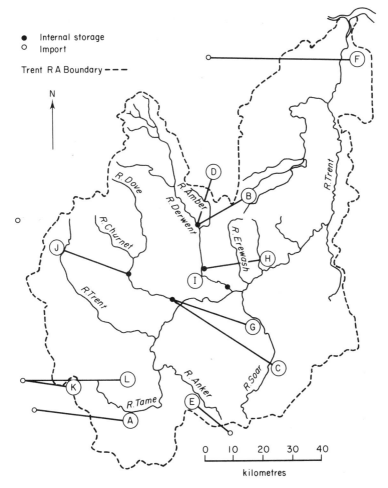

Figure 17.6 Selected optimum development permitting no abstraction from the River Trent

mended by the WRC to produce a potable water from the River Trent. The Centre have emphasized that it is not their responsibility to decide whether treated water is 'wholesome' (e.g. in the context of the Water Act of 1973) and it remains uncertain whether Trent water so treated would be regarded as totally acceptable for public supply (Miller and Short, 1972; WRB, 1972a). Accordingly the Water Resources Board recommended that the relatively clean resources of the basin continue to be developed by the construction of a reservoir at Carsington by 1978 (WRB, 1973). The Board also recommended further research on the reclamation of the River Trent with a view to re-assessment of the options before 1985, by which time a second major resource development is expected to be required.

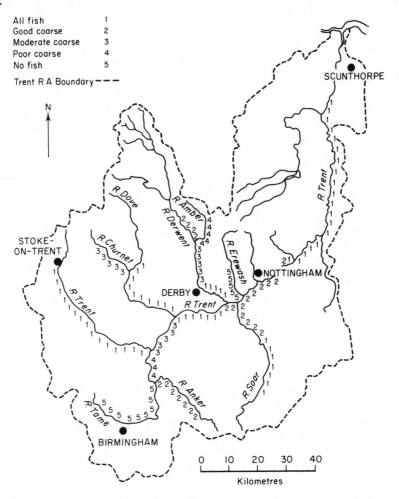

All fish   1
Good coarse   2
Moderate coarse   3
Poor coarse   4
No fish   5

Trent R A Boundary ---

Figure 17.7 Estimated fisheries in the Trent basin in 2001. For solution using the Trent as a source of potable water

17.8 FUTURE WORK

Research is continuing in particular in the fields of water treatment, artificial recharge, and improvement of the quality of the River Tame via a river retention lake (WRB, 1972a). The water quality section of the Trent model has been applied to the River Great Ouse in the context of a study of the water quality aspects feasibility of constructing storage reservoirs in the Wash (Warn, 1973) and is being developed for use in a study of the future effluent disposal problems of the River Ouse (Fawcett, 1974).

Meanwhile the quality of the Trent continues to improve (Lester, 1971; TRA, 1959–72), and the Trent Model will continue to be used to direct research

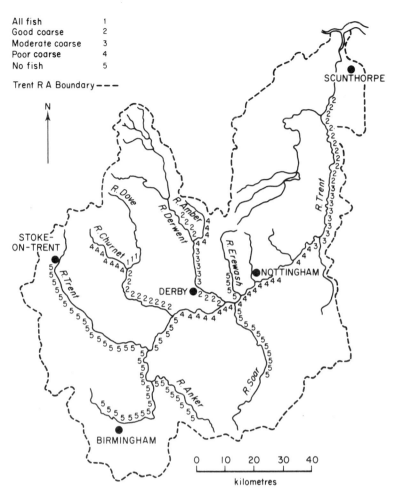

Figure 17.8 Estimated fisheries in the Trent basin in 2001. For solutions not using abstractions from the Trent

into the key areas, and to consider the impact of the results of such research in the planning of the future use of the Trent system.

Acknowledgement

The author thanks the Director of the Water Resources Board for permission to present this paper. The views expressed are those of the author and not necessarily those of the Board.

17.9 REFERENCES

Alabaster, J. S., Garland, J. H., Hart, I. C., and Solbe, J. F. de L. G. (1972). 'An approach to the problem of pollution and fisheries', *Symposium Zool. Soc. London*, **29**, 87–104.

374

Bellman, R. E. (1957). *Dynamic Programming*. Princetown University Press, New Jersey, USA.

Bowden, K., Green, J. A., and Newsome, D. H. (1971). 'A mathematical model of the River Trent System', *IWPC Nottingham Symposium on the Trent Research Programme*, 94–108.

Brewin, D. J., Chang, M. S. T., Porter, K. S., and Warn, A. E. (1972). 'Trent Mathematical Model: (b) Development', *IWE Birmingham Symposium on Advanced Techniques in River Basin Management*, 169–192.

Brown, V. M. (1968). 'The calculation of the acute toxicity of mixtures of poisons to rainbow trout', *Water Research*, **2**, 723–733.

Fawcett, A. (1974). *Third Semi-Annual Report on the Progress with the Steady State Model*. Steering Group for the Bedford Ouse Water Quality Model. Great Ouse River Division, Anglian Water Authority.

Garland, J. A., and Hart, I. C. (1971). 'Water quality relationships in the River Trent system', *IWPC Nottingham Symposium on the Trent Research Programme*, 37–55.

Gibson, J. G. (University of Birmingham), and Porter, K. S. (Water Pollution Research Laboratory) (1972), Unpublished work.

Hart, I. C. (Water Pollution Research Laboratory), and Warn, A. E. (Water Resources Board) (1972). Unpublished work.

Kavanagh, N. J., and Gibson, J. G. (1971). 'Measurement of fishing benefits on the River Trent', *IWPC Nottingham Symposium on the Trent Research Programme*, 26–36.

Lester, W. F. (1971). 'The River Trent and the economic model research programme', *IWPC Nottingham Symposium on the Trent Research Programme*, 10–22.

Lester, W. F., Woodward, G. M., and Raven, T. W. (1971). 'The effect and cost of river purification lakes', *IWPC Nottingham Symposium on the Trent Research Programme*, 74–93.

Miller, D. G., and Short, C. S. (1972). 'Treatability of Trent Water', *IWE Birmingham Symposium on Advanced Techniques in River Basin Mamagement*, 43–74.

Newsome, D. H., Bowden, K., and Green, J. A. (1972). 'Trent Mathematical Model: (a) Construction', *IWE Birmingham Symposium on Advanced Techniques in River Basin Management*, 151–167.

Porter, K. S., and Boon, A. G. (1971). 'Cost of treatment of waste water with particular reference to the river system of the Trent area', *IWPC Nottingham Symposium on the Trent Research Programme*, 56–73.

Satchell, R. L. H., and Edworthy, K. J. (1972). 'Recharge of the Nottinghamshire Bunter Sandstone', *IWE Birmingham Symposium on Advanced Techniques in River Basin Management*, 75–102.

Trent River Authority. '*River Water Quality. Triennial Statistics*', 1959–61, 1962–64, 1965–67, 1968–71.

Trent River Authority (1968). *Water Resources—A Preliminary Study*.

Trent River Authority (1971). *Water Resources—A First Development*.

Warn, A. E. (1973). *Wash Feasibility Study. Water Quality Forecasts for the Great Ouse River System*. Central Water Planning Unit, Reading.

Water Resources Board (1970). *Water Resources in the North: Northern Technical Working Party Report*. HMSO, London.

Water Resources Board (1971). *Water Resources in Wales and the Midlands*. HMSO, London

Water Resources Board (1972). Trent Research Programme (Published in 12 volumes)
  (a) Water Resources Board. *Report by the Water Resources Board*.
  (b) Trent River Authority, Water Resources Board. *Water demands and river flow augmentation*.
  (c) Porter, K. S., and Boon, A. G. *Costs of waste water treatment*.
  (d) Garland, J. A., and Hart, I. C. *Effects of pollution on river quality*.
  (e) Miller, D. G., and Short, C. S. *Costs of river water treatment*.

(f) Lester, W. F., Woodward, G. M., and Raven, T. W. *River purification lakes.*

(g) Satchell, R. L. H., and Edworthy, R. J. *Artificial recharge: Bunter sandstone.*

(h) Broadhead, J. A., and Mackey, P. G. *Artificial recharge: river gravels.*

(i) Jackson, K. J. *Dual Supply systems.*

(j) Kavanagh, N. J., and Gibson, J. G. *Recreation benefits: angling.*

(k) Bowden, K., Green, J. A., Phillips, G. W., and Renold, J. *The Trent Mathematical Model.*

(l) Warn, A. E., Brewin, D. J., Chang, M. S. T., and Porter, K. S. (1975) *Evaluation of the mathematical model.*

Water Resources Board (1973). *Water Resources in England and Wales*, HMSO, London.

18

A Hierarchical Approach to Water-Quality Modelling

D. G. JAMIESON

18.1 INTRODUCTION

In the analysis of large-scale, interconnected systems such as regional water-resource schemes, it is usually found that there are many alternative ways in which the system can be made to meet the requirement placed on it. Not all of these alternatives are equally attractive in terms of costs, and one problem is to identify the least-cost solution. Often this leads to a conflict between the number of alternative solutions that should be analysed and the amount of detail that can be included in each analysis. The water-quality aspect of water-resource systems is no exception. A compromise is not always a reasonable solution if there are a large number of possible alternatives and if the processes involved in the analysis are complex. This is of course the situation in water-quality systems and therefore it is unlikely that any single model will be capable of meeting all user requirements.

Instead of a single all-embracing water-quality model, a hierarchy of specific types of models for different purposes would seem to be one way in which large numbers of alternative solutions could be evaluated without necessarily sacrificing detail. This type of hierarchical structure would act as a filter allowing only the near-optimal solutions to pass for further examination in greater detail. A three-tier structure is proposed for water-quality models which is compatible with the water-quantity models (Figure 18.1). For the purpose of identification, these stages will be referred to as planning, design, and operational.

The purpose of a planning model is to evaluate a large number of potential solutions and isolate a small number of near least-cost solutions. The amount of detail that could be accommodated would be limited in order to keep computing costs within bounds (Figure 18.2). In the design stage, each of the alternative configurations recommended by the planning stage would be re-analysed in considerably more detail than was previously possible. The purposes of a water-quality design model are to assess the long-term performance of the system in terms of the probability of various water-quality levels arising at critical points in the system, and to verify or amend some of the assumptions made in the planning stage. At the operational stage, only the system to be

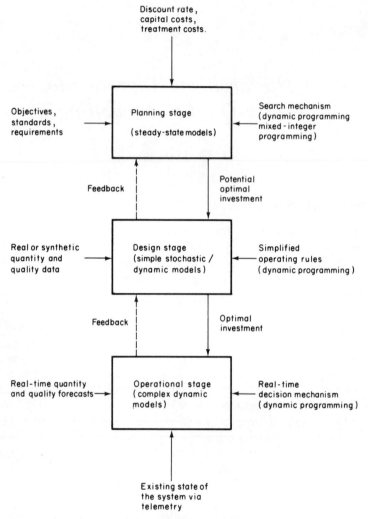

Discount rate,
capital costs,
treatment costs.

Objectives,
standards,
requirements

Planning stage

(steady-state models)

Search mechanism
(dynamic programming
mixed-integer
programming)

Feedback

Potential
optimal
investment

Real or synthetic
quantity and
quality data

Design stage
(simple stochastic /
dynamic models)

Simplified
operating rules
(dynamic programming)

Feedback

Optimal
investment

Real-time quantity
and quality forecasts

Operational stage
(complex dynamic
models)

Real-time
decision mechanism
(dynamic programming)

Existing state of
the system via
telemetry

Figure 18.1 A hierarchical structure for classifying both water quantity
and water quality models

implemented (presumably the least-cost solution) is considered. An operational
model defines how the system should be managed to achieve optimal performance. Some feedback to earlier stages may be necessary if some of the
assumptions made previously are not substantiated when studied in more
detail. However, in general a hierarchy of models which complement each
other enables efficient use to be made of computing resources.

Assigning different types of models to the purpose for which they were
formulated, rather than how they were formulated, seems an appropriate way of
classifying water-quality models. Each class has certain distinctive characteristics. For the purpose of this review chapter, rather than trying to summarize

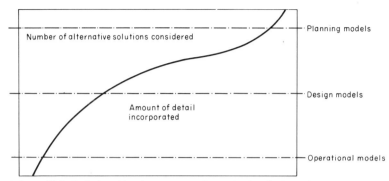

Figure 18.2 Relationship between scale and detail for different classes of models

a rapidly expanding field of research, an example of each type of model has been chosen to illustrate its principal feature. All three examples are taken from research currently being undertaken in this country.

18.2 PLANNING MODELS

Water-quality planning models are characterized by their ability to evaluate large numbers of potential investment programmes. Their inputs usually comprise the requirement of the system in terms of forecast demands and so forth, together with capital costs, pumping costs, and water treatment costs for the various components to be considered. In order to identify optimal or near-optimal least-cost solutions of fulfilling the requirements, a search technique is necessary for large systems. Recently mixed-integer programming has been used for water-quantity models (O'Neill, 1972) and dynamic programming has been used in conjunction with water-quality models (Bowden, Green, and Newsome, 1971).

Planning models are necessarily restricted to using simple evaluation routines in order to consider a large number of alternative solutions without incurring exorbitant computing costs. Therefore in water-quality planning models, only average annual values of water quality parameters are used in the evaluation routine. These simple steady-state models are completely deterministic and since they are based on average conditions, cannot account directly for any natural fluctuations in water-quality parameters.

The Trent Mathematical Model is a well-known example of a water-quality planning model. The details of its formulation have been presented in Chapter 17, and only the salient features that characterize planning models are highlighted here. The River Trent is regarded as a means of disposing effluent, a potential source of water for public supply, and a habitat for fish; it also has amenity and recreational uses. The objective was to identify the minimum net present cost investment programme for a whole range of river quality states to meet these requirements. The identification process has five phases namely calibration, prediction, search, allocation, and investment.

The calibration phase is mainly an accounting procedure for each of the 16 water-quality parameters considered. Average annual quantities of effluent discharge and their associated concentrations of pollutants are compared with the corresponding average concentrations of pollutants in the river system. If measurement errors are assumed to be small, then the degree of natural purification in the various reaches of the river can be deduced.

In the prediction phase, estimates are made of the abstractions and effluent returns for a specified future date. It is assumed that the natural river flow and purification processes remain unchanged. For each effluent return, one of three different standards of treatment is permitted, each standard having its own associated capital and operating costs. Similarly at each abstraction point on the river there is an associated range of capital and operating costs depending upon the quality of the river water at that point. The quality of the river water at each abstraction point is predicted using the steady-state model with upstream effluent discharges as input. Thus for any predefined combination of effluent treatment standards and river abstractions, a monetary cost can be attributed to that solution. Any incidental monetary benefits accrued from improved fisheries, recreation, and so on, resulting from changes in the quality of the river water are included to offset the cost of that solution.

When it is realized that there are 80 major effluent inputs in the Trent system, it will be appreciated that this can generate a prodigious number of alternative solutions. The search phase uses dynamic programming to select and carry forward a maximum of thirty alternatives covering the complete range of river quality states to the next river stage immediately downstream. Anything other than the least-cost solution of reaching that river stage for each river quality state is discarded.

Rather than assume that the requirement of each demand centre can be met from only one point in the system, alternative internal links are considered as well as imports of water from neighbouring catchments. The purpose of the allocation phase is to select the most appropriate pattern of links by deleting those which have the largest associated costs without impairing the supply to the demand centre. The deletion of a link causes repercussions in the remainder of the system and therefore necessitates an automatic reevaluation of the preceding phases.

The final step is the investment phase. Instead of just considering a single time frame, the whole process of prediction, search, and allocation is repeated at discrete time increments over the planning period. The investment phase examines the resulting discounted or net present costs for the optimal schemes at different future dates in order to formulate a development programme. Alternatively these least-cost solutions can form the input to a national or regional resources-allocation model of the type developed by O'Neill, 1972.

18.3 DESIGN MODELS

Following isolation of a small number of near-optimal least-cost solutions at the planning stage, the selected configurations require verification in considerably more detail to enable the optimal system to be identified. For the past decade water-quantity design models have been used to define the capacities of reservoirs required to meet forecast demands for water with a given reliability. Recently this type of model has been extended to incorporate water-quality parameters so that estimates can be made of the frequency or probability that certain water-quality levels will arise at specified points within the proposed water-resource system. Instead of being confined to using average values of flows and water-quality characteristics, design models must accommodate natural flow and water-quality fluctuations, on a daily basis if necessary. The simplified operation rules by which the system is to be managed should cater not only for quantity constraints such as maintained flows but also water-quality constraints.

Design models are concerned primarily with estimating the long-term performance of a water-resource system and at present to be confident about the prediction requires a large amount of real or synthetic input data. Consequently the models have to be relatively simple as a single iteration is generally expensive in terms of computer time. The type of model usually adopted is therefore an unsophisticated, deterministic simulation. Lag and attenuation effects observed as an element of water passes round the system are preserved, but the stochastic fluctuations are solely the result of the input data.

Research into this class of water-quality models was undertaken at the Water Resources Board (Page and Warn, 1973). The Welland and Nene river system comprising two pumped-storage reservoirs, Empingham and Pitsford, and a confined aquifer, the Lincolnshire Limestone, has been simulated (Figure 18.3). The effluent from one of the major demand centres is returned to the river upstream of the intake pumps to Empingham Reservoir, thereby creating an appreciable degree of recirculation. Initially only water-quantity was modelled, but this was gradually extended to include a conservative water-quality parameter (chloride) and a degradable water-quality parameter (biochemical oxygen demand, BOD). Whereas chloride is simply cumulative, BOD was also allowed to degrade exponentially with time.

Currently synthetic flow data generation is used in the hydrological design of water-resource systems. To retain this facility when the model includes water-quality parameters requires that the flow and quality data are compatible and any natural relationship between them is preserved. Preliminary analyses indicate that flow rate and water-quality are only weakly correlated naturally. For example, regression of chloride concentration in the River Welland with the corresponding flow rate accounts for only 12% of the variance. A slight improvement is achieved with the use of logarithmic flow values. Attempts have been made to isolate any seasonal component in the chloride record using Fourier analysis, but again the variance explained was only 17%. The variance explained increased to 22% when a linear flow term was superimposed on the seasonal variation. Unfortunately the chloride record, which comprises one sample per fortnight over a period of 5 years, is not adequate to identify any autoregressive component. Even with improved records, there is likely to be a

382

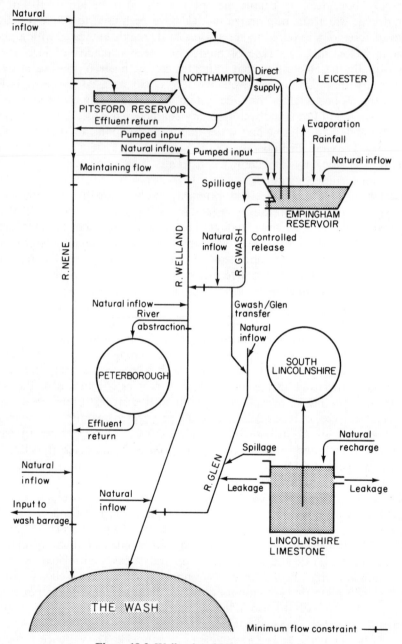

Figure 18.3 Welland and Nene subsystem

large portion of unexplained variance after the removal of the deterministic components. However, providing the mathematical distribution of the residuals can be identified, a flow-compatible chloride record can be synthesized in a similar way to the flow record itself. Even if the mathematical distribution of the residuals cannot be identified, the sample cumulative density function can be used directly to generate the random residuals.

By using the model to monitor the fluctuations of chloride and BOD over a long period of record at strategic points within the water-resource system, a histogram of water-quality levels can be derived (Figures 18.4 and 18.5) which enables the probability of various concentrations arising to be estimated. If the level of chloride or BOD is unacceptably high, it should be possible to reduce the extreme values by improving the quality of the effluent returned. Alternatively it may be possible to achieve an improvement in water quality at the expense of water quantity by modifying the operating rules.

To illustrate the interaction between water quality and quantity, a suggestion by the former river authority to change the minimum flow constraint below the Empingham intake pumps on the River Nene has been evaluated. Increasing the value of the flow on the Nene below which abstraction would not be permitted was considered as a means of reducing the amount of effluent that would be recirculated through the reservoir. For the data record used and the forecast 2001 demands, the model estimated that increasing the existing minimum flow constraint by 50% would reduce the median chloride concentration in Empingham Reservoir by 11%, and the maximum concentration predicted by 13%. Similarly the median BOD concentration in the reservoir would be reduced by 14% but the maximum concentration predicted would be increased by 141%. This large increase in the maximum BOD concentration was the consequence of the reservoir being drawn down more severely during times of water shortage, which in turn was the logical outcome of increasing the minimum flow constraint. The more severe drawdown not only meant that the contents of the reservoir were less and therefore the BOD of the inflow had a more dominant effect on the average BOD of the reservoir, but it also indicated that water was not remaining in the reservoir as long as usual thereby further increasing the extreme values. The actual values quoted are tentative and not important, but the example does show that the repercussions of including water-quality aspects are not always immediately obvious.

The techniques described are essentially for evaluating the design of a water-resource system rather than being a design procedure. Ideally it should be possible to specify an acceptable reliability and let the model itself derive the appropriate reservoir capacities, pump-sizes, and so forth, to meet the water-quantity requirements and water-quality constraints. This would require the inclusion of automatic optimization which although theoretically possible is beset by practical difficulties. At present the cost of one interation of the model parameters is extremely expensive and therefore ways of making this type of simulation more efficient are being examined. In anticipation that the cost of a single evaluation can be reduced to a reasonable level, research

Each character represents a frequency of 105

Int	Cum	Frequency	
1	0	0	I
2	0	0	I
3	0	0	I
4	0	0	I
5	0	0	I
6	0	0	I
7	0	0	I
8	0	0	I
9	0	0	I
10	0	0	I
11	0	0	I
12	0	0	I
13	0	0	I
14	0	0	I
15	0	0	I
16	0	0	I
17	0	0	I
18	0	0	I
19	0	0	I
20	0	0	I
21	0	0	I
22	0	0	I
23	0	0	I
24	0·0	31	I
25	0·1	116	I •
26	0·1	117	I •
27	0·2	155	I •
28	0·3	118	I •
29	0·5	379	I ••••
30	0·8	611	I ••••••
31	1·2	617	I ••••••
32	1·7	942	I •••••••••
33	2·5	1441	I •••••••••••••
34	3·5	1982	I ••••••••••••••••••
35	4·5	1719	I •• ••••••••••••••
36	5·8	2486	I ••••••••••••••••••••••••
37	8·9	3884	I ••••••••••••••••••••••••••••••••••••
38	10·4	4831	I •••
39	13·9	7378	I ••
40	17·9	7378	I •••
41	22·7	8809	I •••
42	27·8	9169	I ••
43	33·2	9976	I ••
44	38·7	9934	I ••
45	44·4	10445	I ••
46	49·6	9609	I ••
47	54·6	9124	I ••
48	59·3	8424	I ••
49	63·6	8325	I ••
50	67·6	7338	I ••
51	71·2	6208	I •••
52	74·5	6018	I ••
53	77·3	4989	I •••
54	79·6	4153	I ••••••••••••••••••••••••••••••••••••••
55	81·6	3699	I ••••••••••••••••••••••••••••••••••
56	83·5	3438	I ••••••••••••••••••••••••••••••••
57	85·9	2816	I ••••••••••••••••••••••••••
58	86·5	2669	I •••••••••••••••••••••••••
59	87·7	2226	I •••••••••••••••••••••
60	88·9	2188	I ••••••••••••••••••••
61	90·1	2135	I ••••••••••••••••••••
62	91·1	1898	I ••••••••••••••••••
63	92·0	1595	I •••••••••••••••
64	92·7	1398	I •••••••••••••
65	93·4	1265	I ••••••••••••
66	94·0	1100	I ••••••••••
67	94·6	1039	I ••••••••••
68	95·2	1179	I •••••••••••
69	95·8	1063	I ••••••••••
70	96·3	949	I •••••••••
71	96·8	823	I ••••••••
72	97·2	753	I ••••••••
73	97·6	716	I •••••••
74	97·9	586	I ••••••
75	98·2	547	I •••••
76	98·5	491	I •••••
77	98·7	324	I •••
78	98·8	327	I •••
79	99·0	253	I ••
80	99·2	309	I •••
81	99·3	323	I •••
82	99·5	227	I ••
83	99·6	309	I •••
84	99·7	323	I •
85	99·8	227	I •
86	99·8	199	I •
87	99·8	133	I •
88	99·8	151	I •
89	99·9	109	I •
90	99·9	141	I •
91	99·9	93	I
92	99·9	88	I
93	100·0	68	I
94	100·0	17	I

Figure 18.4 Histogram of concentration of Chloride in Empingham Reservoir

Each character represents a frequency of 52

Int	Cum	Frequency
1	0·0	86
2	0·2	271
3	0·4	362
4	0·5	278
5	0·6	252
6	0·7	181
7	0·8	119
8	0·9	120
9	0·9	109
10	1·0	95
11	1·1	119
12	1·2	331
13	1·6	597
14	2·3	1328
15	3·0	1381
16	3·5	1353
17	5·0	2282
18	6·3	2261
19	7·6	2450
20	9·2	2922
21	11·3	3772
22	13·4	3763
23	15·2	3341
24	17·2	3642
25	19·2	3723
26	21·4	3959
27	23·7	4248
28	25·9	3981
29	28·2	4179
30	30·4	4100
31	32·7	4182
32	35·0	4221
33	37·3	4093
34	39·6	4238
35	41·0	4040
36	44·0	4030
37	46·2	4028
38	48·3	3766
39	50·3	3653
40	52·2	3423
41	53·8	2980
42	55·1	2805
43	56·7	2250
44	57·7	2011
45	58·8	1982
46	59·9	1938
47	61·1	2285
48	62·7	2865
49	64·4	3013
50	66·0	2984
51	67·5	2848
52	69·0	2594
53	70·3	2466
54	71·5	2183
55	72·7	2127
56	73·7	1913
57	74·7	1745
58	75·6	1623
59	76·5	1643
60	77·3	1431
61	78·0	1391
62	78·7	1327
63	79·4	1178
64	80·0	1106
65	80·6	1011
66	81·0	905
67	81·5	873
68	82·0	798
69	82·4	851
70	82·9	878
71	83·5	863
72	83·8	755
73	84·2	763
74	84·5	669
75	84·9	649
76	85·3	665
77	85·6	600
78	85·9	584
79	86·2	566
80	86·5	545
81	86·9	571
82	87·1	511
83	87·4	490
84	87·7	509
85	87·9	484
86	88·1	399
87	88·4	437
88	88·6	418
89	88·8	378
90	89·0	344
91	89·2	357
92	89·4	366
93	89·6	380
94	89·8	394
95	90·0	304
96	90·2	304
97	90·3	330
98	90·5	326
99	90·7	298
100	90·8	273
101	91·0	304
102	91·1	242
103	91·3	240
104	91·4	212
105	91·5	223
106	91·7	272

Figure 18.5 Histogram of concentration of BOD in Empingham Reservoir

was undertaken on the development of a tailor-made stochastic search technique for optimizing this specific class of models.

18.4 OPERATION MODELS

The purpose of operational models is to assist in the efficient management of water-resource systems. They can be distinguished by their ability to forecast the future state of a system over a short period ahead. An operational model uses data relating to the past and present state of the system as input in order to produce flow and water-quality forecasts on which management decisions can be taken. The management decision itself can be an intrinsic part of the model. Recently such a model has been used as the framework of an automated control scheme for a multi-purpose reservoir system (Jamieson and Wilkinson, 1972). The proposed scheme included a small digital computer whose function was firstly to ascertain the existing state of the system using a telemetry scheme, secondly, forecast the future state of the system by means of a simulation, and finally decide on the optimal releases from storage. The decision-mechanism was based on a dynamic-programming formulation. The investigation did not specifically consider water-quality aspects but provision was made for their inclusion at some future date. It was recognized that there would be some form of trade-off possible between a short-term gain in water-quality and a long-term deficit in water quantity.

Research is already in progress at University of Cambridge with the objective of developing an automated control scheme to maintain the levels of water-quality parameters within tolerable bounds (Young, Beck and Singh, 1973). They recognize that the degree of mixing with the recipient streamflow provides a convenient means of distinguishing between the various operational water-quality models currently available.

(i) *No mixing*: a first-order partial differential equation model represented by

$$\frac{\partial \bar{c}(z,t)}{\partial t} = \mathbf{f}_1 \left[\frac{\partial c(z,t)}{\partial z}, \mathbf{c}(z,t) \right] \tag{18.1}$$

$$z_0 \leqslant z \leqslant z_1 \; ; t_0 \leqslant t \leqslant t_f$$

(ii) *Partial mixing*: a second-order partial differential equation model represented by

$$\frac{\partial \mathbf{c}(z,t)}{\partial t} = \mathbf{f}_2 \left[\frac{\partial^2 \mathbf{c}(z,t)}{\partial z^2}, \frac{\partial \mathbf{c}(z,t)}{\partial z}, \mathbf{c}(z,t) \right] \tag{18.2}$$

$$z_0 \leqslant z \leqslant z_1 \; ; t_0 \leqslant t \leqslant t_f$$

(iii) *Complete mixing*: a first-order ordinary differential equation with delay T_d to allow for distributed effects between spatial points z_0 and z_1

$$\frac{\mathrm{d}\mathbf{c}(z_1,t)}{\mathrm{d}t} = \mathbf{f}_3 \left[\mathbf{c}(z_1,t), \mathbf{c}(z_0, t - T_d) \right] \tag{18.3}$$

$$t_0 < t < t_f$$

In these equations t is the independent variable of time, z is the independent variable of distance, c is a vector of dependent concentration variables, and f_1, f_2, and f_3 are vector functions in the variables as shown. z_0 and z_1 represent the spatial boundaries of the system, while t_0 and t_f indicate the operational time interval.

While theoretically equation (18.2) would seem the most appropriate model to represent the system dynamics, Whitehead and Young (1974) have shown that a simpler lumped parameter model based on equation (18.3) gives a reasonable representation of the observed temporal variations of dissolved oxygen (DO) and BOD in a single reach of the River Cam.

To be useful as an operational tool, the degree of complexity may well be constrained by the size of the computer that will eventually control the system. Moreover there are techniques for improving the forecasting ability of a model without necessarily restructuring the basic model. For example, if the differences between the observed values and the forecast values are examined they are seldom found to be purely random, which suggest that there is a residue of information not accounted for by the model. Generally these residuals can be modelled using time—series analysis (Box and Jenkins, 1970). The resulting mathematical model of the discrepancy is then merged with the deterministic forecast to account for the composite effect of imperfections in the structure of the deterministic model. The technique has already been used successfully for river-flow forecasting (Jamieson, Wilkinson, and Ibbitt, 1972).

18.5 DATA REQUIREMENTS

Obviously any results or recommendations derived from a water-quality model are no more reliable than the input data used. In the past the lack of input data has generally been cited as the cause of inconclusive results. In the specific case of water-quality models the excuse is probably justified in most parts of the country. There are very few rivers where water-quality measurements are made which would warrant anything more than a steady-state model at present.

Each class of model defined has different input requirements for historic water-quality data in terms of length of record and frequency of sampling. For a planning model where only average annual values of water-quality parameters are necessary, as long an historic record as possible is desirable even though it may have a coarse sampling frequency of anything up to one month. The historic data will always be a very small sample, but obviously confidence in how representative it is of its parent population increases with length of record. On the other hand, a design model requires not only a long historic record to identify trends and seasonal effects, but also more frequent sampling than that necessary for a planning model. Measurements on a daily or five-daily basis would enable any autoregressive component that might be present in the observed fluctuations to be isolated from the random perturbations. To some extent the ideal sampling frequency may be dependent upon what types and sizes of storage are to be considered in the proposed water-resource system. The data requirements for an operational model are quite different. Only a few years of historic record may be necessary to calibrate and test the model. However the frequency of sampling required would probably be impractical without automated data collection and processing.

Acknowledgement

This paper was presented with the permission of the Director of the Water Resources Board. The views expressed are those of the author and not necessarily those of the Board.

18.6 REFERENCES

Bowden, K., Green, J. A., and Newsome, D. H. (1971). 'A mathematical model of the River Trent system', *IWPC Nottingham Symposium on the Trent Research Programme*, 94–102.
Box, G. E. P., and Jenkins, G. M. (1970). *Time-series analysis, forecasting and control.* Holden-Day, San Fransisco, 150–151.
Jamieson, D. G., and Wilkinson, J. C. (1972). 'Dee Research Programme: 3 A short-term control strategy for multi-purpose reservoir systems', *Water Resources Research*, **8** (4), 911–920.
Jamieson, D. G., Wilkinson, J. C. and Ibbitt, R. P. (1972). 'Hydrological forecasting with sequential deterministic and stochastic stages', *International Symposium on Uncertainty in Hydrologic and Water Resource Systems*, Tucson, Vol. 1, p. 177–185.
O'Neill, P. G. (1972). 'A mathematical-programming model for planning a regional water resource system', *Journ. Inst. Water Engrs.*, **28**, 47–61.
Page, C., and Warn, A. E. (1973). 'Water-quality considerations in the design of water-resource systems', *Proceedings of the 7th International Conference of Int. Assoc. on Water Pollution Research*, Paris.
Whitehead, P., and Young, P. C. (1974). 'A dynamic-stochastic model for water quality in part of the Bedford Ouse River system', *Proceedings of the First IEIP Working Conference on Modelling and Simulation of Water Resources*, Ghent.
Young, P. C., Beck, M. B., and Singh, M. (1973). 'The modelling and control of pollution in a river system', *IFAC Symposium on Control of Water Resources Systems*, Haifa.

19

Systems Analysis and Modelling Strategies in Ecology

J. N. R. JEFFERS

19.1 INTRODUCTION

The chapters in this book are an expression of the considerable interest in the use of mathematical models as a strategy for research. Such interest has been expressed in many areas of scientific research, but it is perhaps especially in ecology that the role of systems analysis and mathematical models has been emphasized in recent years. Such emphasis is not without its critics. Gifford (1971) suggests that two developments in ecological thought have debased our values in ecology in the name of quantification of the ecosystem—ecological energetics and the use, or rather the misuse, of the computer model for whole-systems analysis. He suggests that the measurement of energy transfer between organisms provides us with virtually no insight into what limits useful production in a system, and pursuit of this pernicious methodology has deflected able minds from consideration of the real problems which surround us and which demand our urgent attention.

Clearly, however, the use of mathematical models in ecology is not confined to the measurement of energy transfer between organisms, and it is a gross over-simplification to regard systems analysis as being solely concerned with energetics in ecology. Living organisms and their inter-relationships are of interest not only as chemical mixtures, physical forces, and biological phenomena. They are influences that shape our environment and determine our perception of environmental quality, and they have a significant effect on the availability of renewable and non-renewable resources. For renewable resources, particularly, the understanding of the structure, functioning, and dynamics of living systems and of the effects of our impact on those systems is vital to our survival on this planet.

We must develop a science, or combine existing sciences, to provide the basis for a rational use of our natural resources. While the development of any science proceeds through observation, measurement, hypothesis, prediction, and control, further progress necessarily depends on the development of major concepts capable of carrying the science beyond the present limits. The major developing concept of ecology at the present time is that of the ecosystem, a concept possessing a generality which is universal and providing a framework for the examination of the essential relationships between the component parts of natural systems. Ecosystems, as the basic units in resource management, have been recognized for some time, and measurements have been made of variables and components of many ecosystems in isolated studies. However, our understanding of such systems is incomplete, and we lack the ability to control them or to predict their future. To use resources without totally destroying them we must be able to predict the consequences of our management of ecosystems, and the current emphasis on mathematical models of ecological systems is based on the belief that we can construct valid models of such systems and use these models to predict the consequences of our manipulation.

In this chapter, I am attempting to do no more than to outline the role of systems analysis models in ecology and the necessary stages in a modelling strategy for ecology, and to examine the inter-relationships between scientists and decision-makers in the use of natural resources. If such a chapter should seem superficial or too much concerned with generalities, my justification is that, while many scientists are concerned with the construction of mathematical models of individual ecosystems or components of ecosystems, I believe that too little attention is being given to the necessary impact of modelling strategies upon the organization of research, and upon the inter-relationship between the research scientist and the resource manager. Nevertheless, some examples of the application of modelling strategies to the solution of ecological problems are given at the end of the chapter.

Much of what I have to say has been influenced by my participation in an Expert Panel in the Role of Systems Analysis and Modelling Approaches in the programme on Man and the Biosphere, which is currently being formulated by Unesco. The Man and Biosphere programme (MAB) is an inter-disciplinary inter-governmental programme of research. The general objective of the programme is to develop the basis within the natural and social sciences for the rational use and conservation of the resources of the biosphere and for the improvement of the global relationship between man and the environment. Briefly, the programme is intended to:

(1) identify and assess changes;
(2) examine the structure, functioning, and dynamics of ecosystems;
(3) study the inter-relations between ecosystems and socio-economic processes;
(4) develop means for measuring environmental changes;
(5) increase global coherence of environmental research;
(6) promote simulation and modelling as tools for environmental management;
(7) promote environmental education.

Each of these objectives is interpreted in the context of man, the biosphere, and the reciprocal interactions of the one upon the other. Those responsible for the MAB programme have come to the conclusion that integration and coordination of scientific activity are needed to bring about greater coherence of ecological research, and to ensure that the significance of the MAB programme is greater than the sum of its constituent parts. Systems and modelling approaches can help to bring about this greater coherence, and mathematical models can play an important role in the understanding and optimal management of natural resources, and particularly renewable resources. These discussions and the report of the Panel of Experts (Unesco, 1972) have necessarily influenced this chapter.

19.2 THE ROLE OF MODELS IN ECOLOGY

Natural systems are essentially complex, and, when faced with complex and highly interacting systems, human judgement and intuition may lead to wrong decisions and to irreversible damage. In recent years, there has been extensive study of the behaviour of less complex interacting systems in such fields as engineering, physiology, and economics. This diverse body of experience has enabled progress to be made over the past ten years in the development of methods for understanding the dynamics of ecosystems and the impact of changes upon them, including changes generated by man. Such studies are frequently referred to as systems ecology, and the studies are based on the assumption that the state of an ecosystem at any particular time can be expressed quantitatively, so that changes in the system can be described by mathematical expressions.

The use of mathematical models in ecology, or indeed in any other branch of science, is not new. The transfer of chemical and physical concepts to thinking about the inter-relationships between biological organisms, and the relationships between such organisms and their environment, have provided the basis for much of the progress in ecological research. The translation from physical models to abstract models in which the variables and properties of living systems are formulated as mathematical relationships is, however, more recent. Mathematical models differ from physical analogues in the extent of their abstraction, and in their ability to handle complex relationships, so that what is new in the present emphasis on quantitative models is the increased use of mathematical techniques of interpretation.

It is perhaps necessary to emphasize the distinction between static and dynamic mathematical models of ecological relationships. The former are concerned solely with the state of a particular ecosystem at one point in time, and are essentially descriptive. Dynamic models, in contrast, are concerned with the development of the model in time. They may be purely descriptive, but, if correctly formulated, they may also be predictive. More complex formulations of the model may even be capable of indicating some optimum decision about the future form of the management of the ecosystem.

In the formulation of dynamic models, there is a general trend towards the adoption of the terminology used by Derusso and colleagues (1967). This terminology makes a distinction between state variables, driving variables, and output variables as follows:

(1) The *state variables* are defined as the measurable properties of the system, and include such variables as biomass, numbers of organisms, concentration of mineral or nutrient elements, water content, and so on. The seasonal dynamics of the state variables reflect the changes taking place in the system and the values of these variables at any particular time themselves depend on changes within the system.

(2) The *driving variables* are not affected by processes internal to the system but act upon the system from outside. In the management of natural or semi-natural systems, the most important driving variables are usually the major climatological or meteorological factors influencing the system, but biotic variables may also be considered as driving variables, e.g. the number of animals grazing on the vegetation.

(3) The *output variables* are the quantities that the model is required to predict. Sometimes these variables are a subset of the state variables, but, in addition, they will usually include quantities calculated from the state variables. Clearly, the output variables are the quantities of most direct interest to the manager who proposes to use the model.

(4) Many ecological, physiological, and physical processes may be simulated in the model. These processes result in changes in the state variables, and the rates at which processes take place are determined by the current or previous values of these and other state variables and of the driving variables. In this way, they constitute the mechanism by which these variables are coupled. Clearly, in such a concept of a dynamic model, there is no restriction to energetics.

Much of the early work on the application of mathematical models to biological problems has depended on the concept of differential equations in which the rates of change in the state variables of a system have been expressed in terms of the factors influencing them. It is relatively simple, at least in concept, to build up suitable sets of equations expressing the supposed inter-relationships between variables in this way. With modern computing techniques, it is possible to find numerical solutions to sets of equations over specified time intervals and to record the new values of the variables describing the state of the system. In a real sense, this is a simulation of an ecosystem, a representation of the ecosystem, and the processes within it, so that the behaviour of the system as a whole in time can be studied. Such simulations are obviously abstractions or simplifications of the total system, but they are usually more comprehensive and precise than the mental models which underlie our understanding of complex systems.

Figure 19.1 shows how mathematical models of different kinds may help to bridge the gap between ecological theory, scientific research, and management. By systems analysis, it is possible to incorporate data and information from a wide array of studies into a single inter-related model. The sets of

393

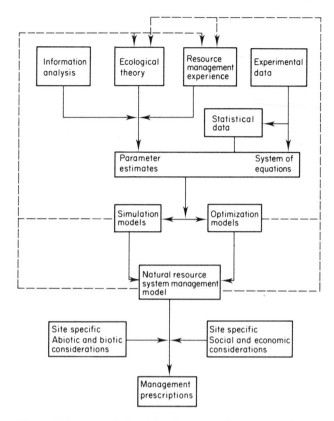

Figure 19.1 Inter-relationships between theory, experience, data, statistical analysis, simulation, and optimization in natural resource management
(Modified version of figure from Report of Expert Panel on the Role of Systems Analysis and Modelling Approaches in the Programme on Man and the Biosphere, Unesco, 1972)

coupled differential or difference equations, when solved, predict the response of the system to induced changes. Alternatively, it may be possible to find the particular set of changes which will produce a response which is optimum in some defined way, consistent with constraints which have to be placed on the management alternatives. It may even be possible to couple simulation and optimization models so that whole sets of management strategies may be evaluated. If simulation and optimization models can be structured to give the necessary output, then such an approach has important implications for the management of natural resources.

However, mathematical models of biological systems which are expressed in the form of differential or difference equations lack one essential property of all living systems, i.e. the property of variability. For this reason, increasing attention is being given to probabilistic systems of modelling which contain stochastic elements capable of describing, or predicting, the variability which is so characteristic of living organisms. Again, such models may be formulated as a simulation of the variable system, or they may be formulated so as to seek the particular combination of parameters which leads to an optimum solution. In general, the mathematical background of stochastic models is more complex, although, paradoxically, the arithmetic of the solutions may be similar. The multivariate nature of many biological problems, however, places an additional level of complexity on stochastic models and we have so far, little experience in the construction of successful models of this kind. Nevertheless, both the mathematics and the computing power exist for such models, and it requires only the successful fusion of ecological experience with mathematical expertise for the successful achievement of mathematical models which are capable of solving complex biological problems.

Isolated sets of experimental and survey data may be utilized in constructing both simulation and optimization models, whether these are of the deterministic or probabilistic kind, but, ideally, data should be derived from integrated research programmes. Where experimental data and information are not available, guesses may have to be made by quantifying the knowledge of experienced scientists and managers. In this way, models may be constructed for areas in which data are not readily available, and the output from the models may first be judged for the extent to which it is consistent with experience. If consistent, then experiments may be conducted on the model itself by systematically varying the values of the coefficients in the model or by varying the connection between the parts of the model. Interestingly enough, there is a return to many of the methods of design of experiments and surveys, now regarded as outdated by younger mathematicians, for the purpose of exploring the implications of models, and it is possible to test a great variety of management approaches in a relatively short time, and to select those that appear promising, before evaluating them under field conditions.

To summarize, the role of mathematical models in ecological research is to test the effects of the modification of biological systems, as this can be done far more quickly and cheaply than in the real-life system, and without the risk that unexpected effects will have embarrassing and possibly irreversible consequences. If it is desired to manage the ecosystem to obtain some definable goal, such as maximum yield, minimum loss of surface soil or nutrients, or maximum biological production, the simulation model can be incorporated into a computer program designed to test all possible combinations of natural management practices, and hence to find the desired optimum. The importance of this approach is clear, and mathematical modelling has great practical potential in the management of natural resources. Wherever alternative land use and management policies are under discussion, choice between them can be greatly facilitated by a reliable dynamic model, but it must not be supposed that a single model of any system will serve all purposes, and it is important to determine which practical problems require the simulation of a total ecosystem, and which can be solved by simulation of only part of the system.

Perhaps the most important consequence for this approach, however, is that, in the face of a shortage of well-trained managers and scientists, systems analysis techniques, although themselves requiring the scarce expertise of mathematical ability, may help to evaluate, with minimum cost and effort, many management alternatives that would otherwise be too costly and time-consuming to determine. For mathematical models to be used in this way, however, managers and scientists must have a general understanding of the use and limitations of modelling and systems approaches, and have available to them the necessary data, as well as the generalized mathematical models required to simulate and analyse the systems. I do not know of any country where this situation has been achieved.

19.3 THE NECESSARY STAGES IN A MODELLING STRATEGY

The examination of the use which has so far been made of systems analysis in ecological research suggests that there are some necessary stages in research projects, and that it is necessary to emphasize these stages. Many problems in the application of modelling strategies have arisen from the attempt to eliminate particular stages of the process of investigation with the result that the research has been badly planned or executed.

(1) Setting of objectives and preliminary synthesis

The first phase in any research project should result in the setting of objectives and the creation of an initial synthesis of existing information. The objectives of the project should specify:

- (a) the population about which inferences are to be made as a result of the investigation
- (b) the experimental manipulation, modification, or disturbance, if any, to which the population will be exposed
- (c) the variables that will be measured.

Other aspects may be included in the objectives where human interaction with the project is important, as, for example, in the perception of environmental quality.

Once the preliminary objectives have been set, it is then possible to begin to assemble relevant existing knowledge. Some of this knowledge will be in the form of publications, some in the form of unpublished reports and field data, and some will exist only in the minds of experienced scientists. All the information from the various sources will require to be incorporated into written summaries, and to be quantified in as far as this is possible.

A preliminary synthesis may well consist of a 'word model', in which an attempt is made to describe the model in purely descriptive and verbal terms. Such a model may well seem trivial in concept, but experience has already suggested that it is surprising how little agreement there may be between the workers from various disciplines cooperating in a multi-disciplinary research

396

project on the model that they have in mind. The expression of the mental concepts in terms of a verbal model will lead to an important exchange of information, identification of areas where different concepts are held, and, hopefully, agreement on the final model. Only when an agreed 'word model' has been achieved, will it usually be worthwhile to express the model in quantitative terms.

Once a preliminary synthesis has been attempted, it may then be necessary to reexamine the objectives. The operation of assembling data will have helped to clarify ideas in the minds of the scientists concerned with the project. Some of the objectives will be capable of refinement, some will need to be omitted, and new objectives may be added. Frequently, it will be recognized that the scope of the project had originally been defined too broadly or too narrowly. It may be necessary to redefine the population about which inferences are to be made, or, alternatively, for which representative samples can be derived. There may also be some revision of the list of the state, driving, and output variables to be included.

There is an important distinction to be made between objectives and the criteria by which the success of the objectives can be achieved. Unless the criteria by which the achievement of the objectives can be assessed are also defined, it may be difficult to recognize the extent to which the project has been successful, or even to define the objectives in such a way that they are anything more than statements of broad intent. The combination of defined objectives and statements of the criteria by which the success of the project can be judged will usually greatly clarify the aims of the project. It will also provide a basis for decisions which will have to be made in later phases of the investigation.

(2) Experimentation phase

The second phase of research projects usually involves experimentation, both in the field and in the laboratory, as well as with the mathematical models of the preliminary synthesis. The nature of the experimentation naturally depends upon the level of sophistication of the project. Broadly, however, it will be necessary to collect data upon which the validity of the model of the preliminary synthesis can be tested. Simulation models, for example, usually require the collection of sets of coordinated driving variables and of sets of time-series observations of state variables. The driving variables may then be used to derive the equation systems of the model, and output from the model tested against the time-series observations. In some instances it may be necessary to test the model against the expectations of panels of experts but, where possible, such experts should be different from those used to develop the preliminary synthesis.

Adjustment or revision of models is greatly accelerated by the formulation of procedures for the rapid transmission of information on the structure and operation of the models between different projects or parts of the same project. Experiments conducted on models lead to new ideas on the management of systems, and help to promote a dialogue between the scientists involved in the research and those responsible for management decisions.

However, perhaps the most useful by-product is the development of simulation and optimization models which may be used as the basis for management games. In the sense that the word 'game' is used here, it is more an approach to the use of the models than a separate theory in itself, such as has been derived in the statistical theory of games. The development of models of natural resources, for example, and the use of such models in management games is, in effect, a form of management education.

(3) The management phase

Once the initial validation of the model has been achieved, and a certain degree of feedback from the experimental phase to the synthesis phase has taken place, it is appropriate to begin the third phase in a modelling strategy, i.e. the consideration of the uses of outputs from the research in the practical management of resources. Small-scale, prototype management plans and manipulations can often be defined and undertaken as a result of the early work on synthesis and experimentation with models. Such small-scale studies of the application of results can be conducted in cooperation with management agencies. After appropriate testing on a small scale, more comprehensive management schemes can frequently be undertaken, again with consultation and feedback between the scientific groups developing the synthesis, experiments, and models.

The emphasis of this phase on the continuing cooperation between the scientist and the manager differs markedly from the usual practice in the conduct of scientific research, even in the western European and North American communities. The traditional approach, by which the scientist produces a report or paper, and by which the results are incorporated, or ignored, in management action, usually without any further consultation with the scientist, places no emphasis upon the need for a dialogue between the manager and the model, and between the manager and the scientist responsible for developing the model. In this situation, it is not surprising that many of the models that have so far been produced for the management of natural resources have been markedly deficient in reflecting management needs, or that managers have frequently failed to understand the implications of well-conceived models of natural systems. Indeed, as has been stressed earlier in this paper, one of the major values of a modelling strategy in ecology is the bringing together of the sceintific and managerial expertise in developing a joint approach to the problems of resource management.

(4) Evaluation phase

The fourth phase, like the third, is frequently omitted from ecological research programmes. However, it is important that an evaluation should be made for each project of the possible impact of the results upon the perception of environmental quality. The research scientist and the manager should together evaluate the effects of the management changes proposed as a result of the research on the structure, functioning, and stability of the relevant ecosystems, and consider how these effects may influence the quality of life. Procedures should then be developed to evaluate the change in public attitudes in relation to the use of resources. Such information may be vital in suggesting policies for management, use, and allocation of resources.

(5) Final synthesis

The end of the project is reached when a final synthesis has been derived in the form of a model, either as a simulation or optimization of the management decisions, and a prediction of the effect of the decisions upon the natural system. The sense in which any synthesis can be regarded as final is, of course, arbitrary. However, we may regard a working model, agreed jointly by the scientists and the managers, and for which it is possible to simulate the effects of management decisions upon the natural systems, or, hopefully, to optimize these effects in some defined way, as a temporary stopping point in the research activity. It may subsequently be necessary to refine the model and to derive an even closer approximation to the operation of the system. Alternatively, it may be more important to turn the attention of the research scientist to systems for which even a first approximation is not currently available. Such decisions depend upon important problems of the allocation of scarce resources of research expertise.

At the phase of final synthesis, it is important that all the information derived from the project should be summarized, evaluated, and published. Much of the information will be held in the form of computer models, and the computer itself serves an important function in providing an unambiguous algorithm of the models, a function which may, indeed, be regarded as one of the most important advantages brought to science by the computer. An algorithm, especially if written in a high-level language, is readily available for examination, and, if necessary, for alteration, so that anyone wishing to make use of the model has only to implement the computer program on whatever computer may be available or suitable. In this way, subsequent dependence upon the scientific expertise, necessarily scarce, which was necessary for the development of the model is minimized. It is important, however, that, even if the 'final' synthesis is recognized as being a temporary stopping point in the investigation, the necessary documentation and evaluation associated with the algorithm should be completed.

19.4 THE INTER-RELATIONSHIP BETWEEN SCIENTISTS AND DECISION-MAKERS

As a consequence of the development of computers and computer languages, new methodologies have been designed to describe complex systems, and, as has been suggested earlier in this chapter, it is now possible to design research strategies for ecological investigations having many interacting components. Interdisciplinary groups are increasingly bridging the gaps between disciplines and methodologies in many universities and research institutes. These groups are often directed towards the solution of problems which are likely to be of critical importance in the long term, but which are of limited practical importance in the short term. On the other hand, management agencies are continually embarrassed by the need for short term, limited solutions to immediate problems, and have therefore evolved a strong bias towards pragmatism. Synthesis and integration of research and management attitudes are needed to combine the precision and the detachment of the research institute or university with the pragmatism and immediacy of the management agency, in order to help develop new approaches to problem-solving.

Experience of the problem-solving process shows that such tasks as the development of systems analysis, the formulation of hypotheses and collection of data, and the planning and implementation of pilot studies and management plans have very often been undertaken by different groups and by different agencies. Inadequate integration of these tasks has resulted in loss of communication: (a) between data experimentation and model development; (b) between simulation models, the overall systems analysis, and the implementation of models in management practice; (c) between the examination of predictions from systems analysis and the implementation of models in management practice; (d) between management practice and development of new hypotheses; and (e) between the implementation of results from pilot studies and the development of new hypotheses.

The development of models often follows a somewhat standardized pattern, in which there is a progressively detailed breakdown of each component into modules that can be easily translated for input into the computer. After a year or so of efforts of this kind, some of the following conclusions can usually be drawn:

(a) The quality of the available data and the understanding of causal pathways, especially as they relate to human ecology, are generally poor.
(b) Systems analysis and data collection must develop a mutual feedback from which the manager can draw the maximum benefit.
(c) Training in systems analysis is valuable for stressing a broad interdisciplinary, problem-orientated philosophy of research.
(d) Systems models can be improved only by building them and striving to correct their weaknesses.
(e) Systems analysi teams must be broadly interdisciplinary.
(f) Systems models often demand large quantities of high-quality data and they may consequently be expensive.

In return for the investment, however, one hopes and expects to obtain a powerful tool for prediction and planning, and a set of procedures for formally applying logical processes as the basis for data collection and management.

19.5 SOME EXAMPLES

Having talked rather superficially about the use of models in ecology, the necessary stages in modelling strategies, and the inter-relationships between scientists and decision-makers, it may seem desirable to give some examples of the application of these ideas to real problems. The examples I will give are drawn from the work which is being undertaken by research workers in my own institute, and are parts of currently continuing research programmes. This inevitably means that the phase of final synthesis has not yet been reached, but it may be helpful to describe them briefly nevertheless.

Much of the current emphasis on the use of modelling strategies in ecological research derives from initiatives developed in the course of the International Biological Programme. This programme aims to provide an understanding of the biological basis of productivity and human welfare. It is divided into a number of sections examining, for example, production in terrestrial, freshwater, and marine communities. In the United Kingdom, studies on production in terrestrial communities are concentrated on woodland, moorland, and grassland. Studies on main sites for woodlands (Meathop Wood) and moorlands (Moor House) are coordinated from the Merlewood Research Station. In both projects, measurements of primary and secondary production have been made, and the main patterns of energy and nutrient transfer in each ecosystem described.

Meathop Wood is a mixed deciduous woodland about 3 km from the Merlewood Research Station near the shore of Morecambe Bay. It lies on a drift-covered outcrop of carboniferous limestone and has a recorded history of woodland management, mainly for charcoal production, since 1684. Botanically, it is typical of the small deciduous native woodlands on calcareous sites in southern Lake District, and the ecosystem shows features related to its intensive exploitation and the low nutrient status of the drift-derived soil.

The aims of the research are, in general, to study a complex deciduous woodland community as a complete ecosystem, and specifically:

(a) to estimate and describe the net primary production, energy flow, nutrient circulation, and water balance of a selected hectare of the wood for several years;
(b) to identify the key organisms at various trophic levels and to give quantitative estimates of their activity;
(c) to compare the results with those from other woodland IBP studies and to formulate hypotheses concerning the factors affecting primary and secondary production.

Thirty workers have contributed to the research, about half from the Merlewood Research Station and half from universities and other institutes. Some preliminary results from the programme, in its final synthesis stage and due to terminate in 1974, were presented at a symposium of the British Ecological Society, notably the results of studies of primary and secondary production, and description of the main patterns of the energy and nutrient transfers. The final results of the synthesis will be published in a separate volume. However, it has already been possible to locate gaps in the information available and to assess the effect of the error terms in the component estimates on the balance between the input and output of the energy budget. The computations indicate that the budget is likely to show an excess of energy input over output and that there might be undetected energy accumulation in the system, possibly as soil organic matter. The synthesis also suggests that the flow of energy through the fungal population may be second only to that through the oak trees, and that the data currently available from the literature are inadequate for estimating fungal population metabolism. Figure 19.2 gives an outline of the general model which is the basis of the synthesis.

The importance of studies of this kind cannot be over-emphasized. Far from being confined to the measurement of energy transfer between organisms and providing virtually no insight into what limits the useful production in a system, as has been alleged, the studies have enabled scientists from a wide range of disciplines to investigate the inter-relationships between the activities of a wide range of different organisms. The underlying model of the ecosystem is useful as a means of (a) locating inconsistencies in the data; (b) revealing the illogicalities that arise if the data are extrapolated, thus indicating the additional information needed for predictive purposes; and (c) giving a general sense of proportion about the ecosystem. The model may also help to indicate ways in which the future management of the wood might influence the development of the ecosystem.

For example, the discontinuous compound interest rates for tree and shrub stems and branches in Meathop Wood imply that, amongst the existing species, sycamore, ash, and birch will grow larger relative to oak and hazel. The wood is known to have a management history of coppicing on a 15-year rotation with sales of oak bark and other oak products important to the economy. Selection for oak may therefore be expected to have been a feature of this form of management. Hazel would perhaps have been favoured by the reduced light competition resulting from frequent thinning of the canopy. The present composition of the stand reflects its history, and there is every indication that the stand composition is changing as it approaches an equilibrium. Observations on the structure of the canopy indicate that the mean canopy height of ash is greater than that of oak, and this is consistent with the idea of the fast-growing species, e.g. ash, birch, and sycamore, over-topping the oak at a certain stage of the wood's recovery from coppicing. As the canopy closes and competition for light increases, the relative suppression of oak may be expected to reach a maximum. This is likely to be temporary, however, for when the mature ash, birch, and sycamore die, their successors will have to develop under or between the longer-lived oak.

The nutrient status of the site is high relative to other forest sites studied in IBP, with the exception of phosphorus which is extremely low. Analysis of tree stems, bark, and roots show, however, that the phosphorus concentration

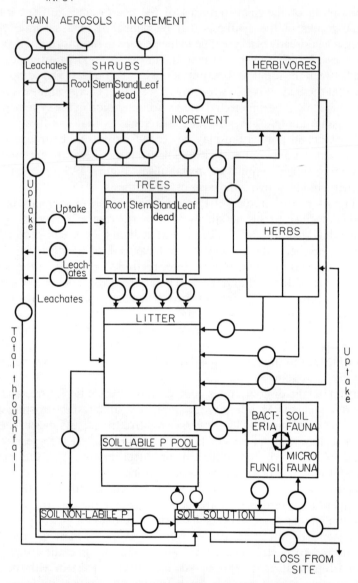

Figure 19.2 Diagrammatic representation of the Meathop Wood synthesis

in ash, which is growing fast, is higher than in oak, which is growing more slowly. This implies that, although the two species may be competing for nutrients, their growth is not obviously limited by an absolute nutrient deficiency. The pattern of tree growth in the mixed deciduous woodland is patently the outcome of a complex of interacting processes and the problem is to separate some aspects which are conceptually discrete and feasible. Essentially, what is being aimed at is a model which will predict the future composition and behaviour of the wood if it is left unmanaged, and what the possible consequences might be of various forms of management.

The second example from the International Biological Programme is concerned with the studies being undertaken at the Moor House National Nature Reserve. The marginal agricultural land of the northern Pennines is dominated by a blanket of peat with vegetation composed of Calluna, Eriophorum and Sphagnum species. The Moor House NNR is a sample of this landscape and the IBP studies have been designed to measure, or to estimate, the production by plants and animals, and the circulation of energy within this type of moorland ecosystem. The research programme has concentrated on the blanket bog but has also used for comparison available information from adjacent grassland areas. Because of its cold climate and the dominance of dwarf shrubs, Moor House has formed an integrated part of the Tundra Biome projects of IBP, in which a comparison is being made of the productivity and production processes of a range of polar and alpine ecosystems.

The Moor House programme has completed its field and laboratory studies and is now in the stage of the final synthesis of the results. The international comparisons are also in the stage of final synthesis and, like the Meathop Wood programme, was terminated in 1974.

The productivity of the main organisms in the bog ecosystem have been measured by the individual research workers contributing to the project. When those results are combined, they show the main pattern of distribution and transfer of dry matter in the ecosystem, and this is summarized in its simplest form in Figure 19.3. The results are expressed in grams of dry weight per square metre and are average annual values, for a depth of 20 centimetres, which represents the main activity zone of the ecosystem. The values are subject to natural variation and methodological errors, e.g. the grouse and psyllid population vary tenfold between years, the microflora standing crop estimate varies one hundredfold depending on the use of direct or viable population estimates. In addition, some estimates are derived from others by addition or subtraction or by assumption of efficiency values derived from literature. Despite the large range associated with many of the estimates, the general pattern remains the same.

This extremely simplified summary of the final synthesis shows that, in the blanket bog, about 1% of the annual primary production is consumed by vertebrate and invertebrate herbivores. Of the dead plant remains in the decomposer cycle, the soil fauna assimilate about 5% and pass through their digestive system about 50%. All this activity occurs mainly in the upper three centimetres of the bog. It is estimated that about 10% of the annual input of dead plants passes below 20 cm in the peat, so that the remaining 85% which is decomposed is converted by the microflora. The high primary production from a large standing crop, with low utilization by herbivores and slow decomposition by fauna and microflora, reflects the low nutrient concentration in the plant tissues and the storage of nutrients by the perennial vegetation. This contrasts with adjacent

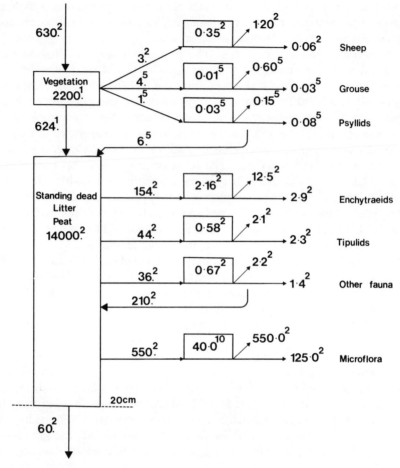

Figure 19.3 Estimated standing crops (gm^{-2}) and transfers $(gm^{-2}\ yr^{-1})$ of organic matter in the surface 20 cm of blanket bog at Moor House NNR. Variability in estimated values, through sampling and systematic errors and between year and between site variation, are indicated by superscripts as:

$$^1 = 0.9 - 1.1 \times \text{mean} \qquad ^5 = 0.2 - 5.0 \times \text{mean}$$
$$^2 = 0.5 - 2.0 \times \text{mean} \qquad ^{10} = 0.1 - 10.0 \times \text{mean}$$

(Reproduced by permission of The Royal Society from *Phil. Trans.*, B, 1976)

grassland on mineral soils where annual primary production is lower, and produced from a smaller standing crop, but the herbivore consumption, soil fauna production, and rates of microbial activity are higher than on the bog. This probably results from the nutrient concentrations in live and dead plants being two or three times higher on the grassland.

Computer simulation, both at Moor House and in the Tundra programme generally, provides one way to study the compatibility of data collected in discrete projects. It also allows an explicit statement of the assumptions involved in the manipulation of the data. In this study production and decay of the three major blanket bog species were used to simulate peat accumulation on two blanket bog sites of different peat depths. The simulation was run for some 2500 years, the period over which the three codominant species had been present and dominant, and for which the depth and weight accumulated in the field had been calculated. Additional runs for 15 years have been made to check with datings obtained using changes in $^{14}C/^{12}C$ ratios due to atom bomb tests. Variation in litter production, litter densities, decay rates, and effect of depth on these decay rates have also been simulated. Such modelling exercises are particularly valuable in helping the individual research workers to understand their data and the ecological system from which the data are derived, and the value to the research worker as an adjunct to critical biological thinking needs to be emphasized. Again, it cannot be over-emphasized that a modelling strategy is of great value in investigations such as that at Moor House, and in the linking of biological comparisons between different polar and alpine sites in the Tundra Biome of the IBP. Because recognition of the value of modelling did not develop until late in the programme, the potential value of this tool has often not been fully exploited. Merlewood has been fortunate in being one of the centres for this type of work, with the establishment of an international microbiological and decomposition data bank, and both resident staff and visiting fellows to work on various aspects of the international synthesis. In this way, it is possible to devise a research programme which is more than the sum total of the scientific papers which are produced by individual research workers.

My third example is a systems study of the uplands of northern England and southern Scotland, undertaken jointly between the Merlewood Research Station and the International Systems Corporation of Lancaster, a consultancy company associated with the Department of Systems Engineering, Lancaster University. The uplands of northern England and southern Scotland form a fairly homogeneous geophysical area with a limited number of land-using activities. Hill farming, forestry, and recreation are the main uses, water provision, military uses, mining and quarrying are lesser, more localized uses. The land, though poor, has some value: an economic value for its products of wood, meat and timber, a social value to local populations for employment and services, and to visitors for amenity and wilderness; and, more debatably, a 'conservation' value for its potential in terms of wildlife, genetic reserves and bioproductivity.

There are many groups of interests involved. There are, for example, tenant farmers and owner occupiers, and local populations such as farm, forestry, and service workers, wives and children. There are government land-owning bodies such as the Forestry Commission, large private owners, charities like the National Trust, and syndicates of business men interested in land, such as those represented by the Economic Forestry Group. There are administrative bodies, such as the Ministry of Agriculture, Fisheries and Food, local planning authorities and the Countryside Commission, and there are non-official organizations—on the one hand, commercial firms, and on the other, represent-

406

ative lobbying organizations, such as the Country Landowners' Association or the National Farmers' Union.

Present trends include a possible long-term decline in soil fertility, a general increase in forestry, and a sometimes unfavourable market trend for agricultural products, rural depopulation, and rapidly increasing demands from outside the area for water and recreation. In the long term, there are prospects of world shortage of primary products and increasing external population pressure. In the relatively unspoiled uplands, there is still a chance for genuine environmental planning, in contrast with the rescue work which has to be attempted in other areas. The present time is crucial, because of entry to the European Economic Community, local government reorganization, and increased public awareness of environmental matters. Each of these developments calls for an overall understanding of the uplands and of the interactions between activities before planning decisions and policy commitments are agreed; otherwise there is little chance of any lasting improvements being obtained.

A pilot study has therefore been undertaken to provide a basis, and justification, for a major study which would be capable of contributing substantially, in a practical way, to the long-term aim. In particular, the pilot study was intended to investigate the applicability of systems analysis to this type of environmental management problem.

The primary conclusion which has been drawn from this pilot study is that a major study is both feasible and worthwhile. The objective of the project is defined as the provision of a comprehensive and agreed systems analysis of the uplands in sufficient detail to enable useful deductions to be made about the changes in structure and policies which can be introduced to ensure improved environmental management. In broad terms, the project is seen to have the structure shown in Figure 19.4. The development of quantitative models is an important but time-consuming part of this project, and the continuing dialogue with the many bodies (both public and private) with interests in the uplands is also a time-consuming, but key activity. Without such work, not only would it be difficult to assemble a complete analysis, but no feeling of commitment would be developed, so that implementation of any possible changes might be strongly resisted.

Three main types of outcome are foreseen and are illustrated by Figure 19.4. These are:

(a) A definition, by reference to a clearly presented systems framework, of possible changes in management structure which are seen to be beneficial for the community generally and acceptable to the majority of interests, and which are seen to arise from the qualitative systems analysis.
(b) A computable model which would indicate the outcome in the long term (i.e. 50–100 years) of pursuing particular management policies. By allowing senior policy-makers to explore the range of outcomes which result from present-day decision, a more secure and objective basis

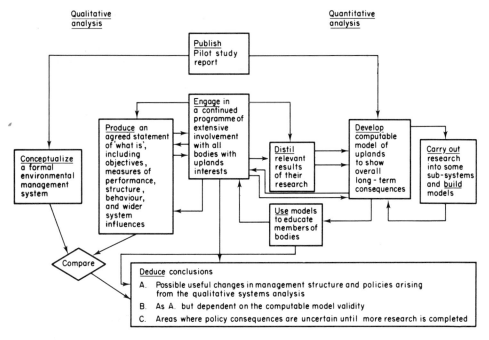

Qualitative
analysis

Quantitative
analysis

Publish
Pilot study
report

Engage in
a continued
programme of
extensive
involvement
with all
bodies with
uplands
interests

Produce an
agreed statement
of 'what is',
including
objectives,
measures of
performance,
structure,
behaviour,
and wider
system
influences

Conceptualize
a formal
environmental
management
system

Distil
relevant
results
of their
research

Develop
computable
model of
uplands
to show
overall
long - term
consequences

Carry out
research
into some
sub-systems
and build
models

Use models
to educate
members of
bodies

Compare

Deduce conclusions

A.   Possible useful changes in management structure and policies arising
     from the qualitative systems analysis
B.   As A. but dependent on the computable model validity
C.   Areas where policy consequences are uncertain until more research is completed

Figure 19.4 Project structure for uplands systems study (reproduced by permission of International Systems Corporation of Lancaster, University of Lancaster)

should be available for such decisions. Further, since all interested parties can engage in exploring the long-term dynamic behaviour, all can gain a fuller understanding of how the uplands as a whole would develop in response to individual actions.

(c) Although a number of research studies have been completed, or are now in hand, in the uplands, the study may be expected to show where gaps in research programmes lie, as well as to reveal where duplication occurs. Further, and perhaps more important, the systems analysis will provide a wider systems framework to which future research can relate, so that sub-systems can then be isolated for more detailed study. In particular, the opportunity for multiple use of land could be examined more efficiently.

We began the study of upland systems with the firm conviction that the major problem would be in modelling the various ecosystems which go to make up the uplands. The preliminary studies have been sufficient to show that we probably already have enough information on the ecology of the area to be able to provide the necessary models of the response of such systems to management policies, but that the main gap is in the understanding of the inter-relationships between the various management practices and policies which have an impact on the area as a whole. The primary objective, therefore, has shifted from conventional ecology to the decision-making process and the impact of decision-making on ecology. It seems likely that similar findings will result from many such investigations, and will quickly lead to a redefinition of priorities in ecological research.

In this chapter, I have tried to summarize the present views of myself and of my staff towards the use of ecological models in resource management. If I have dwelt at length upon the formulation of research strategies, and on the inter-relationship between the scientists and managers, at the expense of detail of ecological models, it is because I believe that the main difficulties for the future lie in creating an efficient interface between the manager and the research scientist. Our ability to model ecosystems to any degree of detail that is required is limited only by the lack of ecologists with sufficient mathematical knowledge to make use of the techniques which already exist for undertaking this task. This is a challenge to which the ecologist is rapidly responding, and I have little doubt that we will see a quick increase in the ability of ecologists and biologists generally to make use of the powerful tools which have been made available by the conjunction of the electronic computer with modern mathematics. The problems lie less in terms of the mathematics than in terms of the basic ecological thinking, but these problems are minute by comparison with those of ensuring an effective use of the finished models by the manager of the resources. Equally, we need to ensure that the models that are produced are relevant to the actual problems of management, and this represents the other facet of the interface.

As I have said in another paper (Jeffers, 1973), I do not believe that the new frontier lies in research *per se*, but in the joining of research to the decision-making process so that the result of research is not a document or a paper to a learned society, but an actual contribution to the decision that is ultimately made, couched in terms that the decision-maker can understand and which leave him in no doubt of the consequences of alternative decisions. As scientists, it is our job to forge the multidisciplinary teams which will be needed to tackle real problems and to see that these teams are equipped both mentally and physically to undertake tasks of sufficient magnitude to make a positive contribution to the continuing crisis of resource management. As managers, it is our job to see that the priority is given to those tasks of sufficient magnitude to make a positive contribution to the continuing crisis of resource management. As managers, it is our job to see that the priority is given to those tasks which are likely to have the greatest impact upon the decisions that have to be made. In return, however, the administrator must modify, if not abandon, the now outmoded method of resource management which arrives at decisions by subjective 'weighting' of undefined assumptions. We have the techniques and much of the information needed to improve the way in which our resources are managed, and the use of models and systems analysis represents an important strategy for future research.

19.6 REFERENCES

P. M. Derusso, R. J. Ray, and C. M. Close (1967). *State variables for engineers*. Wiley, New York.

D. Gifford (1971). *Comment. Bulletin, British Ecological Society*, **2** (2) 2.

J. N. R. Jeffers (1973). 'Systems modelling and analysis in resource management', *Journal of Environmental Management*, **1** (1), 13–28.

Unesco (1972). *Expert Panel on the Role of Systems Analysis and Modelling Approaches in the Programme on Man and the Biosphere*. Final Report. MAB Report Series No. 2.

Subject Index

Abyssal communities, 207
Acetate
  as substrate, 276
  removal of, 270
Acetic acid, 284
  in digesters, 291
  ionization of, 286
Activated sludge, 3, 7, 27, 32–34, 248–251, 265–279, 299–322
Activated sludge process
  activity, 266, 270, 271, 273, 277
  bacterial concentration, 266–271, 275
  MLSS, 274
  model, 265–279
  nitrification, 266
  oxygen demand, 299
  performance, 265
  plant loading, 265
  respiration rate, 275
  retention time, 267, 273
  sludge age, 267, 271
  sludge wastage rate, 266, 272, 273
Advection, 44, 143, 195
  horizontal, 208
  in sea, 215
  vertical, 208
Aeration systems, 154
Aeration tank, 232, 233, 234, 243, 272, 322, 324
  model, 96–100
  oxygen transfer, 326
  retention time, 324
Aerosols, as source of nutrients, 402
Air pollution, 8
Algae, 107–135, 145, 147, 148, 154, 157, 158, 160–164
  concentration of, 108, 115, 127
  death of, 160, 162
  growth of, 112–134, 158
  production by, 108–135
  problems of, 108
  respiration, 160
  sedimentation of, 160

  yield of, 127
Algebraic equations, 19, 47, 48, 59, 60
  linear, 48, 60
Algorithms, use of, 398
Algorithm methods, 81, 92
  non-sequential, 81
  sequential, 81
Alkalinity, effect on digestion, 282, 284, 292, 293, 294, 295, 296, 300
Allocation model, 355, 356, 365, 366, 367, 368
Allylthiourea, 171, 176
Ammonia
  as substrate, 276
  removal of, 270
Ammonium, in digestion, 288, 295
Anaerobic decomposition, 249
Anaerobic digestion, 5, 7, 12, 17, 26, 27, 33, 250, 251, 281–302, 322, 324
  batch, 292
  failure, 281, 282, 285, 286, 292–294, 296–298, 300
  gas scrubbing, 297, 298
  hydraulic overload, 282, 286, 292–294, 297, 299, 300
  inhibition, 282, 283
  inhibition function, 282, 285, 286, 287, 300
  loading frequency, 293, 296
  organic overloading, 282, 285, 286, 292–296, 300
  organism recycle, 300, 301
  retention time, 292, 293–295, 300, 324
  sludge recycle, 292, 293, 296, 299–301
  sludge thickening, 296, 300
  stability, 282, 293–296, 298, 300
  toxic overloading, 282, 286, 292, 293, 297, 299, 300
Analogue models, 254
Analytical methods, 19, 47
Annelid worms, in filter beds, 311, 312, 313, 314
Ash, 401, 402

Assimilation, by zooplankton, 221
Autocorrelation analysis, 74
Autoregression analysis, 381

Bacteria, 181
  biochemical activity, 266, 276
  in activated sludge, 268–273
  in bottom muds, 149
  in filter beds, 309–317
  in rivers, 188, 190
  in soil, 402
  lysis, 275
  metabolic rate, 309, 313
  zoogleal, 309
Bacteria beds, 249
Bacterial growth, 175, 177–179, 181, 182,
  265–279
  rate, 268, 269, 271–273, 275, 282, 284,
  285
  specific growth rate, 265, 266, 271, 272,
  275, 276, 285, 286, 300
  yield, 175, 176, 179, 180, 265, 267, 268,
  273, 284
Bellman's Principle, 92
Benthic bacteria, 179–182, 184
Benthic communities, 207
Benthic processes, 179, 180, 181, 183, 184
Bessel functions, 49
Beta transform, 77
Bicarbonate, in digestion, 287, 288, 290,
  291, 293, 300
Binomial theorem, 43, 55
Biochemical oxygen demand, 8, 11, 13, 15,
  44, 50, 67, 70, 173–178, 184, 189,
  228, 235, 239, 241, 244, 245, 267,
  274, 315, 317, 325, 328, 381, 383,
  385, 387
  carbonaceous, 171–174, 176, 189
  decay coefficient, 45, 46, 145, 148, 153,
  172, 173, 174
  interaction with DO, 137–164
  loading, 251, 323
  nitrogenous, 172–174, 176, 177
  production, 162
  removal, 96–98, 233, 235, 246, 266, 270
  river water, 358, 359
  statistical analysis of data, 71, 72
  UOD, 172
Biochemical processes, 207
Biological models, 215–222
Biological processes, 3, 7
  in sea, 215
Birch, 401, 402
Bottom muds, 145, 146, 147, 148, 149

Branch and bound methods, 81, 84
Buoyant jet, 208, 209, 210, 212
  entrainment coefficient, 212

Cadmium, in river water, 358, 359, 360
Calluna, 403
Carbon, total organic, in river water, 358,
  359
Carbon 14: Carbon 12 ratios, 405
Carbon: Nitrogen ratio, in wastes, 309
Carbon dioxide, in digestion, 283, 284,
  287, 288, 290, 291, 292, 293, 295,
  296, 297, 300
Chemical treatment, 233
Chi-squared test, 76
Chloride, 381, 383, 384
  in river water, 358, 359
Chlorination, 14
Chlorophyll concentration, 110–134
Chromium, in river water, 358, 359, 360
Ciliata, in filter beds, 310, 311, 312, 313,
  314
Coagulation, 232
COD removal, 266
Coke oven wastes, 168
Coliforms, 13
Computers, 21, 34, 35, 70, 227, 228, 232,
  233, 281, 398, 399
  analog, 21, 294
  arithmetic, 60
  control, 33, 36
  digital, 9, 21, 33, 294, 346
  hybrid, 21, 27, 282, 398
  languages, 21–25, 398
  optimization, 345, 346
Computing, 53, 54, 195, 227–229, 232,
  233, 235, 239, 241
  costs of, 379, 381
  interactive, 232, 246
  library programs, 58
  simulation, 36 (*see also* Simulation)
  step size, 232, 233, 246
  techniques, 392
Conservative substances, 208
Continuous culture, 268, 282
Continuous systems modelling program,
  21–23, 36
Control, 30–32
  algorithm, 11, 29–34, 36
  automatic, 11, 34
  computer, 32
  dynamic, 137
  engineering, 137, 138
  feedback, 11, 29, 34

412

loop, 29, 30
strategies, 32, 35, 281–302
systems, 33, 36
theory, 4, 16, 36
Copepods, in sea, 221
Copper, in river water, 358, 359, 360
Correlation coefficients, 184, 186, 190, 258
Correlogram analysis, 76
Costs, 34, 35, 84–87, 90, 91
  analysis programme, 228, 233
  of sewers, 228, 229, 233, 237, 238, 241
  of waste treatment, 228, 232, 233, 234, 241, 242
  sand filters, 347–350
  surface, 93, 99, 100
  treatment plants, 12–15
  waste treatment plants, 96–100, 319–337
  water resource development, 355, 361, 365–373, 377–388
  water treatment, 361, 367, 370
Created response surface technique, 91, 92
Cross correlation, 78
Currents, 207
  tidal, 215
Cutting plane methods, 81, 83
Cyanide, in wastes, 169

Decay constants, 169, 170, 171, 184, 187, 188, 190
Densimetric circulation, 199
Densimetric effects, 207, 209, 212, 213
Design, 43
  of sand filters, 339
  of treatment plants, 9, 12, 14, 19, 28
  of wastewater treatment plants, 319–337
  process, 81–104
Deterministic models, 227
Deterministic processes, 75, 138, 139, 154, 155, 156, 159
Detergents, removal of, 266, 267, 270
Diatoms, 118
Differential equations, 18, 19, 22, 27, 40, 41, 45, 48, 52, 53, 55, 57, 58, 60, 63, 107, 142, 185, 315, 316, 317, 392, 393
  boundary conditions, 45–47, 49–5., 58–60, 316
  first order, 143, 149, 158, 386
  initial conditions, 53, 54, 64, 315
  parabolic, 50
  partial, 50, 51, 53, 62, 64
  second order, 44, 47, 143, 146
Diffusion, 20, 46, 50, 51, 52, 53, 193–204

coefficient, 316
  in sea, 212–214, 219
  molecular, 194
  turbulent, 194, 195, 198, 199
  velocities, 214
Dispersion, 44, 147, 207, 209, 210, 211, 212, 213, 214, 215
  coefficients, 193
  horizontal, 208
  longitudinal, 193, 195, 198, 200, 201
  vertical, 208
Dissolved oxygen, 9, 67, 68
  consumption, 144
  in river water, 358, 359
  mass balance, 171, 173, 174
  production of, 144, 154
  river models, 79, 137–264
Dry weather flow, 237, 238, 239, 245, 321
Dyes, 195
  as tracers, 200, 201
Dynamic behaviour, 4, 14, 25, 27, 29–31
Dynamic models, 16, 18, 22, 29, 33, 36, 137, 139, 141, 150, 151, 153, 158, 162, 163, 164
  validation of, 164
Dynamic programming, 16, 81, 361–373, 378, 379, 380, 386

Ecological investigation, 107, 389–409
Ecology, 389–408
  woodland, 400–402
Economic analysis, 150
Economic models, 245
Ecosystems dynamics, 107, 395–409
Enchytraeids, 404
Eriophorum, 403
Estuaries, 7, 168, 193–204, 210, 215
  stratified, 200
Estuarine discharges, 207, 210, 211, 213, 216, 217
Experimental design, 70, 71

Farming, 405, 406
Feedback, 397
  controllers, 163
  signals, 299
Ferric hydroxide floc, 341
Fick's Law, 315
Finite difference methods, 85, 146, 213, 218, 342
Fish, marine, 207
Fisheries, 355, 356, 357, 359, 360, 361, 364, 370, 372, 373, 380
  coarse, 168

Flagellata, in filter beds, 310, 311, 312
Flocculation, 232, 248
Flocculent particles, 248, 249, 250
Floods, 181
Flow
  steady, 198, 201, 202
  synthetic data generation, 381
  unsteady, 200, 201, 202
  turbulent, 203
Forcing functions, 24, 27, 151
  generator, 26
  impulse, 24, 25
  pseudo-random binary sequence, 151
  pulse, 24, 25, 151
  pulse train, 26
  ramp, 24, 25
  step, 24, 25, 26, 151
Forestry, 405, 406
FORTRAN, 227, 245
Fourier, analysis, 381
Fourier, series, 149
Frequency distributions, 69, 72, 213, 383
  log-normal, 189
Frequency response analysis, 27
Friction velocity, 196
Froude number, 209
Functional analysis, 40
Fungi
  in filter beds, 309, 312
  in soil, 401, 402

Gamma transform, 77
Gas composition, from digestion, 282, 284,
  292, 296, 299, 300
Gas flow rate, in digestion, 282, 284, 292,
  296, 299, 300
Gas transfer rates, 288, 292
Gaussian distribution, representing concen-
  tration, 213
Glucose
  as substrate, 276
  removal rate, 270
Graph theory, 16
Grazing, 108, 120, 126, 132
  of phytoplankton, 220
  threshold, 220
Gradient methods, 233, 246
Grass plot treatment, 322, 324
  loading rate, 324
Grouse, 404
Growth
  of attached plants, 145
  of bacteria (see Bacterial growth)
Games theory, 16, 397

Harmonic analysis, 68, 72, 73, 74, 75, 77
Hazel, 401, 402
Heavy metals, effect on settlement, 249
Henry's Law, 288
Herbivores
  concentration of, 108, 132
  in peat, 403
  in sea, 220
  in soil, 402
  mortality of, 221
Hydraulic modelling, 254
Hydrographs, of storm water, 229, 230,
  231, 235, 236, 237, 239, 240
Hydrology, 257

Industrial wastes, 357
  cost of treatment, 361
  discharge to sea, 207
  effect on sewage treatment, 321
  settlement of, 249
  treatment by filters, 308
Infiltration, 228
Information flow, 10, 11, 12, 29
  theory of, 20
Insect larvae, in filter beds, 311, 312, 313,
  314
Inter-tidal zone, 207
Investment model, 355, 356, 368, 269

Kalman filter, 158
Kinematic viscosity, 209, 247, 341
Kinetics
  first order, 169, 175, 183, 184, 185, 189
  Michaelis–Menten, 189, 270, 276, 285,
    316
  model, 149, 180, 182, 189
  Monod, 180, 181
Kolmorgoroff–Smirnov test, 76
Kozeny–Carman equation, 341

Lagrangian multipliers, 89, 90
Laminar flow, 247, 248
Laplace transforms, 41, 64
Leaching, of nutrients, 402
Lead, in river water, 358, 359, 360
Light, 109–135
  Beer–Lambert Law, 110
  effect on photosynthesis, 109–135
  euphotic depth, 114
  in sea, 220
  vertical attenuation, 110–130, 162
  wavelength, 110
Lime, use in digesters, 296

Linear programming, 81, 82
Linear regression, 73
Linearity, 41, 85, 90
Longitudinal dispersion, 143, 146, 157

Manning roughness, 196
Marine pipelines, 207–211, 213
  diffusers, 208, 209
Marine pollution, 207–222
Markov model, 75, 76
Mass balance, 167, 168, 169, 171, 179, 194
  nutrients in seawater, 218
  on digester, 289, 290
Mass transfer, 51
Mathematical models, 32, 36, 39, 107, 138, 140, 141, 163, 207, 227, 237, 265–279, 319, 320 (*see also* Models, modelling)
  in ecology, 389–408
  of sedimentation, 251–263
  of Trent river system, 355–374
  sensitivity analysis, 334
  validation, 162, 164
Mathematical techniques, 39–65
Mechanical aerators, 163
Metals, in seawater, 210
Methane, 281, 283, 287, 296
  production in digesters, 299
Method of characteristics, 149
Method of maximum likelihood, 154, 155
Michaelis–Menten coefficient, 220
Michaelis–Menten function, 189, 270, 276, 285, 316
Microorganisms, 41, 42
  growth of, 40–44 (*see also* Bacterial growth)
Microstrainers, 322, 324
  loading rate, 324
Mixed-integer, programming, 378–379
Mixing length, 197
Modelling of sewerage systems, 227–246
Modelling strategies, 389–408
Models, modelling, 68, 207–222 (*see also* Mathematical models)
  black box, 40, 41, 77, 140, 141, 150, 152, 154, 157
  calibration of, 380
  chloride in rivers, 141, 150
  design, 377, 379, 380, 381, 387
  deterministic, 19, 24, 251, 381, 383
  dispersion, 213
  dynamic, 281–302, 391, 392, 394
  ecological, 303–318

economic, 18
estuarine, 195
fundamental, 265, 276
heuristic, 58
hierarchical, 377–388
hydrodynamic, 208
information flow, 291
internally descriptive, 140–144, 152, 153, 155, 157
linearity, 42, 45, 47, 48, 55, 56, 60
mathematical, 4, 13, 17, 18, 20, 21, 24, 31, 42, 77
multiple regression, 188
of tertiary treatment, 339–351
operational, 378, 379, 386
optimization, 393, 394, 397
planning, 377, 379, 387
procedural, 17
prototype, 320
physical, 17, 23
real time, 378
river temperature, 78, 141
simulation, 393, 397, 399
steady state, 12, 18, 19, 41, 42, 43, 44, 54–58, 68, 282
sensitivity, 66
stability, 42, 43, 54, 56–58
statistical, 13, 77
stochastic, 381, 394
structure, 152
time series, 141
verification, 24
water quality, 137, 146, 151, 377–388
water quantity, 377, 381
Monod function, 161, 265, 270, 276, 282, 285, 286
Moorland, 403, 404, 405
Multiple regression, 77
Multiple regression models, 79

Nematoda, in filter beds, 310, 311, 312, 313, 314
Network analysis, 16
Nickel, in river water, 358, 359, 360
Nitrate, loads in river waters, 163
Nitrification, 145, 148, 167–191
  dynamics of, 167, 178, 180
  inhibition of, 171, 176
  model of, 169, 180, 182–190
Nitrifying
  bacteria, 169, 175, 177, 179
  regime, 169
  yield coefficient, 176, 179, 180
Nitrobacter, 175, 176, 188, 190

Nitrogen
  ammoniacal, 167–172, 174, 176–184, 186, 188, 189, 190, 211, 266, 325, 328, 358, 359, 360
  balance, 211
  budget, 169, 170
  concentration in wastes, 308
  cycle, 145, 162
  in seawater, 215, 217, 218, 219
  inorganic, 167, 169
  nitric, 185, 187
  nitrous, 176, 184–188, 190
  organic, 211
  oxidized, 167, 184, 210, 211, 358, 359
  removal of, 235
Nitrosomonas, 175, 176, 177, 178, 179, 183, 184, 186, 189, 266, 271, 276
Non-linear iterative methods, 81, 82, 84
  constrained, 82, 89, 92
  gradient, 82, 84, 85, 86, 87
  search, 82
Non-linear programming, 81, 82, 92, 96
  dynamic programming, 82, 92, 93, 96–100, 361–373, 378–380, 386
  dynamic search, 82, 95, 96
  geometric programming, 82, 93, 95
Normal distribution, 69, 70, 76
Numerical methods, 21, 22, 41, 47–49, 53, 58, 60, 107, 146, 190, 392
  divergence/convergence, 48, 49, 59, 62
  finite difference, 47, 85, 146, 213, 218, 342
  integration, 59, 60, 274
  linearity, 60–62
  Runge–Kutta, 47, 59
  series solution, 47–49
  stability, 53, 56, 57, 60
  step size, 47, 53, 58, 60
  trial and error, 170, 233, 246, 294
Nutrients, 126–130
  availability, 132
  concentrations, 127–129, 220, 404
  deficiency, 401, 402
  enrichment, 215–219
  excretion, 222
  limitation, 127, 133
  model, 128
  requirement, 127, 130
  transfer, 400–403
  uptake, 129, 219, 220, 222
  upwelling of, 215, 218, 219

Oak, 401, 402
Optimization, 15, 16, 23, 31, 36, 81–104, 233–246, 281
  of rapid filtration, 339–351
  of water resource development, 355–373
  of water resources systems, 383, 386
  of waste treatment cost, 319–337
  of wastewater treatment, 250
  routines, 239
  techniques, 4, 16
Organic matter, in soil, 401
Oxygen consumption, 171, 174
Oxygen resources, 167, 168, 171
Oxygen uptake rate, 299

pH, effect on digestion, 282, 283, 284, 286, 287, 292, 296, 297
Parameter estimation, 147–164
Particle size
  distribution, 248
  non-flocculent, 252
Paviani's method, 89, 92
Peat, 403, 404
Pelagic communities, 207
Penalty functions, 89, 90
Percolating filters, 50–53, 303–318, 322, 324
  design, 303, 305, 317
  dissolved oxygen, 308
  distributors, 305
  dosing, 304, 305, 315
  film accumulation, 305, 306, 308, 316, 317
  filter community, 309–312
  functional analysis, 313–317
  grazing fauna, 308, 311–314
  high rate, 305
  hydraulic conditions, 304, 305, 317
  hydraulic loading, 324
  media, 303, 304
  mixing patterns, 305
  model, 314–317
  retention time, 50, 304, 305, 315
  standard rate, 305
  temperature effects, 306–308
Phenols, in river water, 358, 359
Phosphorus, 32, 183, 232, 235, 246, 401, 402
  in seawater, 210, 216, 217, 219
  in wastes, 308
Photosynthesis, 7, 107–135, 145, 154, 157
  light relationship, 110–135
  rate of, 146–148
  self-shading, 112
Physical modelling, 254
Phytoplankton, in sea, 218

Pipeline hydraulics, 237
    coefficients, 237, 238
Plankton, marine, 207
Plug flow tanks, 251, 252
Potassium bromide 82, 184
Power spectrum analysis, 74
Pragmatic models, 265, 276
Predation
    curves, 221
    in filter beds, 311, 313, 314, 317
    in sea, 222
Primary clarifier, 232, 233, 234, 242, 243
Principal component analysis, 79
Probability density function, 257
Probability distribution, 76
Probability theory, 17, 18
Process kinetics, 232
Production
    primary, 222, 400, 403, 404
    sea, 215, 222
    secondary, 222, 400
Program evaluation review technique, 17
Propionic acid, as substrate, 276
Protozoa, 310, 311
    in bottom muds, 149
Psyllids, 404
Pulses, sinusoidal, 210, 212
Pumping stations, 229

Quadratic functions, 86

Rain, as a source of nutrients, 402
Rainfall
    intensity, 228, 229, 230, 232, 237, 238,
        239, 240
    runoff, 230, 231, 237
Random numbers, generator, 26
Raw water quality, 242
Reaction kinetics, 20, 40, 41, 189, 190
Reactors
    continuous stirred tank, 18, 19, 22, 24,
        26, 27, 36, 40, 41, 42, 55–58, 140,
        143, 157, 217, 266, 269, 273, 288,
        292
    kinetics, 42
    mass balance, 289
    plug flow, 27, 269, 292
    retention time, 27, 40, 41
    tubular, 45
Reaeration, 144, 146, 153, 173
    capacity, 172
    constant, 172, 174
    escape coefficient, 174
Recharge of aquifers, 357, 372

Recirculation, in sedimentation tanks, 253
Recreation, 405, 496
    benefits from, 361, 362, 364, 370
Recreational uses, 379, 380
Regression, 257, 258
    analysis, 209, 261, 262
Respiration, 145
    algal, 108–135
    rate of, 146, 147, 148
Reservoirs, 107–135
    circulation, 127
    storage, 355, 365, 366, 370
    stratification, 134
    water quality, 383, 386
Resource management, 390, 406
Retention time
    in reactors, 27, 40, 41
    in rivers, 146, 180, 182
Reynolds analogy, 196, 198
Reynolds number, 209
Richardson number, 198, 202
River
    basins, 4, 5, 7, 14
    models, 355, 356, 357, 360, 363, 364,
        366, 367
    purification lakes, 357, 361, 372
Rotifera
    in filter beds, 310, 311, 312
Runoff, 145, 146
    coefficient, 228, 237, 239

Salinity, near estuaries, 211
Sampling, 68, 404
    design of, 67
    variance, 69
Sand filters, 322, 326, 339–351
    anthracite, 347
    backwashing, 348
    cost, 347–350
    depth, 339, 342–348, 350, 351
    design, 339, 342
    filter coefficient, 340
    filtration velocity, 340, 343, 344, 348,
        350
    grain size, 340–347, 351
    head loss, 339, 341, 342, 347, 348
    media, 339, 340
    multilayer, 345, 347
    porosity, 340, 341
    upflow, 345
Sarcodina, in filter beds, 310, 311, 312
Scour, 145
Scale up, 23
Seawater, density of, 207

Secondary clarifier, 232, 234, 242, 243
Sedimentation, 24, 30, 145, 247–263
    of algae, 109, 120
    primary, 57, 249–251, 255, 256
    rate of, 146
    secondary, 37, 249, 250, 257
Sedimentation tanks, 249–263, 322, 324,
        328–331
    BOD loading, 260
    density currents, 249, 252
    design, 247, 249, 250
    efficiency, 255, 256
    hindered settlement, 248
    hydraulic loading, 257
    operation, 247, 250
    overflow rates, 248, 249, 256, 260, 261
    retention time, 249, 252, 255, 256, 259,
        260, 262, 324, 326, 328, 331
    sludge pumping rates, 324
    solids loading rate, 324
    theory, 247, 248
    velocity gradients, 249
Sensitivity analysis, 23, 237, 238, 246
    of costs, 349
Self-purification, 137, 167, 171, 359, 380
Septic tanks, 4
Serial correlation, 75
Settleable solids, 233, 248, 251, 257
    removal of, 257
Settling column tests, 249, 255
Settling velocity, 247, 248, 252
Sewage
    composition, 9, 11, 12, 321
    crude, 169, 248, 249
    discharge to sea, 222
    effect on settlement, 250
Sewage works, 153, 168, 169, 228, 241,
        242, 319, 320, 322, 357
    costs, 361
    model, 96–100
    optimum design, 96–100
    steady state conditions, 321
Sewage treatment
    preliminary, 320
    primary, 322, 326
    secondary, 322, 326
    tertiary, 322, 326
Sewerage, 7
    systems, 168, 227–246
Sewer network, 138, 228, 229, 230, 240,
        241, 245, 246
    model, 227–231, 234–236, 245
    storage in, 231
Shear, 194, 199, 200

stress, 196
    transverse, 199, 202
    velocity, 195
    vertical, 198, 199, 202
Sheep, 404
Simplex
    linear, 81, 82, 83
    non-linear, 87, 88, 89, 92, 96, 97, 98, 99,
        103, 104
Simulation, 13, 16, 17, 19, 21, 24, 31, 67,
        68, 77, 79, 138, 149, 162, 183, 184,
        213, 228, 245, 246, 249, 281, 284,
        290, 292, 295, 296, 303
    of water resource development, 366–368
Simultaneous equations, 86, 94
Skewness, 77
Sludge
    age, 268, 269
    density, 30
    drying beds, 325
    loading rate, 325
    dumping, 207
    growth rate, 268
    loading rate, 233
    production rate, 233
    thickening, 322, 324
    treatment and disposal, 232, 320, 322,
        324, 325, 327
    treatment plant model, 227, 233, 235,
        245
Soda ash, use in digesters, 296
Soil fertility, 406
Soil microfauna, 402
Solid wastes, 8
Solids, total dissolved in river water, 358,
        359
Solids thickening, 250
Sphagnum, 403
Stability analysis, 16
Standards, water quality, 12, 13, 163
Statistical analysis, 29, 34
Statistical instability, 79
Statistical methods, 16
Statistical stability, 78
Statistical synthesis, 77
Statistical techniques, 67–79
Statistics, 156
Stationarity, 77
Steady state conditions, 109
Steady state models, 139, 147, 149, 150,
        234, 235
Steady state processes, 137, 146, 150
Stearate, removal of, 270
Stochastic disturbances, 138, 141

Stochastic effects, 142
Stochastic errors, 138, 141
Stochastic models, 19, 24, 150, 255, 257–263
Stochastic processes, 16, 20, 68, 75, 138, 151, 163, 259, 260, 378
Stochastic systems, 293, 394
Stokes's law, 247, 248
Storage tanks, for storm water balancing, 228, 232, 243, 244
Storm water, 241
    overflows, 163, 228, 232, 235, 236, 237, 244
    runoff, 227, 228, 229, 235, 236, 239, 240, 241, 243, 244, 245
    tanks, 322
Stratification, in sea, 207
Stream flow, 12, 67
    augmentation, 13
Stream velocities, 184, 193, 194, 195, 196, 199
Streeter–Phelps formulation, 138, 145, 172
Substrate utilization, 265–279
Sulphur, in wastes, 169
Sum of squares, 70
Suspended solids, 233, 243, 247, 248, 249, 251, 255, 260, 261
    in rivers, 358, 359
    measurement, 248
    removal, 97, 98, 235, 256
Suspensions, 249, 250, 252
    characteristics, 257, 262
    types of, 248
Sycamore, 401, 402
Systems, 3, 4, 7, 8, 17, 18, 21, 27, 30, 35
    analysis, 137, 138, 139, 149, 150, 153, 163, 389–408
    control, 4, 5, 8–10, 13–15, 19–21, 27, 29–31
    engineering, 4, 6, 281
    first order, 27
    linear, 26, 27
    management, 75
    oscillatory, 27
    second order, 27
    sensitivity, 78
    states, 139

Telemetry, 386
Temperature, river water, 10, 12, 72, 73, 358, 359
Tidal effects, 200, 202, 203
Tidal waters, 209
Time delay, 161

Time of travel, in rivers, 173, 183, 187
Time series analysis, 72, 77, 78, 150, 152, 387
Time series models, 150
Tipulids, 404
Toxicity tests, 360
    median lethal concentrations, 360
Tracers, 18, 19, 23, 24, 26, 27, 36, 182, 190
    pulse, 18, 24–26
    radioactive, 169, 183, 184
Transmission of water, costs of, 365, 366, 367
Transportation delay, 140, 143
Transfer functions, 30
Transient response analysis, 4, 12, 24, 29, 36
Trial and error methods, 170, 233, 246, 294
Trickling filters (see Percolating filters)
Trophic network
    in sea, 219
    levels, 400
Trout, rainbow, 359, 360
Turbidity, 69, 77, 134, 162
    in sea, 222
Turbulence, 182, 193–204
    isotropic, 197, 198

Velocities
    in sea, 209, 210, 212–214, 219
    in sewers, 228
Vertical mixing, in sea, 220
Viscosity
    kinematic, 209, 247, 341
    vertical turbulent, 196
Volatile acids, 282, 283, 284, 290, 292, 296
    inhibition by, 282
    unionized, 286
Von Karman's constant, 196

Water, potable, 355, 356, 360, 361, 369, 372
Water abstraction, 137, 369, 371, 373, 383
Water demands, 405, 406
Water distribution, 16
Water quality, 137, 138, 167, 168, 355, 356, 358–360, 364, 365, 372, 379, 381, 383, 386
    management, 153
    standards, 12, 13, 163
    statistical relationships, 359
    surveillance, 178

Water quality models, 137, 146, 151,
377–388
allocation phase, 380
investment phase, 380
optimization phase, 381
Water resources systems, 79, 377, 381,
386
design of, 383
development of, 355–373
management of, 168
Water supply, 138, 167
Water temperature, data analysis, 74, 75,
76
Water treatment, 6, 15, 138, 356, 357, 360,
364, 372
coagulation, 360
costs, 379
desalination, 360
filtration, 360
model, 77, 79
storage, 360
softening, 360
Water velocities
eddy velocities, 219
in sea, 209, 210, 213, 214, 219

Wastewater
domestic, 9, 11, 12, 169, 208, 248, 249,
321
flow, 230, 231, 243, 245
industrial, 8, 13, 168, 208
production, 228
Wastewater treatment, 3, 5–7, 9, 13,
15–17, 19, 23, 24, 27, 28, 32–35,
138, 168, 250, 281, 319–337, 361, 362
control of, 72
costs of, 367, 370, 380
efficiency of, 232, 235
models, 227, 228, 231–234, 239, 245
Wave action, effect on dispersion, 198
Waves, 207
Woodland ecology, 400–402, 403

Zero-one method, 81, 84
Zinc, in river water, 358, 359, 360
Zooplankton
excretion, 222
in sea, 220
metabolism, 221, 222
respiration of, 221
reproduction of, 221